BIOPROCESSING TECHNOLOGY IN FOOD AND HEALTH

Potential Applications and Emerging Scope

BIOPROCESSING TECHNOLOGY IN FOOD AND HEALTH

Potential Applications and Emerging Scope

Edited by

Deepak Kumar Verma
Ami R. Patel, PhD
Prem Prakash Srivastav, PhD

AAP APPLE
ACADEMIC
PRESS

Apple Academic Press Inc.	Apple Academic Press Inc.
3333 Mistwell Crescent	9 Spinnaker Way
Oakville, ON L6L 0A2 Canada	Waretown, NJ 08758 USA

© 2019 by Apple Academic Press, Inc.

First issued in paperback 2021

Exclusive worldwide distribution by CRC Press, a member of Taylor & Francis Group
No claim to original U.S. Government works
ISBN 13: 978-1-77463-167-6 (pbk)
ISBN 13: 978-1-77188-688-8 (hbk)

Library and Archives Canada Cataloguing in Publication

Bioprocessing technology in food and health : potential applications and emerging scope / edited by Deepak Kumar Verma, Ami R. Patel, PhD, Prem Prakash Srivastav, PhD.

Includes bibliographical references and index.
Issued in print and electronic formats.

ISBN 978-1-77188-688-8 (hardcover).--ISBN 978-1-351-16788-8 (PDF)

1. Functional foods. 2. Food--Microbiology. 3. Food industry and trade--Technological innovations. I. Patel, Ami, editor II. Srivastav, Prem Prakash, editor III. Verma, Deepak Kumar, 1986-, editor

QP144.F85B56 2018	613.2	C2018-903696-6	C2018-903697-4

Library of Congress Cataloging-in-Publication Data

Names: Verma, Deepak Kumar, 1986- editor. | Patel, Ami R., editor. | Srivastav, Prem Prakash, editor.

Title: Bioprocessing technology in food and health : potential applications and emerging scope / editors, Deepak Kumar Verma, Ami R. Patel, Prem Prakash Srivastav.

Description: Toronto ; New Jersey : Apple Academic Press, 2019. | Includes bibliographical references and index.

Identifiers: LCCN 2018027606 (print) | LCCN 2018028139 (ebook) | ISBN 9781351167888 (ebook) | ISBN 9781771886888 (hardcover : alk. paper)

Subjects: | MESH: Food Microbiology | Functional Food--microbiology | Fermented Foods--microbiology | Food-Processing Industry

Classification: LCC RM218.7 (ebook) | LCC RM218.7 (print) | NLM QW 85 | DDC 613.201/579--dc23

LC record available at https://lccn.loc.gov/2018027606

Apple Academic Press also publishes its books in a variety of electronic formats. Some content that appears in print may not be available in electronic format. For information about Apple Academic Press products, visit our website at **www.appleacademicpress.com** and the CRC Press website at **www.crcpress.com**

ABOUT THE EDITORS

 **Deepak Kumar Verma** is an agricultural science professional and is currently a PhD Research Scholar in the specialization of food processing engineering in the Agricultural and Food Engineering Department, Indian Institute of Technology, Kharagpur (WB), India. In 2012, he received a DST-INSPIRE Fellowship for PhD study by the Department of Science & Technology (DST), Ministry of Science and Technology, Government of India. Mr. Verma is currently working on the research project "Isolation and Characterization of Aroma Volatile and Flavoring Compounds from Aromatic and Non-Aromatic Rice Cultivars of India." His previous research work included "Physico-Chemical and Cooking Characteristics of Azad Basmati (CSAR 839-3): A Newly Evolved Variety of Basmati Rice (*Oryza sativa L.*)". He earned his BSc degree in agricultural science from the Faculty of Agriculture at Gorakhpur University, Gorakhpur, and his MSc (Agriculture) in Agricultural Biochemistry in 2011. He also received an award from the Department of Agricultural Biochemistry, Chandra Shekhar Azad University of Agricultural and Technology, Kanpur, India. Apart from his area of specialization in plant biochemistry, he has also built a sound background in plant physiology, microbiology, plant pathology, genetics and plant breeding, plant biotechnology and genetic engineering, seed science and technology, food science and technology etc. In addition, he is member of different professional bodies, and his activities and accomplishments include conferences, seminar, workshop, training, and also the publication of research articles, books, and book chapters.

Ami R. Patel, PhD, is an Assistant Professor in the Division of Dairy and Food Microbiology, Mansinhbhai Institute of Dairy & Food Technology-MIDFT, Dudhsagar Dairy Campus, Gujarat, India. Professor Patel has expertise in specialized areas involve isolation, screening, and characterization of exopolysaccharides from potential probiotic cultures and employing them for food and health applications. In addition, she is engaged with teaching undergraduate, postgraduate, and research students in food microbiology, microbial biotechnology, food biotechnology, food science and technology, clinical microbiology and immunology, etc. She has authored a number of peer-reviewed papers and technical articles in international and national journals as well as book chapters, books, proceedings, and technical bulletins. She has received a number of awards and honors, including the receiving the Vice Chancellor Gold Medal for her PhD work, being selected as a *BiovisinNxt*11 fellow to attend an international conference at Lyon, France, and receiving a *Erasmus Mundus* scholarship from the European Union for three years to visit Lund University, Sweden, as a guest researcher. She is also serving as an expert reviewer for several scientific journals. She is a member of several academic and professional organizations, including the Indian Dairy Association (IDA) and the Swedish South Asian Network for Fermented Foods (SASNET).

She earned her BSc (Microbiology) and her MSc (Microbiology) degree from Sardar Patel University, Gujarat, and received her doctorate degree in Dairy Microbiology from the Dairy Department of SMC College of Dairy Science, Anand Agricultural University, Gujarat, India.

Prem Prakash Srivastav, PhD, is Associate Professor of Food Science and Technology in the Agricultural and Food Engineering Department at the Indian Institute of Technology, Kharagpur (WB), India, where he teaches various undergraduate-, postgraduate-, and PhD-level courses and guides research projects. His research interests include the development of specially designed convenience, functional, and therapeutic foods; the extraction of nutraceuticals; and the development of various low-cost food processing machineries. He has organized many sponsored short-term courses and completed sponsored research projects and consultancies. He has published various research papers in peer-reviewed international and national journals and proceedings and many technical bulletins and monographs as well. Other publications include books and book chapters along with many patents. He has attended, chaired, and presented various papers at international and national conferences and delivered many invited lectures at various summer/winter schools. Dr. Srivastav has received several best poster paper awards for his presentations. He is a member of various professional bodies, including the International Society for Technology in Education (ISTE), the Association of Food Scientists and Technologists (India), the Indian Dairy Association (IDA), the Association of Microbiologists of India (AMI), and the American Society of Agricultural and Biological Engineers, and the Institute of Food Technologists (USA).

CONTENTS

LIST OF CONTRIBUTORS

Iuliana Aprodu
Faculty of Food Science and Engineering, "Dunărea de Jos" University of Galati,
Domnească Street 111, 800201 Galati, Romania, Tel.: +40336130183, Mob.: +40744963478,
Fax: +40236460165, E-mail: Iuliana.Aprodu@ugal.ro

Carlos Eduardo Barão
Federal Institute of Paraná (IFPR), Paranavaí PR 87703-536, Brazil, Tel: +55 44 34820110
E-mail: carlos.barao@ifpr.edu.br

Andrei Sorin Bolocan
APC Microbiome Institute, University College Cork, College Road, Cork, Ireland,
Tel.: +353214901781, Mob.: +353851529997
E-mail: andrei.bolocan@ucc.ie, andrei.s.bolocan@gmail.com

Luminița Ciolacu
Department for Farm Animal and Public Health in Veterinary Medicine, Institute of Milk Hygiene,
Milk Technology and Food Science, University of Veterinary Medicine Vienna, Veterinärplatz 1,
1210 Vienna, Austria, Tel: +43 1 25077 3510 E-mail: luminita.Ciolacu@vetmeduni.ac.at

Adriano Gomes da Cruz
Instituto Federal de Educação, Ciência e Tecnologia do Rio de Janeiro (IFRJ), Mestrado
Profissionalem Ciência e Tecnologia de Alimentos (PGCTA), Maracanã, Rio de Janeiro
PR 20270-021, Brazil, Tel: +55 21 22641146 E-mail: food@globo.com

Lorraine Draper
APC Microbiome Institute, University College Cork, College Road, Cork, Ireland,
Tel.: +353214901781 E-mail: l.draper@ucc.ie

Kimmy G.
Department of Food Engineering and Technology, Sant Longowal Institute of Engineering and
Technology Longowal, Sangrur, Punjab 148106, India, Mob.: +00–91–8699271602
E-mail: kishorigoyal09@gmail.com

Goksen Gulgor
Department of Food Engineering, Faculty of Agriculture, Uludag University, Gorukle 16059,
Bursa, Turkey, Tel.: +90 224 294 15 06 E-mail: goksengulgor@uludag.edu.tr

Colin Hill
APC Microbiome Institute, University College Cork, College Road, Cork, Ireland,
Tel.: +353214901781 E-mail:c.hill@ucc.ie

Suellen Jensen Klososki
Federal Institute of Paraná (IFPR), Paranavaí PR 87703-536, Brazil, Tel.: +55 44 34820110
E-mail: suellen.jensen@ifpr.edu.br

Nevzat Konar
Food Engineering Department, Siirt University, 56100 Siirt, Turkey, Tel.: +90 5322714611
E-mail: nevzatkonar@hotmail.com

Mihriban Korukluoglu
Department of Food Engineering, Faculty of Agriculture, Uludag University, Gorukle 16059 Bursa, Turkey, Tel.: +90 224 294 14 97 E-mail: mihriban@uludag.edu.tr

Dipendra Kumar Mahato
Agricultural and Food Engineering Department, Indian Institute of Technology Kharagpur, West Bengal 721302, India, Mob.: +91–9911891494 E-mail: kumar.dipendra2@gmail.com

Alaa Kareem Niamah
Department of Food Science, College of Agriculture, University of Basrah, Basra City, Iraq, Mob.: +00–96–47709042069 E-mail: alaakareem2002@hotmail.com

AncaIoana Nicolau
Department of Food Science and Engineering and Applied Biotechnology, Faculty of Food Science and Engineering, Dunarea de Jos University of Galati, Galati, Romania, Mob.: +40 755746227

Sirin Oba
Department of Food Processing, Amasya University Suluova Vocational School, Amasya, Turkey, Tel.: +90 5545826755 E-mail: sirin.oba@amasya.edu.tr

Elena-Alexandra Oniciuc
Department of Food Science and Engineering and Applied Biotechnology, Faculty of Food Science and Engineering, Dunarea de Jos University of Galati, Galati, Romania, Mob.: +40 744998122
E-mail: elena.Oniciuc@ugal.ro

Ibrahim Palabiyik
Food Engineering Department, Namik Kemal University, 59030,Tekirdağ, Turkey, Tel.: +90 5542313361 E-mail: ipalabiyik@nku.edu.tr

Ami Patel
Division of Dairy and Food Microbiology, Mansinhbhai Institute of Dairy and Food Technology-MIDFT, Dudhsagar Dairy Campus, Mehsana, Gujarat 384002, India, Mob.: +00–91–9825067311, Tel.: +00–91–2762243777 (O), Fax: +91–02762–253422 E-mail: amiamipatel@yahoo.co.in
E-mail: anca.nicolau@ugal.ro

Tatiana Colombo Pimentel
Federal Institute of Paraná (IFPR), Paranavaí PR 87703-536, Brazil, Tel: +55 44 34820110
E-mail: tatiana.pimentel@ifpr.edu.br

Derya Genc Polat
Tayas Food, Gebze, Kocaeli, Turkey, Tel: +90 5366754219 E-mail: deryagenc@tayas.com.tr

Gabriela Râpeanu
Faculty of Food Science and Engineering, "Dunărea de Jos" University of Galati, Domnească Street 111, 800201 Galati, Romania, Tel.: +40 336130183, Mob.: +400742038288, Fax: +40236460165, E-mail: Gabriela.Rapeanu@ugal.ro

Michele Rosset
Federal Institute of Paraná (IFPR), Colombo PR 83403-515, Brazil, Tel.: +55 41 35351835
E-mail: michele.rosset@ifpr.edu.br

Osman Sagdic
Food Engineering Department, Yildiz Technical University, Istanbul, Turkey
E-mail: osagdic@yildiz.edu.tr

Nihir Shah
Division of Dairy and Food Microbiology, Mansinhbhai Institute of Dairy and Food Technology-MIDFT, Dudhsagar Dairy Campus, Mehsana,Gujarat 384002, India, Mob: +00–91–9925605480, Tel.: +00–91–2762243777 (O), Fax: +91–02762–253422 E-mail: nihirshah13@yahoo.co.in

Pratibha Singhal
Department of Bioscience and Biotechnology, Banasthali University, Vanasthali, Rajasthan 304022, India, Mob.: +00–91–9643850855 E-mail: pratibhasinghal89@gmail.com

Prem Prakash Srivastav
Agricultural and Food Engineering Department, Indian Institute of Technology Kharagpur, West Bengal 721302, India, Tel.: +91 3222281673, Fax: +91 3222282224
E-mail: pps@agfe.iitkgp.ernet.in

Nicoleta Stănciuc
Faculty of Food Science and Engineering, "Dunărea de Jos" University of Galati, Domnească Street 111, 800201 Galati, Romania. Tel.: +40336130183, Mob.: +40729270954, Fax: +40236460165, E-mail: Nicoleta.Stanciuc@ugal.ro

Pooja Thakkar
Department of Life Sciences, Gujarat University, Ahmedabad, India, Mob.: +00–91–9909804647
E-mail: poojathakkar16603@gmail.com

Omer Said Toker
Food Engineering Department, Yildiz Technical University, Istanbul, Turkey, Tel: +90 5053124247
E-mail: stoker@yildiz.edu.tr

Deepak Kumar Verma
Agricultural and Food Engineering Department, Indian Institute of Technology Kharagpur, West Bengal 721302, India, Tel.: +91 3222281673, Mob.: +91 7407170259, Fax: +91 3222282224
E-mail: deepak.verma@agfe.iitkgp.ernet.in,rajadkv@rediffmail.com

Rishika Vij
Department of Veterinary Physiology and Biochemistry, COVAS, Chaudhary Sarwan Kumar Himachal Pradesh Krishi Vishvavidyalaya, Palampur, Himachal Pradesh 176062, India, Mob.: +00–91–9466493086 E-mail: rishikavij@gmail.com

Martin Wagner
Department for Farm Animal and Public Health in Veterinary Medicine, Institute of Milk Hygiene, Milk Technology and Food Science, University of Veterinary Medicine Vienna, Veterinärplatz 1, 1210 Vienna, Austria, Tel.: 0043 1 25077 3500, Fax: 0043 1 25077 3590
E-mail: martin.wagner@vetmeduni.ac.at

LIST OF ABBREVIATIONS

α-la	α-lactalbumin
AAB	Acetic acid bacteria
AAD	Adenylic acid deaminase
Ab	*Acetobacter*
AF	Aflatoxin
AFB	Aflatoxin B
AFB_1	Aflatoxin B_1
AFs	Aflatoxins
AISI	American Iron and Steel Institute
ALDC	α-acetolactate decarboxylase
AMG	Amyloglucosidase
AMP	Adenosine monophosphate
ANS	8-anilinonaphthalene-sulfonic acid
APs	Aminopeptidases
Asn	Asparagine
Asp	Aspartic acid
a_w	Water activity
BA	Biogenic amine
BACs	Biologically active compounds
BC	β-carotene
BOD	Biochemical oxygen demand
BSA	Bovine serum albumin
Bw	Body weight
C	Cytosine
C3G	Cyanidin-3-O-glucoside
Ca	Calcium
$CaCO_3$	Calcium carbonate
CDS	Circular dichroism spectroscopy
CFU	Colony-forming units
CLA	Conjugated linolenic acid
CMC	Carboxymethylcellulose
CO_2	Carbon dioxide
CP	Chlorpyrifos
CSE	Carotenoids extract from sea buckthorn
CT	Conching time
Cy3glc	Cyanidin-3-glucoside

CYR	Cyanidin 3-rutinoside
Cys	Cysteine
D3G	Delphinidin-3-O-glucoside
DNA	Deoxyribonucleic acid
DP	Degree of polymerization
DZ	Diazinon
EFSA	European Food Safety Authority
EPS	Exopolysaccharides
EPS	Extracellular polymeric substance
EU	European Union
Exo-PG	Exo-polygalacturonase
FDA	Food and Drug Administration
FMC	Food microbial community
FNT	Fenitrothion
FOS	Fructooligosaccharide
FPC	Fermentation-produced chymosin
FSIS	Food Safety and Inspection Service
FTIR	Fourier transform infrared spectroscopy
FUM	Fumonisin
G	Guanine
G-(-)	Gram-negative
G-(+)	Gram positive
GABA	γ-aminobutyric acid
GC-content	Guanine-cytosine content
GF	Gel filtration
GHP	Good hygiene practices
GI	Glucose isomerase
GIT	Gastrointestinal tract
Glu	Glutamic acid
GMOs	Genetically modified organisms
GMP	Good manufacturing practices
GO	Glucose oxidase
GOS	Galactooligosaccharides
GRAS	Generally Recognized as Safe
H_2O	Water
H_2O_2	Hydrogen peroxide
HAMLET	Human alpha-lactalbumin made lethal to tumour cells
HePS	Heteropolysaccharides
HHP	High pressure homogenization
HIPEF	High intensity pulsed electric field
His	Histidine
HoPS	Homopolysaccharides

HPH	High-pressure homogenization
HSA	Human serum albumin
IA	Itaconic acid
IARC	International Agency for Research in Cancer
ICT	Immobilized cell technology
IEC	Ion-exchange chromatography
IgA	Immunoglobulin A
IgG	Immunoglobulin G
IgM	Immunoglobulin M
Ile	Isoleucine
ILSI	International Life Sciences Institute
IMP	Inosine monophosphate
ITC	Isothermal titration calorimetry
JECFA	Joint FAO/WHO Expert Committee on Food Additives
K_a	Binding constant
KI	Potassium iodide
K_q	Biomolecular quenching rate constant
K_{sv}	Stern–Volmer constant
LA	Lactic acid
LAB	Lactic acid bacteria
Lc	*Lactococcus*
LDL	Low-density lipoprotein
Leu	Leucine
LF	Lactoferrin
Lys	Lysine
MLF	Malolactic fermentation
MPT	Methylparathion
MT	Malathion
NMR	Nuclear magnetic resonance
NSLAB	Non-starter lactic acid bacteria
O_2	Oxygen
°D	Dornic degrees
OpdB	Organophosphorus hydrolase
OPPs	Organophosphorus pesticides
OTA	Ochratoxin A
P3G	Pelargonidin-3-O-glucoside
PAGE	Polyacrylamide gel electrophoresis
PAHs	Polycyclic aromatic hydrocarbons
PCR	Polymerase chain reaction
PE	Pectinesterase
PEA	Phenylethylamine
PFGE	Pulsed-field gel electrophoresis

PFU	Plaque-forming units
PG	Peptidoglycan
PG	Polygalacturonase
PHAs	Polyhydroxyalkonate
Phe	Phenylalanine
PHs	Protein hydrolysates
PLP	Pyridoxal 5'-phosphate
Pro	Proline
PSD	Particle size distribution
PSL	Prebiotic substance at various levels
PSs	Polysaccharides
PT	Parathion
QAC	Quaternary ammonium compounds
RDA	Recommended daily allowance
rDNA	Recombinant DNA
ROS	Reactive oxygen species
RTE	Ready-to-eat
SCFs	Supercritical fluids
SEM	Scanning electron microscopy
SSA	Specific surface area
ST	Surface tension
TAs	Teichoic acids
*Tre*S	Trehalose synthase
Trp	Tryptophan
Tyr	Tyrosine
UHT	Ultra-high temperature
UHT	Ultra-high temperature processing
US	United States
USDA	United States Department of Agriculture
USFDA	United State Food and Drug Administration
USFDA	United States Food and Drug Administration
UV	Ultraviolet
WHO	World Health Organization
WPC	Whey protein concentrate
α-LA	α-lactalbumin
β-LG	β-lactoglobulin
γ-PGA	Poly-γ-glutamic acid
ΔG	Gibbs free energy
ΔH	Enthalpy change
ΔS	Entropy change
σ_0	The lifetime of fluorophore in the absence of the quencher

PREFACE

Bioprocessing technology is the systematic use of technology of biological materials such as living microbial cells, organelles, or biomolecules (such as enzymes, vitamins, bacteriocins, antibiotics, and vaccines) to carry out a process for commercial, medical, or scientific reasons. Recent technologies have helped us to expand our knowledge and understanding of how microorganisms are related to food and human health. Microorganisms are small entities with immense potential applications and have a beneficial as well as detrimental role to play in our life.

The properties of each individual microbial species are often unique; results from experiments using even closely related bacteria often are different. A number of industrially important compounds, such as enzymes, bioactive peptides, exopolysaccharides, bacteriocins, and so forth, are obtained from food-grade microorganisms. Simultaneously, the complex microbiota in the gastrointestinal tract assists in food digestion, metabolism, and diseases resistance.

This volume, *Bioprocessing Technology: Emerging Scope and Prospective Applications in Food and Health*, is divided into three main parts. The functional foods market represents one of the most fascinating areas of investigation and innovation in the food sector, as suggested by the increasing number of scientific literature globally. In this regards, chapters compiled in Part I: Functional Food Production and Human Health discusses the newly emerged bioprocessing technological advances in the functional foods (chocolates and whey beverages) with their prospective health benefits. Food safety is one of the frontier areas of research in today's world. Thus, Part II: Emerging Applications of Microorganism in Safe Food Production of the book covers recent breakthroughs in microbial bioprocessing that can be employed in the food and health industry, such as, instance, for prospective application of microbial-derived exopolysaccharides and enzymes derived from genetically modified microorganisms. Another chapter comprehensively describes the role of different food microbial communities (including beneficial, spoilage-type, and harmful/pathogenic microorganisms). In Part III: Emerging Scope and Potential Application in the Dairy and Food Industry, enhancing the stability and functionality of whey proteins through microencapsulation and potential of food-grade microbes for biodegradation of toxic compounds such as mycotoxins, pesticides, and polycyclic hydrocarbons are the innovative concepts discussed.

We have bring together a group of outstanding international contributors in the forefront of bioprocessing technology, and together they have produced an outstanding reference book that is expected to be a valuable resource for researchers; teachers; students; food, nutrition, and health practitioners; and all those working in the dairy, food, and nutraceutical industry.

We extend our sincere thanks to all the authors who have contributed with dedication, persistence, and cooperation in completing their chapters in a timely manner to this book and whose cooperation has made our task as editors a pleasure. We hope that the reader will find this book informative and stimulating.

—**Deepak Kumar Verma**
Ami R. Patel
Prem Prakash Srivastav

PART I
Functional Food Production and Human Health

CHAPTER 1

IMPORTANCE OF CHOCOLATES IN FUNCTIONAL FOODS: FORMULATION, PRODUCTION PROCESS, AND POTENTIAL HEALTH BENEFITS

NEVZAT KONAR[1,*], OMER SAID TOKER[2], IBRAHIM PALABIYIK[3], DERYA GENC POLAT[4], SIRIN OBA[5], and OSMAN SAGDIC[2,6]

[1]*Food Engineering Department, Siirt University, 56100 Siirt, Turkey, Tel.: +90 5322714611*

[2]*Food Engineering Department, Yildiz Technical University, Istanbul, Turkey, Tel.: +90 5053124247, E-mail: stoker@yildiz.edu.tr*

[3]*Food Engineering Department, Namik Kemal University, 59030, Tekirdağ, Turkey, Tel.: +90 5542313361*

[4]*Tayas Food, Gebze, Kocaeli, Turkey, Tel.: +90 5366754219, E-mail: deryagenc@tayas.com.tr*

[5]*Department of Food Processing, Suluova Vocational School, Amasya University, Amasya, Turkey, Tel.: +90 5545826755*

[6]*E-mail: osagdic@yildiz.edu.tr*

Corresponding author. E-mail: nevzatkonar@hotmail.com

1.1 INTRODUCTION

Nutrition is essential for maintenance, growth, reproduction, and health of an organism. In recent years, the relationship between health/disease and nutrition has come to light, which induces changes in the expectations of consumers from food materials. Consumer trends have tended toward consumption of food products with health-beneficial effects. In this respect, functional food concept first aroused in Japan in 1994 (Pappa-lardo and Lusk, 2016). Functional foods are defined as the food products which are fortified by ingredients having positive physiological effects (Kubomara, 1998). After understanding the relationship between health and diet, consumer demand is currently increasing for healthy and natural food products. The increase in cardiovascular disease and obesity and other diet-related illnesses has led consumers to buy healthier, functional, and natural food products (Konar et al., 2016). Therefore, researchers and producers have made an effort to discover novel functional ingredients and to improve novel functional foods. For this, the most known food ingredients used for fortification purposes in food products are presented in Figure 1.1.

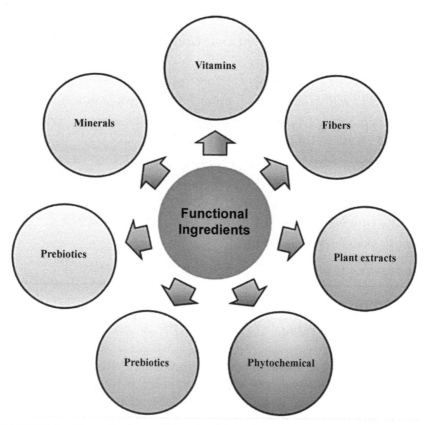

FIGURE 1.1   **(See color insert.)** Functional ingredients used in the food industry.

These ingredients were used for increasing functionality of many food products and in the market, there are many different commercial products such as mineral/vitamin-enriched products, prebiotic/probiotic/synbiotic food products are present. Functionality of food products is not the only driving force for consumer acceptability of the products. Sensory properties, especially taste and aromas, are also important crucial factors playing an important role in the preference of such products. Therefore, functional ingredient type and its level should be attentively selected to produce food products which are similar to conventional products in terms of desired quality characteristics. In order to maintain product quality, the enrichment process should be evaluated by two aspects. Regarding ingredients, the functional ingredient used in the formulation must not negatively affect the quality characteristics of the food products such as color, taste, odor, aroma, and other quality parameters including textural characteristics. Considering food products, aroma intensity, consumer prevalence, and group, other health-beneficial effects become prominent during the improvement of functional food products.

When considering these, chocolate is one of the probable products suitable for enrichment or delivering of functional ingredient since it has strong aroma and taste due to cocoa present in the formulation. Chocolate is loved by most of the people due to its attractive characteristics such as shiny gloss of the surface, snap during breaking, smooth texture, and melt-in-mouth characteristics (Jeyarani et al., 2015). Moreover, chocolate has many health-beneficial effects because cocoa is rich in polyphenols (Macht and Mueller, 2007). Due to unique aroma of the chocolate, it is consumed by people of all ages around the world.

The main aim of this chapter is to present possibilities related to the use of chocolates as a delivery agent of probiotic organisms and prebiotic substances. The biggest advantage of chocolate for consumption prevalence is its sensorial properties and consumer perception of the product. However, the attitudes and behaviors of consumers have changed significantly, especially since the last 25 years and functional and healthy foods have gained prominence in the market, with a large number of products being developed. Consumers of chocolate have functional product expectations at a significant level. Among the main bioactive compounds, probiotics and prebiotics may be used with this aim. Therefore, health benefits of chocolate and production of prebiotic, probiotic, and synbiotic chocolates are discussed in this chapter.

1.2 FORMULATION AND PRODUCTION PROCESS: AN OVERVIEW

Chocolate is the suspension of cocoa mass and sugar mixture in a cocoa butter matrix (Andarea-Nightingale et al., 2009). During chocolate manufacturing, various chemical and physical operations take place with the addition of additives to obtain chocolates with desired physical and organoleptic attributes (Jovanovic and Pajin, 2004). Depending on the contents of cocoa solids, milk fat, and cocoa butter in the formulation, chocolate categories are known as dark, milk, and white (Afoakwa, 2010). All chocolate categories are obtained by performing same process conditions and different parameters. Chocolate production principally consists of five stages as depicted in Figure 1.2 (Schumacher et al., 2009). Quality of the final product is greatly affected by the production conditions (Cidell and Alberts, 2006). Conching and refining steps cause specific sensory and textural properties by influencing suspension viscosity and particle size (Afoakwa et al., 2008a, 2008b).

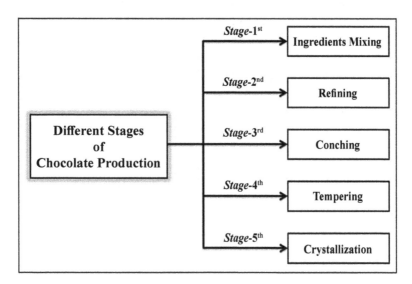

FIGURE 1.2   Different stages for production of chocolate (Schumacher et al., 2009).

Proper particle size should be obtained after the refining process (Bolenz et al., 2003). Apart from using the right ingredients, processing technology is also essential for obtaining desirable attributes. Smooth texture can be achieved by reducing particle size using a three- or five-roll refiner in the refining step of chocolate (Lucisano et al., 2006; Torres-Moreno et al., 2012). Organoleptic and rheological properties of milk chocolate are critically affected by terminal particle size (Afoakwa et al., 2007; Konar et al., 2014a; Torres-Moreno et al., 2012).

Conching step includes agitation of chocolate at elevated temperatures (>40°C) to obtain satisfactory texture of chocolate mixture (Konar et al., 2014a). Variation in conching temperature can influence consistency, final texture, and sensorial properties of the chocolate (Bolenz et al., 2003; Bolenz et al., 2005; Bordin et al., 2009; Torres-Moreno et al., 2012). Moreover, conching time (CT) affects rheological properties and the production costs (Sokmen and Gunes, 2006). Reduction in particle size during conching guarantees the fluidity of the mass (Schumacher et al., 2009). Processing steps (refining, conching, and tempering) and composition of the chocolate (amount and type of emulsifiers, amount of fat, and particle size distribution [PSD]) equally affect the rheological attributes of molten chocolate (Fernandes et al., 2012). During the manufacturing process, determination of flow properties of chocolate is essential to obtain high-quality products with well-defined texture (Servais et al., 2004).

Regarding the fat requirements for obtaining desirable flow properties, Afoakwa et al. (2008a) and Beckett (1999) concluded that specific surface area (SSA) of solids and the largest particle size are two key parameters. An increase in particle size leads to more spherical particles, causing to a broadening of the PSD with a final reduction in solid loading while the amount of fat increases. The SSA decreases with increasing particle size, as determined in the PSDs of single constituents, which has been reported in previous studies (Afoakwa et al., 2008a; Beckett, 1999; Sokmen and Gunes, 2006). PSD affects flow and textural attributes of tempered and molten dark chocolates, influencing the microstructure, spreadability, tempering and pre-crystallization behavior, hardness, and organoleptic qualities (Afoakwa et al., 2008b).

Tempering is an important step in chocolate production (Dhonsi and Stapley, 2006; Glicerina et al., 2013). This process is to carry out the pre-crystallization in order to have the most stable solid form of cocoa butter having a polymorphic structure in the final product chocolate (Afoakwa et al., 2008a). Cocoa butter can be crystallized in six different polymorphic forms, called forms I–VI or forms sub-α, α, β2', β1', β2, β, according to increasing melting point and stability (Dahlenborg et al., 2015). The tempering process is carried out with the controlled melting and chilling of the chocolate to ensure the formation of form V having the most suitable crystal structure from six different polymorphic forms (Debaste et al., 2008).

The tempering process has four important stages: complete melting (50°C), cooling to crystallization temperature (32°C), crystallization (27°C), and conversion of unstable crystals (29–31°C) (Afoakwa et al., 2008b). Conventional pre-crystallization is carried out by pumping fluid chocolates through different tempering zones (Bolliger et al., 1998). Well-tempered chocolate has a bright surface, stability in storage (especially blooming), whereas bad or improper tempered chocolate has gummy and sandy structure and fat blooming during storage (Debaste et al., 2008; Hartel, 2001).

1.3   POTENTIAL HEALTH BENEFITS OF CHOCOLATE

Consumption of chocolate and cocoa products was supported with the awareness about potential health benefits of cocoa components (Cooper et al., 2008). From very early times until now, chocolate has been used for health benefits such as reducing fever, minimizing diarrhea in childhood, and increasing breast milk (Wilson, 2010). During the 20th century, medical benefits of chocolate were worked on and obtained from conclusive data. However, some products consisting of derived components of cocoa bean gained a unique popularity. Chocolate is also routinely used in many laxative, tonic, tablets, and lozenges. Biomedical science has begun to experimentally recognize the healing and prevention of disease potential of chocolates. When the role of cocoa and dark chocolate in promoting health and happiness has been explored in scientific nutrition researches, especially in the last two decades, the use of cocoa or chocolate as a medicine has been increasing (Paoletti et al., 2012; Desch et al., 2010).

Chocolate products are widely consumed in the human diet depending on the cocoa content (Robbins et al., 2012). Health benefits of chocolate result from multiple elements in cocoa. In recent years, many people have included cocoa and chocolate into their well-balanced diet due to their unique flavor and aroma (Crozier and Hurst, 2014).

More evidence suggests consumption of chocolate and cocoa may improve insulin sensitivity (Grassi et al., 2005) aspects of cognitive function

(Crichton et al., 2016) through enhanced neuronal activity (Scholey and Owen, 2013), visual performance, and cerebral blood flow (Engler et al., 2004), the latter of which could in part be associated to cardiovascular properties previously mentioned (Heiss et al., 2010).

1.3.1 HEALTH-BENEFICIAL COMPOUNDS PRESENT IN CHOCOLATE

The recent researches have focused on detecting the components of chocolate and its antioxidant function which has positive effect on human health. Together with the latest developments on analytic technologies, it has been realized that there are more than 500 components in the cocoa's chemical composition. It has also been found out that several components have pharmacological or physiological importance. These components can be divided as hydrophilic and lipophilic components. In addition to this, lipophilic components exist in cocoa powder and cocoa butter. The lists of the hydrophilic components are getting attention in recent years (Vriesmanna et al., 2011; Crozier and Hurst, 2014).

1.3.1.1 LIPOPHILIC COMPONENTS

Cocoa butter is an expensive raw material that is used in chocolate, confectionery, pharmacy, and cosmetic industries. Cocoa butter usage in ointments, suppositories, and pomades come into prominence gradually more and more by the way of advertisements. At the same time, it has been used for skin massages, moisturizing the chapped lips, and treating retracted nipples. Cocoa butter meets one-third of the chocolate content with its valuable physical and organoleptic features (Beckett, 2008). Cocoa butter consists of 97% triglyceride and the rest 3% includes materials such as free fatty acids, mono and diglycerides, phospholipids, and trace components (Rogers et al., 2008). The taste of the cocoa butter is neutral. About 50–60% calorie content of a typical cocoa bean comes from butter. After the fermentation of cocoa and processes like roasting, it consists of aromas such as pyrazine, thiazoles, pyridines, and short-chained oil acids. Cocoa butter contains approximately 29–38% oleic acid and it is one of the most common oil acids. The ratio of stearic and palmitic acid is approximately 29–38 and 20–26%, respectively. Cocoa oil may also contain a trace of linoleic acid and arachidonic acid (Dimick and Manning, 1987).

1.3.1.2   POLYPHENOLS

They are sometimes called as "phenolic" according to the presence of at least one phenol structure, which is a hydroxyl group on an aromatic ring. Three main groups of the polyphenols are flavan-3-ol or catechins, anthocyanins, and proanthocyanids (Lee et al., 2003). Fresh cocoa bean's main polyphenol compound contains (−) epicatechin and then (−) catechin and these compounds dimmers and trimmers. The complex alteration products of catechin and tannin give brown and purple colors to cocoa bean and responsible for bitter taste. Catechin and epicatechin, the main components in chocolate, represent the two forms of flavanol monomer, which contains saturated bond at hydroxyl group and third carbon atom (Ackar et al., 2013). Main antioxidant activity in cocoa products results from procyanidins. Researchers showed that the genetic characterization affected the cocoa polyphenol content. In addition, the season in which the plant grows also affects the polyphenol content of country's cocoa bean. Processing of the cocoa bean affects its high polyphenol content (Hii et al., 2009).

1.3.1.3   METHYLXANTHINES

The purine alkaloids caffeine (1,3,7-trimethylxanthine) and theobromine (3,7-dimethylxanthine) that are naturally present in chocolate are chemical components that belong to methylxanthine group. Theophylline, which is a methylxanthine, (1,3-dimethylxanthine) is only found at trace amounts in chocolate. The primary methylxanthine in cocoa is theobromine (3, 7-dimethylsxanthine). Theobromine forms the 2% of cocoa powder or cocoa grains in terms of weight. Theobromine is the main alkaloid in cocoa and it has been claimed that it contributes to cocoa's typical bitter taste. The amount of cocoa liquor used in chocolate affects the content of caffeine and theobromine. The structures of these are close to each other and they have similar pharmacological features. In commercial chocolate product, various other factors including the type of cocoa bean, maturity, and fermentation conditions may affect the level of theobromine and caffeine (Kaspar, 2006). Methylxanthine concentrations may differ extensively among chocolate forms. Methylxanthines are quickly absorbed through gastrointestinal (GI) way and metabolized in the liver. After chocolate consumption, 99% of the caffeine is absorbed in the small intestine (Scholey and Owen, 2013). Natural products containing methylxanthines have started to be accepted as functional food. Pharmacologically active components of methylxanthines, namely theobromine and caffeine can be found in cocoa/tea/coffee (Franco et al., 2013).

1.3.1.4 ANANDAMIDE

Anandamide (N-arachidonoylethanolamine) is a brain lipid which is bonded cannabinoid receptors with high affinity and imitating the psychoactive effects of herbal rooted cannabinoid medicines. It increases the sensitivity and enthusiasm by activating the cannabinoid receptors in brain. It has been thought that chocolate and cocoa contain unsaturated N-acylethanolamines and this compound is chemically and pharmacologically related with anandamide. Scientists are studying the combinations of chemicals that are the reason of chocolates happiness effects and isolations of its chemical compounds (Bruinsma and Taren, 1999).

1.3.1.5 BIOGENIC AMINES

Biogenic amines, amino acids are deamination products and they are related with a series of physiological effect such as strong desire for chocolate consuming and migraine pain. Among these components, the ones present in chocolate and drawing the most attention are tyramine and phenylethylamine (PEA). When comparing to the other foods, chocolate has minimal levels of PEA. PEA is brain synapse structural neuromodulator. Various researchers claim that the PEA is an important modulator of mood. In the experiments, it has been indicated that the PEA is the pharmacologically active one and stimulant agent. The PEA, its metabolite, and phenylacetic acid decrease the release of liquids causing depression on biological tissues and the PEA and/or an amino acid suppress depression by replacing with a fore material. The PEA exists at important concentrations in chocolate. According to some experts, certain level of PEA taken with chocolate consumption can regulate the spiritual mood (Bruinsma and Taren, 1999).

1.3.1.6   MINERALS

Cocoa is a rich resource with regard to minerals. It has the potential of providing copper, iron, magnesium, potassium, and zinc minerals in human nutrition at significant amounts (USDA, 2017). Comparing with black tea, red wine, and apples, cocoa and cocoa products contain relatively much more magnesium. Cocoa's iron content is much more than red meat (beef) or chicken liver. However, bioavailability of iron within food is a very important factor. Cocoa's potassium and calcium content is also considerable (Hesse, 1993; Gray, 2005; USDA, 2017).

1.3.2   POSITIVE EFFECTS OF CHOCOLATE ON HEALTH

It has been thought that these compounds included in cocoa and chocolate products are responsible for various physiologic effects. The main nutritive contents of cocoa beans are fat, carbohydrates, and proteins. Additional to these compounds, cocoa beans contain various compounds such as various enzymes, water- and fat-soluble vitamins, dietary fibers, phospholipids, sterols, minerals (Cu, Fe, K, Mg, P), xanthenes (caffeine and theobromine), and polyphenol compounds (flavonoids and phenolic acids). It is claimed that regular chocolate and cocoa consumption is inversely proportional to several disease risks. In the last decade, the potential health effects of cocoa, chocolate, and cocoa-flavored drinks have started be researched with clinical tests.

The study methods of the epidemiological and clinical sciences about diseases have been previously determined in detail. In brief, this method is about a prospective population study of men and women inhabited in a selected area. Data are collected from the patient or non-patient participants in various ways. Some criteria (weight, length, body mass index, education status, social class, physical activity, smoking, alcohol consumption, and individual characteristics, etc.) of participants are important. Such prospective biomedical studies on human are also designed to answer specific questions about the health effects of chocolate or cocoa. Therefore, experimental subjects can consume chocolate to estimate the effect of chocolate components on diseases. In the literature, this kind of epidemiological and clinical studies determined the effects of cocoa/dark chocolate/cocoa beverage consumption on some diseases and function (Kwok et al., 2016). These are shown in Table 1.1.

TABLE 1.1 Study on Effect of Cocoa/Chocolate Consumption on Human Disease and Physiological Function of Body.

Consumption of cocoa/dark chocolate	Effects on human disease and body functions	References
Flavonol-rich cocoa consumption	• Endothelial function improved	Heiss et al. (2007)
Regular dark chocolate consumption	• Decreased total cholesterol	Grassi et al. (2008)
	• Improved β-cell function	
High-procyanidin chocolate	• Anti-inflammatory properties	Schwab et al. (1996)
Dark chocolate containing epicatechin	• Coronary artery diameter and endothelium-dependent coronary vasomotion increased	Flammer et al. (2007)
Procyanidins and epicatechin of dark chocolate	• Oxidative stress	Engler et al. (2004)
Cocoa beverage consumption and potassium, magnesium, and calcium in cocoa or stearic acid present in chocolate	• Platelet inhibition	Pearson et al. (2005)
	• Suppressing leukocyte activation	
Dark chocolate	• Decreasing platelet function	Hermann et al. (2006)
High-flavanol cocoa	• Increased high-density lipoprotein	Baba et al. (2007);
Long-term daily consumption of cocoa powder	• Decreased levels of plasma-oxidized low-density lipoprotein	Mellor et al. (2010)
	• Reduces inflammation	
Dark chocolate bar containing polyphenols	• Decreased blood pressure	Almoosawi et al. (2010); Grassi et al. (2010)
Insulin resistance		
High-flavanol cocoa	• Improves endothelial function	Grassi et al. (2009)
High-flavanol and caffeine cocoa	• Altering glucose metabolism.	Ceriello et al. (2004)
Chocolate	• Reduce the risk of diabetes	Greenberg (2015)

TABLE 1.1 *(Continued)*

Consumption of cocoa/dark chocolate	Effects on human disease and body functions	References
Immune function and carcinogenesis		
Diet consisting of 10% cocoa	• Enhances antioxidant defenses in the thymus	Ramiro-Puig et al. (2007)
Cocoa of procyanidins and catechins/dark chocolate	• Prevent inflammatory and improved DNA resistance	Spadafranca et al. (2010)
Central nervous system		
Flavonoid-enriched cocoa	• Cerebral blood flow	Francis et al. (2006)
	Skin	
Cocoa butter/cocoa flavanols	Moisturizers and protecting skin from damage from UV light	Williams et al. (2009)
Psychoactive effects		
Flavanols or methylxanthine compounds in cocoa/chocolate	• Cravings and cognitive function • Physiological capacity to modulate the central nervous system gastrointestinal (GI) tract inducing increased diuresis	Macht and Dettmer (2006); Osman et al. (2006); Scholey et al., (2010)
Cocoa's methylxanthine compounds, caffeine, and theobromine	• Mood-altering	Smit et al. (2004)

Various possible mechanisms were suggested that explain cocoa compounds' protective effects against many diseases (Katz et al., 2011; Vicente et al., 2016). Clinical studies showed that chocolate consumption decreased several cardiovascular risk factors (Kwok et al., 2016), showed antihypertensive effects (Ried et al., 2012) and reduced body weight and body fat. Moreover, cocoa consumption had beneficial effects on preventing cancer (Martin et al., 2013) and inflammatory diseases (Vicente et al., 2016). Blood pressure was observed to decrease with trial evidence and regular chocolate intake is associated with preventing many diseases (Vicente et al., 2016).

Cocoa, one of the primary ingredients in plain chocolate, is rich in flavonoids, epicatechin, flavanol, procyanidin, and so forth, which could enhance the antioxidant activity in plasma, endothelial, and tissue. Some studies suggest that habitual intake of cocoa/chocolate polyphenols prevents the development of cancers, cardiovascular diseases, diabetes, osteoporosis, and neurodegenerative diseases (Lee et al., 2003; Botelho et al., 2014).

1.4   IMPORTANCE OF CHOCOLATE IN FUNCTIONAL FOOD

Suitability of chocolate/chocolate products for delivering of functional ingredients is mentioned above. Main components of chocolate are cocoa mass and sugar where sugar suspends in a cocoa butter (Andarea-Nightingale et al., 2009). Chocolates are widely consumed throughout the world due to sensory pleasure and positive emotions (El-Kalyoubi et al., 2011), which is one of the important factors of why chocolate is a good alternative for delivering of functional ingredients. According to the *MarketWatch* Portal, as reported by Silvia Ascarelli on July 21, 2015, annual consumption of chocolate in Switzerland was 9 kg per capita as depicted in Figure 1.3 (Ascarelli, 2015). As seen from Figure 1.3, chocolate is among the widely consumed products' group. Moreover, conventional chocolate has also beneficial health effects since it is rich in cacao-based phenolic compounds, minerals, and proteins. Positive effects of chocolate consumption on the cardiovascular system are (Kay et al., 2006; Hooper et al., 2012) reduction in the production of stress hormone cortisol and catecholamines (Martin et al., 2010), skin, cholesterol concentrations, and the release of neurotransmitters anandamide and serotonin (Lamuela-Raventos et al., 2005; Katz et al., 2011; Sokolov et al., 2013). Regarding European and American diet, cocoa solids are considered as a significant source of polyphenols (Vinson et al., 2006) having a beneficial effect due to their antioxidant activity. Regarding health aspect, the main concern about chocolate consumption is its high calorie value due to sugar and fat composition, which can reduce by using sweeteners and bulking agents instead of sucrose to produce sugar-free chocolates.

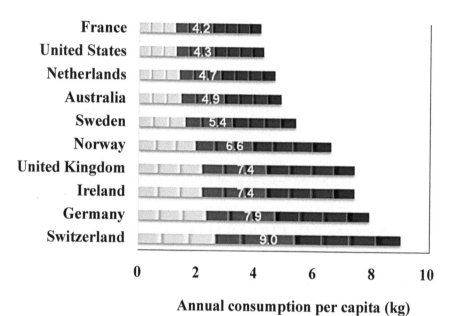

FIGURE 1.3   (See color insert.) Annual chocolate consumption of some countries per capita.

Source: Adapted from Ascarelli (2015).

Regarding confectionery industry, chocolate is one of the products having greatest growth rate, sales percentage of which increased 7% from 2006 to 2007 (about $2.2 billion) (Palazzo et al., 2011; Aidoo et al., 2013). It might have resulted from unique texture, flavor, and eating pleasure (Afoakwa, 2010). As a summary, chocolate or chocolate-based products draw attention in terms of delivering bioactive compounds due to especially three factors mentioned as the following:

- Being widely consumed products
- Intensity of flavor
- Health-beneficial effects of conventional chocolates

Therefore, in recent years, the number of studies about improving the functionality of chocolate or chocolate-based products has increased.

Table 1.2 indicated that studies related to improvement of functionality of chocolate or chocolate products can be collected under different titles as the following:

- Reduction of calorie: decreasing sugar or fat content of the products
- Reduction of saturated fatty acid content
- Enrichment with vitamins
- Enrichment with minerals
- Enrichment with polyphenol-rich fruits

All of the studies mentioned indicated that chocolate matrix is suitable for delivering the targeted bioactive compounds. In Table 1.2, studies about the production of prebiotic, probiotic, and synbiotic chocolates were not mentioned since detailed information about them will be given in the following sections. During the production of functional chocolate, quality characteristics of the products such as sensory properties, texture, rheology, melting profile should be considered to manufacture the product with desired quality. In addition, in order to achieve the desired contribution from bioactive compounds, production process of the chocolate should be known. Addition level of the bioactive compound should be determined considering stability characteristics of the bioactive compounds.

Therefore, in this chapter, we focused on prebiotic, probiotic, and synbiotic chocolate studies and brief information about the production process of the chocolate was mentioned. Knowledge about production process can enable the producers or researchers to achieve maximum benefit from corresponding bioactive material used in the formulation for enrichment purposes.

1.4.1 IN PREBIOTIC

Prebiotics are nondigestible food ingredients that beneficially affect the host by selectively stimulating the growth and/or activity of one or a limited number of bacteria in the colon, and thus improves host health, whereas dietary fiber is the edible part of plants or analogous carbohydrates that are resistant to digestion and absorption in the small intestine of the human with complete or partial fermentation in the large intestine

TABLE 1.2 Examples of Some Studies Conducted for Improving Functionality of Chocolate or Chocolate Products.

Chocolate type	Enrichment agent	Aim	References
Dark and milk chocolate	Dried fruits (prunes, papaya, apricots, raisins, cranberries)	Improvement of antioxidant properties of chocolate, contributing to dietary intake of polyphenolic antioxidants	Komes et al. (2013)
Dark chocolate	Inulin/polydextrose Stevia and thaumatin	Sugar-free chocolate	Aidoo et al. (2015)
Dark chocolate	Sugar alcohols (xylitol, lactitol, maltitol), dietary fibers (inulin, oligofructose), syrups (agave syrup, rice syrup), natural sweeteners (stevioside)	Reduced sugar chocolates	Belšćak-Cvitanović et al. (2015)
Dark	Phytosterol	Reduction of cholesterol	Botelho et al. (2014)
Chocolate spread	Omega-3 fatty acids	Enrichment of omega-3 fatty acids	Jeyarani et al. (2015)
Sucrose-free milk chocolate	Inulin, β-glucan fibers	Reduction of fat	Rezende et al. (2015)
Sucrose-free milk chocolate	Calcium	Fortification of Calcium	Konar et al. (2015)
Chocolate paste	Oleogel prepared by shellac wax/rapeseed oil	Reduction of saturated fatty acid content	Patel et al. (2014)
Dark chocolate	Encapsulated mulberry waste extract	Increasing anthocyanin content	Gültekin-Özgüven et al. (2016)

(Saarela et al., 2006). Recently, dietary fibers have become popular as food additives. These fibers are used because they have low digestive energy content, a high capacity to absorb water, and are said to support the digestive system (Bolenz et al., 2006). Functional food products containing health-promoting cereal fractions and probiotics are already on the market (Saarela et al., 2006) and one of these products is chocolate and chocolate-derived foodstuffs. For instance, Bolenz et al. (2006) studied the use of dietary fibers as bulking agent in chocolate. The ingredients studied, such as wheat fiber, are relatively cheap because these materials originate from sidestream materials of the fruit and vegetable processing industry. Unfortunately, the addition of wheat fiber up to 20% of chocolate had an adverse effect on the flow properties of the molten chocolate.

All carbohydrate prebiotics are fibers but since prebiotics should be utilized by beneficial microbes only, not all fibers are necessarily prebiotic (Saarela et al., 2006). Prebiotics have attracted the interest of researches and the food industry due to their nutritional and economic benefits (Scheid et al., 2013).

Prebiotics can be applied to a variety of foods (Morris and Morris, 2012; Volpini-Rapina et al., 2012). The nutritional value and the possibility of improving some sensory properties of food formulations, enhancing taste to the products, make the use of prebiotic ingredients advantageous (De Morais et al., 2015). Prebiotics could be used in chocolate production depending on their functional properties as sugar replacers, fiber, and prebiotics (Wang, 2009). The quantity of sucrose in chocolate that can be replaced with sidestream materials without obtaining unfavorable high viscosities and resulting change of sensory perception might be limited as a result of the differences in behavior of the sugar and fibers in a hydrophobic matrix, like chocolate. The exact fiber choice will further be dependent on nontechnical issues like price and availability (Bonarius et al., 2014).

In addition to water-soluble prebiotic fibers such as lactose derivatives, fructooligosaccharide (FOS), and galactooligosaccharides (GOS), inulin and polydextrose, largely water-insoluble fibers could have the potential as probiotic protectants (Charalampopoulos et al., 2002). Among the most widely studied and commercially used prebiotics are inulin, polydextrose, FOS, and GOS (Davis et al., 2010).

The biopolymer inulin is a favorable diabetic food ingredient, and an agent to promote the growth of intestinal bacteria (Aidoo et al., 2015). It responds to a variety of consumer demands: it is fiber-enriched, prebiotic, and low sugar. As a dietary fiber, inulin passes through the digestive tract largely undigested (Roberfroid et al., 1998). The extensive use of inulin in the food industry is based on its nutritional and technological properties (Aidoo et al., 2014; Rezende et al., 2015). Inulin products containing mainly short-chain molecules enhance flavor and sweetness and are used to partially replace sucrose (Aidoo et al., 2015). Inulin presents 10% of the sweetness power of sucrose (Franck, 2002), allowing it to partially replace sucrose in certain formulations (De Castro et al., 2009; Wang, 2009; Villegas et al., 2010).

Inulin is the most used and studied compound and it can be a part of chocolate matrix for these purposes. By various studies, inulin affects the textural, rheological, sensorial, melting, microstructure, and other physical properties of different chocolate types, as milk, dark, and white were studied (Konar et al., 2016). The usage of inulin as a bulk sweetener or/and prebiotic substance in chocolate was investigated and favorable outcomes were obtained in terms of similarity of quality of functional chocolate with conventional one (Gomes et al., 2007; Shah et al., 2010; Aidoo et al., 2014; Konar et al., 2014a, 2014b; Rezende et al., 2015). Therefore, it is recognized as an efficient fat replacer for use in chocolate. Moreover, Gomes et al. (2007) and Shah et al. (2010) observed that the addition of inulin results in good acceptance of sucrose-free chocolates (Rezende et al., 2015).

The hardness was used as textural quality parameter of chocolate. Inulin absorbs moisture and this causes hardness of chocolates. Farzanmehr and Abbasi (2009) studied milk chocolates containing inulin. These researchers obtained different hardness values (10.1–15.0 N), depending on the composition of the samples.

Shourideh et al. (2012) reported an increase in moisture content with increase in inulin concentrations in their dark chocolate formulations containing different mixtures of D-tagatose and inulin. The authors attributed this to hydrophilic groups present in inulin which causes increase and preservation of moisture in samples with a high content of inulin. The low concentrations of inulin (<25%) in sugar-free chocolate formulations could have thus resulted in the chocolates falling within acceptable moisture limits.

Inulin has also potential effects on flow properties of chocolate and chocolate products. A decrease in the yield value for an inulin-containing chocolate was observed by Bolenz et al. (2006). Shourideh et al. (2012) indicated that when a high level of inulin was used, it produced an incremental effect in the plastic and apparent viscosity. The increase in the inulin content led to a significant increase in plastic viscosity (Rezende et al., 2015). A low level of inulin improves the flow properties but reduces the apparent viscosity. Moreover, in another study, Aidoo et al. (2015) investigated the rheological properties, melting behaviors, and other physical quality characteristics of sugar-free chocolates processed from inulin (12.0%, w/w) and polydextrose (36%, w/w) mixtures as bulking agents sweetened with stevia and thaumatin extracts. Sucrose replacement with the inulin/polydextrose and stevia/thaumatin extracts resulted in significantly higher Casson viscosity according to the findings of this study (Aidoo et al., 2015).

In the study of Konar et al. (2015), chocolates were investigated that had been prepared according to the composition specified as a result of the previous work (9.00% w/w inulin and 34.0% w/w maltitol). Certain physical (PSD, brightness, chroma, water activity, and hardness) and rheological features of the samples resulting from the addition of calcium carbonate (CaCO) in different quantities (300, 450, 600, 750, and 900 mg CaCO to 100 mg milk chocolate) were studied. Both the Herschel–Bulkley and Casson models were used to investigate the rheological findings. It was determined by comparing certain rheological (rate index, Casson yield stress, and Casson viscosity) and physical (chroma and hardness) parameters that samples containing 409.5 mg calcium (nearly 41.0% of the RDA of calcium) per 100 g chocolate did not show significant differences from samples from the control group. Furthermore, these calcium-containing samples were shown to exhibit positive differences in other physical properties (brightness and water activity) that could be noteworthy and significant with respect to visual quality and shelf life (Konar et al., 2015).

Process conditions may need to be optimized for inulin containing chocolate, especially refining and conching conditions. The effects of substituting fine sugar with isomalt and maltitol in milk chocolate samples that contain inulin (9.0% w/w) and the use of varying conching temperatures (50, 55, and 60°C) in the sample preparation process on their physical (color, hardness, water activity) and rheological properties were examined. Rheological data were obtained using the Herschel–Bulkley model which showed the best fitting for predicting rheology by Konar (2013). It was determined that all properties included within the scope of the study are affected by the use of different bulk sweeteners or varying conching temperatures ($P<0.01$). While color properties, such as brightness (L*), hue angle (h), water activity (a), and rate index properties varied in a narrow range, it was determined that the yield stress and viscosity properties, which are among the important quality parameters of chocolate and can determine effects on sensory properties, manifest variations within a broad range, depending on the conching temperatures and the bulk sweeteners used. It was concluded that maltitol is a more suitable fine sugar substitute in milk chocolates containing inulin (Konar, 2013).

The effects of inulin as a prebiotic substance at various levels (PSL) (0, 60, 90, and 120 g/kg) in milk chocolate, as well as the use of varying CT (3.50, 4.0, and 4.50 h) and refining conditions (20.0, 25.0, and 28.0 μm) as the mean particle size, D [4,3] in the sample preparation process were examined with respect to changes in the physical (color, hardness, water activity) and rheological properties of the samples by Konar et al. (2014a). The SSA and largest particle size (D90) values increased with increasing values of PSL; however, intended D [4,3] values were reached for all PSL values. CT, PSL, and PSD had no significant effect on the brightness and chroma in the inulin-containing milk chocolate; however, it has been found that hardness, water activity, yield stress, and viscosity produced significant changes (Konar et al., 2014a).

One of the factors affecting various features of inulin and quality of the food products was degree of polymerization (DP) of inulin. Particularly, commercial inulin is supplied as low (DP>23) and high DP (DP<10). However, there is a need to investigate DP versus chocolate quality and DP versus bioavailability of inulin interactions (Konar et al., 2016).

Polydextrose is a nondigestible, odorless, white-to-cream amorphous powder with virtually no sweetness, low molecular weight, randomly bonded polysaccharides of glucose and a calorie content of 1 kcal/g. Polydextrose comprises mainly glucose in its highly branched polymer, with small quantities of randomly distributed sorbitol and citric acid. All possible glycosidic bonds with anomeric carbons of glucose are present, α and β 1–2, 1–3, 1–4 and 1–6, among which the 1–6 bond predominates (Konar et al., 2016). It is a functional food additive due to its prebiotic properties (Kolida et al., 2002; Srisuvor et al., 2013). It is well tolerated, and a mean laxative threshold of 90 g/day (1.3 g/kg bw) or 50 g as a single dose has been given (JFECFA, 1985). This compound has an average DP of 12 and an average molecular weight of 200 g/mol (Konar et al., 2014b).

In the study of Konar et al. (2014b), in an effort to obtain a prebiotic milk chocolate, the effects of various levels of polydextrose (0, 60, 90, and 120 g/kg) and the use of varying CT (3.50, 4.0, and 4.50 h) and refining conditions (20.0, 25.0, and 28.0 μm mean particle size, D [4,3]) in the sample preparation process were examined and analyzed with respect to the physical and rheological properties of milk chocolate samples. The use and amount of polydextrose and CT and particle size had statistically significant effects with different magnitudes and directions on the physical and rheological properties of the samples. The samples containing polydextrose significantly differed from the control samples, especially with respect to their rheological properties ($p < 0.01$) (Konar et al., 2014b).

Shah et al. (2010) studied the development of a sucrose-free chocolate sweetened with *Stevia rebaudiana* extract and containing polydextrose as a bulking agent. Gomes et al. (2007) obtained a diet chocolate using various bulking agents as sucrose substitutes. One of the bulking agents in the study was polydextrose (24.14–48.27%). The formulations containing polydextrose, polydextrose, and lactitol and polydextrose and maltitol were evaluated for a sensory analysis due to their good technological performance and adequate machinability of the chocolate mass at different stages of the process. Improved GI function has been demonstrated with a daily intake of 4–12 g polydextrose without adverse effects (Beristain et al., 2006).

Mixtures of prebiotic compounds, especially inulin/polydextrose, were studied in chocolate matrix. Their mixtures were used especially as bulking agents again. Inulin and polydextrose mixtures could be used for sugar-free chocolate manufacture with satisfactory physicochemical properties when sweetened with high-intensity sweeteners (Aidoo et al., 2015). Regardless of the levels of the sugar substitutes used, replacing sucrose with inulin and polydextrose results in darker chocolates (Aidoo et al., 2014). Shourideh et al. (2012) reported darker chocolates for dark chocolate formulations containing 100% inulin. When inulin absorbs moisture, light scattering and lightness decrease, making the chocolate look darker (Shourideh et al., 2012). Bolenz et al. (2006) also reported chocolate samples with 20% inulin has been the most brownish, and had the lowest L* (lightness) value among other texturizing agents in milk chocolate.

Earlier studies reported that chocolate formulations which contain 100% polydextrose show large crystals with dense smaller particles in between the larger crystals and minimal interparticle spaces in comparison to formulations containing 100% inulin which revealed large crystals with more void spaces between the crystals, indicating limited particle–particle interaction strength (Aidoo et al., 2014). A combination of these ingredients will result in chocolates having large crystals with the dense smaller particles of polydextrose filling in the void spaces in the crystal network structure of chocolate formulations with inulin. The end result is chocolate with high solids packing intensity accounting for the low onset values and high peak width at half height for the sugar-free chocolates with 75:25% polydextrose:inulin ratios (Aidoo et al., 2015).

The prebiotic activity of polydextrose was also studied by in vivo methods. Beards et al. (2010) tested the effects of maltitol, polydextrose, and resistant starch addition to chocolate. Forty volunteers consumed reformulated chocolate samples for over a 6-week period. Their results showed that the consumption of samples containing polydextrose–maltitol blend increased the level of *Lactobacilli* and *Bifidobacteria* levels in feces after 6 weeks. An increase in the levels of short-chain fatty acids, that is, propionate and butyrate were observed. Formula development of chocolate provided prebiotic effects to consumers in addition to the decrease in energy values.

Another prebiotic compound used in chocolate is GOS. A study by Winardi-Liem (2011) performed to investigate rheology of prebiotic chocolate bar by using 3.75, 15.4, and 30.4% (m/m) GOS. Overall, the result of this study has shown a positive result of incorporation of GOS in the chocolate system. Although increases in yield stress and apparent viscosity were observed upon GOS addition, it was noted that these increases did not contribute to a noticeable quality/textural change. Therefore, a functional chocolate containing GOS can be produced both by looking at the processing and the quality of the chocolate (Winardi-Liem, 2011). Moreover, the results of the study performed by Suter (2010) demonstrated that a chocolate system can serve as an excellent delivery system for GOS. The GOS maintained concentration and initial profile throughout processing demonstrating stability of these compounds during chocolate manufacturing. By another study, Davis et al. (2010) determined the effect of different doses of GOS on the fecal microbiota of healthy adults, with a focus on *Bifidobacteria* by using chocolate chew including chocolate liquor (7.44%, w/w). This study showed that a high-purity GOS, administered in a confection product at doses of 5 g or higher, was bifidogenic, while a dose of 2.5 g showed no significant effect. However, the results also showed that even when GOS was administered for many weeks and at high doses, there were still some individuals for which a bifidogenic response did not occur.

1.4.2 IN PROBIOTIC

Consumers' sensory and hedonic perceptions could be greatly influenced by the messages highlighted on the front of the packaging, particularly nutrition and health claims for reduced calorie or functional foods (Miraballes et al., 2014). Chocolate lovers want functional chocolate that offers clinically proven physical or emotional health benefits. In the United States, 41% of American consumers are longing for chocolate that help them relax and 38% for "feel-good" chocolate that naturally induces a better mood (Callebaut, 2008). Within the last several years, functional chocolate has gained popularity among American consumers. About 35% of the Americans desire chocolate that strengthen the immune system (Saka, 2011) and this indicates using probiotics for developing functional chocolate and chocolate products. Probiotic has been defined as live microorganisms providing health benefit to the host when existing in adequate level (Konar et al., 2016). They are feed and food supplements. Strain identity is important in order to link a strain to a specific health effect and to enable accurate surveillance and epidemiological studies (Pineiro and Stanton, 2007; Burgain et al., 2011). The term "probiotic" includes a large range of microorganisms, mainly bacteria and also yeasts. As they can stay alive until the intestine and provide beneficial effects on the host health, lactic acid bacteria (LAB), non-LAB and yeasts can be considered as probiotics. LAB is the most important probiotic known to have beneficial effects on the human GI tract. These bacteria are Gram-positive and usually live in a non-aerobic environment but they also can support aerobic conditions (Holzapfel et al., 2001; Anal and Singh, 2007; Burgain et al., 2011; Patel et al., 2017). The species with beneficial properties belong, generally, to *Bifidobacterium* spp. and *Lactobacillus* spp. (Isolauri, 2004).

The consumption of foods such as probiotics that promote wellness, health, and a reduced risk of diseases has grown worldwide. During the past decade, more than 500 new products were introduced to the market (Konar et al., 2016). However, advantages of probiotics for health can only be realized if proper probiotic strain or product selection, and dose guidelines of commercial production, are applied in human food or dietary supplement (Kalliomaki et al., 2010). An international expert group of the International Life Sciences Institute (ILSI) has evaluated the categorized and published evidence of functionality of different probiotics in four areas of human application, namely, metabolism, chronic intestinal inflammatory and functional disorders, infections, and allergy. The ILSI report gives concrete examples demonstrating benefits and gaps, and guidelines and recommendations on the design of next generation of probiotic studies, with the aim to substantiate the current body of information on probiotic benefits (Rijkers et al., 2010).

Different food materials are utilized as a carrier of probiotic microorganisms. They have been incorporated not only in a wide range of foods, including dairy products (such as yoghurt, cheese, ice cream, dairy desserts) but also in nondairy dairy products (such as chocolate, cereals, juices) (Anal and Singh, 2007; Burgain et al., 2011). Yoghurt products accounted for the largest share of sales, representing 36.6%, and scientific development of such products had shown the high sensory acceptance (Coman et al., 2012; Niamah and Verma, 2017).

During improvement of food products functionality, four issues should be taken into consideration: vitality level of probiotics should not decrease during production and storage, novel functional food should be produced by processes which are employed for conventional products as far as possible, the quality of the end product should be similar with the conventional one in terms of sensorial and physical characteristics, and efficiency or suitability of bioavailability level. Therefore, selection of appropriate food matrix depending on probiotic type has become essential (Lalicic-Petronijevic et al., 2015).

For probiotic product development studies, the activity of the initial inoculum, the storage time and the interaction of other ingredients present in the formulations, in particular, should be considered. In general, the food industry has applied the recommended level of 106 cfu/g at the time of consumption for *Lactobacillus acidophilus*, *Bifidobacterium* spp. and other probiotic bacteria (Boylston et al., 2004; Aragon-Alegro et al., 2007). Maragkoudakis et al. (2006) considered probiotic load between 10^5 and 108 cfu/g acceptable for dairy products. Probiotic viability in the food matrix depends on factors, such as pH, storage temperature, oxygen levels, and the presence of competing microorganisms and inhibitors. It is important that the formulation maintains the activity and viability of the probiotic for extended periods of time (Shah, 2007). In addition to the survival of probiotics at the right dose, it needs to be established that a different food matrix does not influence the functionality of the probiotic bacteria (Coman et al., 2012). Vitality level of various probiotics, as *Lactobacillus helveticus, L. rhamnosus, L. paracasei, L. acidophilus, L. brevis, Bifidobacterium longum, B. lactis,* and *Bacillus indicus,* was investigated in the different types of chocolate matrix as white, dark, and milk (Saarela et al., 2006; Aragon-Alegro et al., 2007; Maillard and Landyut, 2008; Coman et al., 2012; Erdem et al., 2014; Champagne et al., 2015; Lalicic-Petronijevic et al., 2015; Konar et al., 2016; Zaric et al., 2016).

Probiotics may exert positive effects on the composition of gut microbiota and overall health. However, in order to be beneficial, the bacterial cultures have to remain alive and active at the time of consumption (Coman et al., 2012). In addition to exhibiting antioxidant activity, chocolate might serve as a better probiotic carrier than dairy products for intestinal delivery (Erdem et al., 2014). Possemiers et al. (2010) claimed that chocolate ensured probiotic survival up to four times higher than milk-containing products. Hence, an important issue is to extend the viability and bioavailability of probiotics in the chocolate matrix. Encapsulation has the potential for this aim. The incorporation of probiotic cells encapsulated by spray-coating technology has been carried out in chocolate (Maillard and Landuyt, 2008). According to these authors, probiotic viability in the small intestine was three times higher when incorporated in chocolate than in dairy product. In this case, probiotic chocolate process was transposed to a larger scale but the challenge here was to obtain a process which is compatible with probiotic survival because high temperatures are required in the usual process (Burgain et al., 2011).

Possemiers et al. (2010) also incorporated encapsulated probiotic cells in chocolate. Results have shown that the introduction of encapsulated probiotic strains into chocolate can be an excellent solution to protect them from environmental stress conditions. In chocolate, the lipid fraction of cocoa butter was shown to be protective for *Bifidobacteria* (Lahtinen et al., 2007, Burgain et al., 2011). Additionally, the viability of probiotics (*B. longum* R0175) was further improved when the microencapsulated cells by spray dryer were incorporated into chocolate particles in the study of Champagne et al. (2015). In addition to the spray-coating technology, it has the potential to be used for this purpose in freeze-drying.

In the last 10 years, some commercial probiotic chocolates were presented to the market. In 2007, Barry Callebaut developed a process to produce chocolate containing encapsulated probiotic cells with the Probio-cap™ technology in partnership with Lal'food. According to Callebaut, the addition of encapsulated probiotic cells has no influence on chocolate taste, texture, and mouthfeel. A consumption of 13.5 g/day of probiotic chocolate seems to be sufficient to ensure the balance of the intestinal microflora (Burgain et al., 2011).

Chocolate has also been used by DSM Food Specialties which produces a bar called Attune launched in the United States in January 2007. The product also contains the prebiotic inulin which supports a healthy digestive function. Attune's innovative product line is found in the refrigerated yoghurt section and advertising of this product highlights more input in calcium, fiber, and less sugar than in most yogurts. After 2 years of collaboration, Balchem Encapsulates and Institut Rosell have developed a stabilized form of encapsulated probiotics. Institut Rosell has incorporated encapsulated probiotic cells into yoghurt-covered raisins, nutrient bars, chocolate bars, and tablets (Burgain et al., 2011; Konar et al., 2016).

1.4.3 IN SYNBIOTIC

The combination of the live bacteria (probiotics) and the food components that stimulate their growth (prebiotics) are defined as synbiotics. Without using prebiotics "growth factor," probiotics does not survive well in the digestive system due to its intolerance for oxygen, low pH, and temperature as well as competitive relationship with other bacteria. A higher advantage to the host was revealed when probiotics and prebiotics had been used together due to their synergistic action (Vitali et al., 2010). Therefore, it is better to produce a "synbiotic" chocolate product to promote the probiotics benefits due to synergies between them. However, compatibility of prebiotics and probiotics and their effects on quality parameters and processability of chocolate should be carefully considered to be able to make high-quality synbiotic chocolate.

Potential usable prebiotics are lactose derivatives, GOS, FOS, inulin, and polydextrose (Charalampopoulos et al., 2002). Concerning the effect of prebiotic selection on the quality of synbiotic chocolates,

Erdem et al. (2014) studied several dietary fibers including maltodextrin, carboxymethylcellulose, inulin, lemon fiber, apple fiber, wheat fiber, each used at 0.05 w/w as the prebiotic materials in the manufacturing of chocolate. After sensory analysis, only maltodextrin and lemon fiber including chocolates got the highest scores on basic sensory attributes of chocolate (appearance, texture, taste, mouthfeel, aroma) from 10 voluntary panelists. Therefore, synbiotic chocolates were manufactured including lemon fiber or maltodextrin and probiotic strains *B. indicus* HU36. It was observed that viable cell counts were not affected by fiber type and all samples contained more than 5 log cfu/g bacteria survival. Regarding the sensorial effects of fibers used together at concentrations of 0.015, 0.035, and 0.055 w/w for each one, remarkably lemon fiber and maltodextrin addition did not result in bloom or off-flavor and caused very high levels of smoothness similar to control sample. Only sweetness, firmness, and adherence features of samples negatively changed among all features compared to control sample. Lemon fiber was found to affect more firmness and maltodextrin was more responsible for sweetness properties of the samples. Concerning the adherence score which is always aimed to be kept at minimum, the concentration of fibers increased adherence of the samples equally and even minimum amount of fiber concentrations caused an increase in adherence.

Patel et al. (2008) investigated the development of a synbiotic chocolate mousse, a kind of aerated dessert, with the addition of *L. paracasei* ssp. *paracasei* and inulin. Three kinds of samples prepared as control, probiotic, and synbiotic sample. After 21 days of storage, viable cell counts were significantly higher for synbiotic samples, indicating the beneficial synbiotic effect which was appeared after 21 days. Regarding the sensorial evaluation of the samples, there are no significant differences observed for fresh samples. After storage, sensory scores decreased for both and probiotic samples; however, interestingly, synbiotic product showed higher acceptability compared to control and probiotic samples as days of storage increased. Acceptability level of inulin in the synbiotic chocolate mousse was also studied and Patel et al. (2008) observed that inulin addition decreased the mean sensory scores. However, there is no significant difference in sensory quality observed among inulin-added synbiotic samples at the concentrations between 9 and 15%.

A similar study was conducted by Aragon-Alegro et al. (2007) by using *L. paracasei* ssp. *paracasei* LBC 82 and inulin in the production of synbiotic chocolate mousse. The difference was the inulin content which was 5% in this study. Control, probiotic, and synbiotic samples were investigated in terms of probiotic strain count during storage at 5°C for up to 28 days in addition to sensorial analysis. Initially, *L. paracasei* ssp. *paracasei* was added approximately 7 log cfu/g in the final product. In both synbiotic and probiotic samples, *L. paracasei* always maintained above 7 log cfu/g, which was higher than the recommended level (106 cfu/g) of probiotic microorganisms in food at the time of consumption to have beneficial effects on the consumer's health. Moreover, there is no significant difference observed between sensorial preferences of the samples prepared by the addition of *L. paracasei* ssp. *paracasei* LBC 82 and inulin. This indicated that the probiotic microorganism and the prebiotic ingredient did not interfere in the sensorial preference of the chocolate mousse.

Mandal et al. (2005) encapsulated *Lactobacilli* and prepared probiotic milk chocolate to improve the stability of probiotics during processing and GI transit. They selected *L. casei* NCDC 298 among the five *Lactobacilli* tested regarding their cell–surface hydrophobicity, ability to utilize inulin, and antimicrobial activity against indicator microorganisms. Intestinal microenvironment of mice was evaluated when fed by milk chocolate containing encapsulated *Lactobacilli*. The survival of the *L. casei* NCDC 298 was found to improve at low pH, high bile salt concentration, and during heat treatments when cells were encapsulated in alginate or κ-carrageenan. Maximum survival of cells was noticed in 4% alginate beads, which was strengthened by incorporating resistant maize-starch (2%) and coating with either stearic acid or beeswax. Although probiotic strains count decreased in synbiotic milk chocolate below the acceptable level during 30 days of storage under ambient conditions, viability of encapsulated *Lactobacilli* was unchanged in refrigerated milk chocolate up to 60 days. Moreover, chocolate with encapsulated *Lactobacilli* was preferred by panelists in terms of sensorial quality. Milk chocolate with encapsulated cells increased the fecal *Lactobacilli*, decreased coliforms and carcinogenic β-glucuronidase enzyme in mice when fed.

Selection of prebiotic and probiotic combination is also essential. Since some of the bacteria have very specific nutrient needs, it is possible to increase the numbers of target bacteria. The currently available synbiotic supplements include combinations of *Bifidobacteria* and FOS;

L. rhamnosus GG and inulins; and *Bifidobacteria* and *Lactobacilli* and FOS or inulins (Verma et al., 2012).

1.5   FUTURE REMARKS

The world market and variability of functional dairy products has been rapidly growing. In the current trend, non-fermented and heat-treated food product categories are emerging probiotic foods apart from the traditional fermented dairy products. Therefore, the addition of probiotics into chocolate which is liked by all age-groups people could offer a good alternative to common dairy products. Moreover, although chocolate is rich in natural antioxidants, polyphenols and minerals, especially potassium, magnesium, copper, and iron, nutritional quality of chocolate can be further enhanced by the incorporation of probiotics and/or prebiotics or dietary fibers. However, the problems appeared with the addition of probiotics and/or prebiotics into chocolate products are their survival and stability during the processing and storage, preservation, and GI transit. The selection of suitable strains for food application is a general method to improve probiotics survivability. However, classical add-and-look approach is very difficult to detect a probiotic strain having all required properties. Since stress responses are strain-specific, the stress adaptation approach is another difficult method to conduct (Gadhiya et al., 2015). Nevertheless, future trend points out the microencapsulation technique which is the most suitable and accessible technology to be used in the manufacturing of probiotic or synbiotic chocolate.

1.6   CONCLUSION

The relationship between diet and health has motived consumers to consume food products with health benefits which are known as functional foods which are produced by incorporating of functional ingredients (vitamins, minerals, dietary fibers, plant extracts, phytochemicals, prebiotics, and probiotics) to conventional food products. During the production of functional food materials, ingredient be used for enrichment purposes should not adversely affect the quality characteristics of the corresponding product, considered to achieve desired consumer acceptability of the products. Consumption level and prevalence of the product are also determinant factors. When considering these, chocolate is a suitable product for delivering bioactive compounds since chocolate and chocolate-derived products are widely consumed around the world. In addition, intensity of the product due to cocoa used in the formulation and many health-beneficial effects of conventional chocolates because of polyphenols present in cocoa can be advantageous in this respect. The studies related with functional chocolate indicated that chocolate has an appropriate matrix for delivering functional ingredients. Desired levels of prebiotics and probiotics can be transported to target parts of body. Selection of the most suitable prebiotic/probiotic combination is essential to improve efficiency. Encapsulated probiotics could be preferred in order to preserve them during production processes where different stresses (heat, pressure, mechanical action) are applied. Addition period of the prebiotic/ probiotic/bioactive ingredient should be cautiously specified considering stability of bioactive ingredients and conditions of production processes. The present studies should be supported by further ones which will be about bioaccessibility characteristics of the corresponding functional ingredient in chocolate matrix.

1.7 SUMMARY

Chocolate is a foodstuff which has been consumed by wide social layers and age-groups worldwide. The biggest advantage of this foodstuff for consumption prevalence is its sensorial properties and consumer perception on the product. However, the attitudes and behaviors of consumers have changed significantly especially for the last 25 years and functional and healthy foods have gained prominence in the market, with a large number of products being developed. When this trend is taken into consideration, the existence of saturated fat and sugar in the ingredients has negative effects on the chocolate and the consumers of chocolate have functional product expectations at a significant level. Among the main bioactive compounds, probiotics and prebiotics may be used with this aim. In this chapter, possibilities related with usage of chocolates as a delivery agent of probiotic organisms and prebiotic substances were discussed.

KEYWORDS

- *Bifidobacterium*
- chocolate
- functional foods
- inulin
- prebiotic
- probiotic
- synbiotic
- bioactive
- conching
- dietary fiber
- flavonoids
- fortification
- fructooligosaccharide
- galactooligosaccharides
- inulin
- *Lactobacillus*
- polydextrose
- polyphenols
- terminal particle size

REFERENCES

Ackar, D.; Lendi, T. K. V.; Valek, M.; Šubari, D.; Mililevi, B.; Babi, J.; Nedi, I. Cocoa Polyphenols: Can we Consider Cocoa and Chocolate as Potential Functional Food? *Hindawi J. Chem.* **2013,** *2013,* Article ID 289392, 7.

Afoakwa, E. O. *Chocolate Science and Technology;* Wiley-Blackwell: Oxford, UK, 2010; p 275.

Afoakwa, E. O.; Paterson, A.; Fowler, M. Factors Influencing Rheological and Textural Qualities in Chocolate—A Review. *Trends Food Sci. Technol.* **2007,** *18,* 290–298.

Afoakwa, E. O.; Paterson, A.; Fowler, M.; Vieira, J. Particle Size Distribution and Compositional Effects on Textural Properties and Appearance of Dark Chocolates. *J. Food Eng.* **2008a,** *87,* 181–190.

Afoakwa, E. O.; Paterson, A.; Fowler, M.; Vieira, J. J. Effects of Tempering and Fat Crystallisation Behaviour on Microstructure, Mechanical Properties and Appearance in Dark Chocolate Systems. *J. Food Eng.* **2008b,** *89,* 128–136.

Aidoo, R. P.; Depypere, F.; Afoakwa, E. O.; Dewettinck, K. Industrial Manufacture of Sugar-Free Chocolates Applicability of Alternative Sweeteners and Carbohydrate Polymers as Raw Materials in Product Development. *Trends Food Sci. Technol.* **2013,** *32,* 84–96.

Aidoo, R. P.; Afoakwa, E. O.; Dewettinck, K. Optimization of Inulin and Polydextrose Mixtures as Sucrose Replacers During Sugar-Free Chocolate Manufacture—Rheological, Microstructure and Physical Quality Characteristics. *J. Food Eng.* **2014,** *226,* 1259–1268.

Aidoo, R. P.; Afoakwa, E. O.; Dewettinck, K. Rheological Properties, Melting Behaviours and Physical Quality Characteristics of Sugar-Free Chocolates Processed Using Inulin/ Polydextrose Bulking Mixtures Sweetened with Stevia and Thaumatin Extracts. *LWT-Food Sci. Technol.* **2015,** *62,* 592–597.

Almoosawi, S.; Fyfe, L.; Ho, C.; Al-Dujaili, E. The Effect of Polyphenol-Rich Dark Chocolate on Fasting Capillary Whole Blood Glucose, Total Cholesterol, Blood Pressure and Glucocorticoids in Healthy Overweight and Obese Subjects. *Br. J. Nutr.* **2010,** *103,* 842–850.

Anal, A. K.; Singh, H. Recent Advances in Microencapsulation of Probiotics for Industrial Applications and Targeted Delivery. *Trends Food Sci. Technol.* **2007,** *18*(5), 240–251.

Andarea-Nightingale, L. M.; Lee, S. Y.; Engeseth, N. J. Textural Changes in Chocolate Characterized by Instrumental and Sensory Techniques. J. Texture Stud. **2009,** *40,* 427–444.

Aragon-Alegro, L. C.; Alarcon Alegro, J. H.; Cardarelli, H. R.; Chiu, M. C.; Saad, S. M. I. Potentially Probiotic and Synbiotic Chocolate Mousse. *LWT-Food Sci. Technol.* **2007,** *40,* 669–675.

Ascarelli, S. The World's Chocoholics. In Americans are Far from the World's Biggest Chocolate Eaters, 2015. http://www.marketwatch.com/story/americans-are-far-from-the-worlds-biggest-chocolate-eaters-2015-07-21. (accessed Dec 15, 2016).

Baba, S.; Natsume, M.; Yasuda, A.; Nakamura, Y.; Tamura, T.; Osakabe, N.; Kanegae, M.; Kondo, K. Plasma LDL and HDL Cholesterol and Oxidized LDL Concentrations are Altered in Normo- and Hypercholesterolemic Humans After Intake of Different Levels of Cocoa Powder. *J. Nutr.* **2007,** *137,* 1436–1441.

Beards, E.; Tuohy, K.; Gibson, G. A Human Volunteer Study to Assess the Impact of Confectionery Sweeteners on the Gut Microbiota Composition. *Br. J. Nutr.* **2010,** *104,* 701–708.

Beckett, S. T. *Industrial Chocolate Manufacture and Use,* 3rd ed.; Blackwell: Oxford, 1999.

Beckett, S T. *The Science of Chocolate,* 2nd ed; The Royal Society of Chemistry Publishing: Cambridge, 2008; 55–57.

Belščak-Cvitanović, A.; Kromes, D.; Dujmović, M.; Karlović, S.; Biškić, M.; Brnčić, M.; Jezek, D. Physical, Bioactive and Sensory Quality Parameters of Reduced Sugar Chocolates Formulated with Natural Sweeteners as Sucrose Alternatives. *Food Chem.* **2015,** *167,* 61–70.

Beristain, C. I.; Cruz-Sosa, F.; Lobato-Calleros, C.; Pedroza-Islas, R.; Rodriguez Huezo, M. E.; Verde-Calvo, J. R. Applications of soluble dietary fibres in beverages. *Revista Mexicana De Ingenieria Quimica,* **2006,** *5,* 81–95.

Bolenz, S.; Thiessenhusen, T.; Schape, R. Fast Conching for Milk Chocolate. *Eur. Food Res. Technol.* **2003**, *218*, 62–67.

Bolenz, S.; Amtsberg, K.; Lipp, E. New Concept for Fast Continuous Conching of Milk Chocolate. *Eur. Food Res. Technol.* **2005**, *220*, 47–54.

Bolenz, S.; Amtsberg, K.; Schape, R. The Broader Usage of Sugars and Fillers in Milk Chocolate Made Possible by the New EC Cocoa Directive. *Int. J. Food Sci. Technol.* **2006**, *41*, 45–55.

Bolliger, S.; Breitschun, B.; Stranzinger, M.; Wagner, T.; Windhab, E. J. Comparison of Precrystallization of Chocolate. *J. Food Eng.* **1998**, *35*, 281–297.

Bonarius, G. A.; Vieria, J. B.; van der Goot; A.; Bodnar, I. Rheological Behavior of Fibre-Rich Plant Materials in Fat-Based Food System. *Food Hydrocolloids* **2014**, *40*, 254–261.

Bordin, A.; Brandelli, A.; Wulf, E.; Carrion, F.; Pieta, L.; Venzke, T. Development and Evaluation of a Laboratory Scale Conch for Chocolate Production. *Int. J. Food Sci. Technol.* **2009**, *44*, 616–622.

Botelho, P. B.; Galasso, M.; Dias, V.; Mandrioli, M.; Lobato, L. P.; Rodriguez-Estrada, M. T.; Castro, I. A. Oxidative Stability of Functional Phytosterol-Enriched Dark Chocolate. *LWT-Food Sci. Technol.* **2014**, *55*, 444–451.

Boylston, T. D.; Vinderola, C. G.; Ghoddusi, H. B.; Reinheimer, J. A. Incorporation of *Bifidobacteria* into Cheeses: Challenges and Rewards. *Int. Dairy J.* **2004**, *14*, 375–387.

Bruinsma, K.; Taren, D. L. Chocolate: Food or Drug? *J. Am. Diet. Assoc.* **1999**, *99*, 1249–1256.

Burgain, J.; Gaiani, C.; Linder, M.; Scher, J. Encapsulation of Probiotic Living Cells: From Laboratory Scale to Industrial Applications. *J. Food Eng.* **2011**, *104*, 467–483.

Callebaut. Barry Callebaut International Consumer Survey Shows: Chocolate Lovers Want Functional Chocolate with Proven Health Benefits. 2008. http://www.barry-callebaut.com/51?release=4030.

Ceriello, A.; Motz, E. Is Oxidative Stress the Pathogenic Mechanism Underlying Insulin Resistance, Diabetes, and Cardiovascular Disease? The Common Soil Hypothesis Revisited. *Arterioscler. Thromb. Vasc. Biol.* **2004**, *24*, 816–823.

Champagne, C. P.; Raymond, Y.; Guertin, N.; Belanger, G. Effects of Storage Conditions, Microencapsulation and Inclusion in Chocolate Particles on the Stability of Probiotic Bacteria in Ice Cream. *Int. Dairy J.* **2015**, *47*, 109–117.

Charalampopoulos, D.; Pandiella, S. S.; Webb, C. Growth Studies of Potentially Probiotic Lactic Acid Bacteria in Cereal-Based Substrates. *J. Appl. Microbiol.* **2002**, *92*, 851–859.

Cidell, J. L.; Alberts, H. C. Constructing Quality: the Multinational Histories of Chocolate. *Geoforum* **2006**, *37*, 999–1007.

Coman, M. M.; Cecchini, C.; Verdenelli, M. C.; Silvi, S.; Orpianesi, C.; Cresci, A. Functional Foods as Carriers for SYNBIO, a Probiotic Bacteria Combination. *Int. J. Food Microbiol.* **2012**, *157*, 346–352.

Cooper, K. A.; Donovan, J. L.; Waterhouse, A. L.; Williamson, G. Cocoa and Health: A Decade of Research. *Br. J. Nutr.* **2008**, *99*(01), 1–11.

Crichton, E. G.; Elias, M. F.; Alkerwi, A. Chocolate Intake is Associated with Better Cognitive Function: The Maine-Syracuse Longitudinal Study. *Appetite* **2016**, *100*, 126–132.

Crozier, S. J.; Hurst, W. J. Cocoa Polyphenols and Cardiovascular Health. In *Polyphenols in Human Health and Disease;* Watson, R. R., Preedy, W. R., Zibadi, S. Eds.; Academic Press: Cambridge, USA, 2014; pp 1077–1085.

Dahlenborg, H.; Millqvist-Fureby, A.; Bergenstahl, B. Effect of Shell Microstructure on Oil Migration and Fat Bloom Development in Model Pralines. *Food Struct.* **2015**, *5,* 51–65.

Davis, L. M. G.; Martinez, I.; Walter, J.; Hutkins, R. A Dose Dependent Impact of Prebiotic Galactooligosaccharides on the Intestinal Microbiota of Healthy Adults. *Int. J. Food Microbiol.* **2010**, *144,* 285–292.

De Castro, F. P.; Cunha, T. M.; Barreto, P. L. M.; Amboni, R. D. D. M.; Prudencio, E. S. Effect of Oligofructose Incorporation on the Properties of Fermented Probiotic Lactic Beverages. *Int. J. Dairy Technol.* **2009**, *62,* 68–74.

De Morais, E. C.; Lima, G. C.; De Morais, A. R.; Blini, H. M. A. Prebiotic and Diet/Light Chocolate Dairy Dessert: Chemical Composition, Sensory Profiling and Relationship with Consumer Expectation. *LWT-Food Sci. Technol.* **2015**, *62,* 424–430.

Debaste, F.; Kegelaers, Y.; Liegeois, S.; Ben Amor, H.; Halloin, V. Contribution to the Modeling of Chocolate Tempering Process. *J. Food Eng.* **2008**, *88,* 568–575.

Desch, S.; Schmidt, J.; Kobler, D.; et al. Effect of Cocoa Products on Blood Pressure: Systematic Review and Meta-Analysis. *Am. J. Hypertens.* **2010**, *23*(1), 97–103.

Dhonsi, D.; Stapley, A. G. F. The Effect of Shear Rate, Temperature, Sugar and Emulsifier on the Tempering of Cocoa Butter. *J. Food Eng.* **2006**, *77,* 936–942.

Dimick, P. S.; Manning, D. M. Thermal and Compositional Properties of Cocoa Butter During Static Crystallization. *J. Am. Oil Chem. Soc.* **1987**, *64,* 1663–1669.

El-Kalyoubi, M.; Khallaf, M. F.; Abdelrashid, A.; Mostafa, E. M. Quality Characteristics of Chocolate e Containing Some Fat Replacers. *Ann. Agric. Sci.* **2011**, *56*(2), 89–96.

Engler, M. B.; Engler, M. M.; Chen, C. Y.; Malloy, M. J.; Browne, A.; Chiu, E. Y. Flavonoid-Rich Dark Chocolate Improves Endothelial Function and Increases Plasma Epicatechin Concentrations in Healthy Adults. *J. Am. Coll. Nutr.* **2004**, *23,* 197–204.

Erdem, O.; Gultekin-Ozguven, M.; Berktas, I.; Ersan, S.; Tuna, H. E.; Karadag, A.; Ozcelik, B.; Gunes, G.; Cutting, S. M. Development of a Novel Symbiotic Dark Chocolate Enriched with *Bacillus indicus* HU36, Maltodextrin and Lemon Fiber: Optimization by Response Surface Methodology. *LWT-Food Sci. Technol.* **2014**, *56,* 187–193.

Farzanmehr, H.; Abbasi, S. Effects of Inulin and Bulking Agents on Some Physicochemical, Textural and Sensory Properties of Milk Chocolate. *J. Texture Stud.* **2009**, *40,* 536–553.

Fernandes, V. A.; Müller, A. J.; Sandoval, A. J. Thermal Structural and Rheological Characteristics of Dark Chocolate with Different Compositions. *J. Food Eng.* **2012**, *116*(1), 97–108.

Flammer, A. J.; Hermann, F.; Sudano, I.; Spieker, L.; Hermann, M.; Cooper, K. A.; Serafini, M.; Luscher, T. F.; Ruschitzka, F.; Noll, G.; Corti, R. Dark Chocolate Improves Coronary Vasomotion and Reduces Platelet Reactivity. *Circulation* **2007**, *116,* 2376–2382.

Francis, S. T.; Head, K.; Morris, P. G.; Macdonald, I. A. The Effect of Flavanol-Rich Cocoa on the fMRI Response to a Cognitive Task in Healthy Young People. *J Cardiovasc. Pharmacol.* **2006**, *47*(Suppl. 2), S215–220.

Franck, A. Technological Functionality of Inulin and Oligofructose. *Br. J. Nutr.* **2002**, *87*(2), S287–229.

Franco, R.; Astibia, A. O.; Martínez-Pinilla, E. Health Benefits of Methylxanthines in Cacao and Chocolate. *Nutrients* **2013**, *5,* 4159–4173.

Gadhiya, D.; Patel, A.; Prajapati, J. B. Current Trend and Future Prospective of Functional Probiotic Milk Chocolates and Related Products—A Review. *Czech J. Food Sci.* **2015**, *33,* 295–301.

Glicerina, V.; Balestra, F.; Rosa, M. D.; Romani, S. Rheological, Textural and Calorimetric Modifications of Dark Chocolate During Process. *J. Food Eng.* **2013,** *119*(1), 173–179.

Gomes, C. R.; Vissotto, F. Z.; Fadini, A. L.; Faria, E. V.; Luiz, A. M. Influencia de diferentes agentes de corponas características reologicas e sensoriais de chocolates diet em sacarose e light em calorias. *Ciencia e Tecnologia de Alimentos* **2007,** *27,* 613–623.

Grassi, D.; Lippi, C.; Necozione, S.; Desideri, G.; Ferri, C. Short-Term Administration of Dark Chocolate is Followed by a Significant Increase in Insülin Sensitivity and a Decrease in Blood Pressure in Healthy Persons. *Am. J. Clin. Nutr.* **2005,** *81,* 611–614.

Grassi, D.; Desideri, G.; Necozione, S.; Lippi, C.; Casale, R.; Properzi, G.; Blumberg, . B.; Ferri, C. Blood Pressure is Reduced and Insulin Sensitivity Increased in Glucose-Intolerant, Hypertensive Subjects After 15 Days of Consuming Highpolyphenol Dark Chocolate. *J. Nutr.* **2008,** *138,* 1671–1676.

Grassi, D.; Desideri, G.; Croce, G.; Tiberti, S.; Aggio, A.; Ferri, C. Flavonoids, vascular function and cardiovascular protection. *Curr. Pharm. Des.,* **2009,** *15,* 1072–1084.

Grassi, D.; Desideri, G.; Ferri, C. Blood Pressure and Cardiovascular Risk: What About Cocoa and Chocolate? *Arch. Biochem. Biophys.* **2010,** *501,* 112–115.

Gray, J. *Cocoa and Chocolate—The Case for Health Benefits,* 1st ed.; Mars Incorporated, 2005; p 8.

Greenberg, J. A. Chocolate Intake and Diabetes Risk. *Clin. Nutr.* **2015,** *34,* 129–133.

Gültekin-Özgüven, M.; Karadağ, A.; Duman, Ş.; Özkal, B.; Özçelik, B. Fortification of Dark Chocolate with Spray Dried Black Mulberry (*Morus nigra*) Extract Encapsulated in Chitosan-Coated Liposomes and Bioaccessibility Studies. *Food Chem.* **2016,** *201,* 205–212.

Hartel, R. W. *Crystallization in Food;* Aspen Publishers Inc.: Gaithersburg, USA, 2001.

Heiss, C.; Finis, D.; Kleinbongard, P.; Hoffmann, A.; Rassaf, T.; Kelm, M.; Sies, H. Sustained Increase in Flow-Mediated Dilation After Daily Intake of High-Flavanol Cocoa Drink Over 1 Week. *J. Cardiovasc. Pharmacol.* **2007,** *49,* 74–80.

Heiss, C.; Keen, C. L.; Kelm, M. Flavanols and Cardiovascular Disease Prevention. *Eur. Heart J.* **2010,** *31,* 2583–2592.

Hermann, F.; Spieker, L. E.; Ruschitzka, F.; Sudano, I.; Hermann, M.; Binggeli, C.; Luscher, T. F.; Riesen, W.; Noll, G.; Corti, R. Dark Chocolate Improves Endothelial and Platelet Function. *Heart* **2006,** *92,* 119–120.

Hesse, A.; Siener, R.; Heynck, H. The Influence of Dietary Factors on the Risk of Urinary Stone Formation. *Scanning Microsc.* **1993,** *7,* 1119–1127.

Hii, C. L.; Law, C. L.; Suzannah, S.; Cloke, M. Polyphenols in Cocoa (*Theobroma cacao L.*). *As. J. Food Ag-Ind.* **2009,** *2*(4), 702–722.

Holzapfel, W. H.; Haberer, P.; Geisen, R.; Björkroth, J.; Schillinger, U. Taxonomy and Important Features of Probiotic Microorganisms in Food and Nutrition. *Am. J. Clin. Nutr.* **2001,** *73*(Suppl. 2), 365S–373S.

Hooper, L.; Kay, C.; Abdelhamid, A.; Kroon, P. A.; Cohn, J. S.; Rimm, E. B.; Cassidy, A. Effects of Chocolate, Cocoa, and Flavan-3-Ols on Cardiovascular Health: a Systematic Review and Meta-Analysis of Randomized Trials. *Am. J. Clin. Nutr.* **2012,** *95,* 740–751.

Isolauri, E. The Role of Probiotics in Pediatrics. *Curr. Pediatr.* **2004,** *14,* 104–109.

Jeyarani, T.; Banerjee, T.; Ravi, R.; Gopala Krishna, A. G. Omega-3 Fatty Acids Enriched Chocolate Spreads Using Soybean and Coconut Oils. *J. Food Sci. Technol.* **2015,** *52,* 1082–1088.

JFECFA. Thaumatin. *Toxicological Evaluation of Certain Food Additives and Contaminants;* Report of the 29th Meeting of the Joint FAO/WHO Expert Committee on Food Additives (JECFA), World Health Organization, Geneva. In WHO Food Additive Series 20. Geneva: WHO, 1985.

Jovanovic, O.; Pajin, B. Influence of Lactic Acid Ester on Chocolate Quality. *Trends Food Sci. Technol.* **2004,** *15,* 128–136.

Kalliomaki, M.; Antoine, J. M.; Herz, U.; Rijkers, G. T.; Wells, J. M.; Mercenier, A. Guidance for Substantiating the Evidence for Beneficial Effects of Probiotics: Prevention and Management of Allergic Diseases by Probiotics. *J. Nutr.* **2010,** *140,* 713S–721S.

Kaspar, K. L. Identification and Quantification of Flavanols and Methylxanthines in Chocolates with Different Percentages of Chocolate Liquor. Master of Science Degree in Food and Nutritional Sciences of University of Wisconsin-Stout. 2006, 3–96.

Kay, C. D.; Kris-Etherton, P. M.; West, S. G. Effects of Antioxidant-Rich Foods on Vascular Reactivity: Review of the Clinical Evidence. *Curr. Atheroscler. Rep.* **2006,** *8,* 510–522.

Katz, D. L.; Doughty, K.; Ali, A. Cocoa and Chocolate in Human Health and Disease. *Antioxid. Redox Signal.* **2011,** *15*(10), 2779–2811.

Kolida, S.; Tuohy, K.; Gibson, G. R. Prebiotic Effects of Inulin and Oligofructose. *Br. J. Nutr.* **2002,** *87*(Suppl. 2), S193–197.

Komes, D.; Belščak-Cvitanović, A.; Škrabal, S.; Vojvodić, A.; Bušić, A. The Influence of Dried Fruits Enrichment on Sensory Properties of Bitter and Milk Chocolate and Bioactive Content of their Extracts Affected by Different Solvents. *LWT-Food Sci. Technol.* **2013,** *53,* 360–369.

Konar, N. Influence of Conching Temperature and Some Bulk Sweeteners on Physical and Rheological Properties of Prebiotic Milk Chocolate Including Containing Inulin. *Eur. Food Res. Technol.* **2013,** *23,* 135–143.

Konar, N.; Ozhan, B.; Artik, N.; Dalabasmaz, S.; Poyrazoglu, E. S. Rheological and Physical Properties of Inulin-Containing Milk Chocolate Prepared at Different Process Conditions. *CyTA J. Food* **2014a,** *12*(1), 55–64.

Konar, N.; Ozhan, B.; Artik, N.; Poyrazoglu, E. S. Using Polydextrose as a Prebiotic Substance in Milk Chocolate-Effects of Process Parameters on Physical and Rheological Properties. *CyTA J. Food* **2014b,** *12*(2), 150–159.

Konar, N.; Poyrazoglu, E. S.; Artik, N. Influence of Calcium Fortification on Physical and Rheological Properties of Non-Sucrose Prebiotic Milk Chocolates Containing Inulin and Maltitol. *J. Food Sci. Technol.* **2015,** *52*(4), 2033–2042.

Konar, N.; Toker, O. S.; Oba, S.; Sagdic, O. Improving Functionality of Chocolate: A Review on Probiotic, Prebiotic, and/or Synbiotic Characteristics. *Trends Food Sci. Technol.* **2016,** *49,* 35–44.

Kubomara, K. Japan Redefines Functional Foods. *Prepared Foods* **1998,** *167,* 129–132.

Kwok, C. S.; Loke, Y. K.; Welch, A. A., Luben, R. N., Lentjes, M. A. H; Boekholdt, S. M.; Pfister, R.; Mamas, M. A.; Wareham, N. J.; Khaw, K. T.; Myint, P. K. Habitual Chocolate Consumption and The Risk Of Incident Heart Failure Among Healthy Men and Women. *Nutr. Metab. Cardiovasc. Dis.* **2016,** *26,* 722e–734

Lahtinen, S. J.; Ouwehand, A. C.; Salminen, S. J.; Forssell, P.; Myllärinen, P. Effect of Starch- and Lipid-Based Encapsulation on the Culturability of Two *Bifidobacterium longum* Strains. *Lett. Appl. Microbiol.* **2007,** *44*(5), 500–505.

Lalicic-Petronijevic, J.; Popov-Raljic, J.; Obradovic, D.; Radulovic, Z.; Paunovic, D.; Petrusic, M.; Pezo, L. Viability of Probiotic Strains *Lactobacillus acidophilus* NCFM® and *Bifidobacterium lactis* HN019 and Their Impact On Sensory and Rheological Properties of Milk and Dark Chocolates During Storage for 190 Days. *J. Funct. Foods* **2015,** *15,* 541–550.

Lamuela-Raventos, R. M.; Romero-Perez, A. I.; Andres-Lacueva, C.; Tornero, A. Review: Health Effects of Cocoa Flavonoids. *Food Sci. Technol. Int.* **2005,** *11,* 159–176.

Lee, K. W.; Kim, Y. J.; Lee, H. J.; Lee, C. Y. Cocoa has More Phenolic Phytochemicals and a Higher Antioxidant Capacity than Teas and Red Wine. *J. Agric. Food Chem.* **2003,** *51*(25), 7292–7295.

Lucisano, M.; Casiraghi, E.; Mariotti, M. Influence of formulation and processing variables on ball mill refining of milk chocolate. European Food Research and Technology, **2006,** *223,* 797–802.

Macht, M.; Dettmer D. Everyday Mood and Emotions After Eating a Chocolate Bar or an Apple. *Appetite* **2006,** *46,* 332–336.

Macht, M.; Mueller, J. Immediate Effects of Chocolate on Experimentally Induced Mood States. *Appetite* **2007,** *49,* 667–674.

Maillard, M.; Landuyt, A. Chocolate: an Ideal Carrier for Probiotics. *Agro Food Ind. Hi-Tech,* **2008,** *19*(Suppl. 3), 13–15.

Mandal, S.; Puniya, A. K.; Singh, K. Value Addition of Milk Chocolate Using Inulin and Encapsulated *Lactobacillus casei* NcDc-298. *National Seminar on Value Added Dairy Products,* Dairy Technology Society of India, National Dairy Research Institute, India, 2005.

Maragkoudakis, P. A.; Miaris, C.; Rojez, P.; Manalis, N.; Magkanari, F.; Kalantzopoulos, G. Production of Traditional Greek Yoghurt Using Lactobacillus Strains with Probiotic Potential as Starter Adjuncts. *Int. Dairy J.* **2006,** *16,* 52–60.

Martin, F. J.; Rezzi, S.; Trepat, E. P.; Kamlage, B.; Collino, S.; Leibold, E.; Kastler, J.; Rein, D.; Fay, L. B.; Kochhar, S. Metabolic Effects of Dark Chocolate Consumption on Energy, Gut Microbiota and Stress-Related Metabolism in Free-Living Subjects. *J. Proteome Res.* **2010,** *8,* 5568–5579.

Martin, M. A.; Goya L.; Ramos S. Potential for Preventive Effects of Cocoa and Cocoa Polyphenols in Cancer. *Food Chem. Toxicol.* **2013,** *56,* 336–351.

Mellor, D. D.; Sathyapalan, T.; Kilpatrick, E. S.; Beckett, S.; Atkin, S. L. High-Cocoa Polyphenol-Rich Chocolate Improves, HDL Cholesterol in Type 2 Diabetes Patients. *Diabet. Med.* **2010,** *27,* 1318–1321.

Miraballes, M.; Fiszman, S.; Gambero, A.; Varela, P. Consumer Perceptions of Satiating and Meal Replacement Bars, Built up from Cues in Packaging Information, Health Claims and Nutritional Claims. *Food Res. Int.* **2014,** *64,* 456–464.

Morris, C.; Morris, G. A. The Effect of Inulin and Fructo-Oligosaccharide Supplementation on the Textural, Rheological and Sensory Properties of Bread and Their Role in Weight Management: A Review. *Food Chem.* **2012,** *133,* 237–248.

Niamah, A. K.; Verma, D. K. Microbial Intoxication in Dairy Food Product. In *Microorganisms in Sustainable Agriculture, Food and the Environment.* as part of book series on *Innovations in Agricultural Microbiology;* Verma, D. K., Srivastav, P. P., Eds.; Apple Academic Press: USA, 2017. Vol. 1, pp 143–170.

Osman, J. L.; Sobal J. Chocolate Cravings in American and Spanish Individuals: Biological and Cultural Influences. *Appetite* **2006,** *47,* 290–301.

Palazzo, A. B.; Carvalho, M. A. R.; Efraim, P.; Bolini, H. M. A. Determination of isosweetness concentration of sucralose, rebaudioside and neotame as sucrose substitutes in new diet chocolate formulations using the time-intensity analysis. *Journal of Sensory Studies,* **2011,** *26,* 291-297.

Paoletti, R.; Poli, A.; Conti, A.; Visioli, F. *Chocolate and Health*; Springer-Verlag: Mailand, 2012; pp 10–39.

Pappalardo, G.; Lusk, J. L. The Role of Beliefs in Purchasing Process of Functional Foods. *Food Qual. Preference,* **2016,** *53,* 151–158.

Patel, P.; Parekh, T.; Subhash, R. Development of Probiotic and Synbiotic Chocolate Mousse: A Functional Food. *Biotechnology* **2008,** *7,* 769–774.

Patel, A. R.; Cludts, N.; Sintang, M. D. B.; Lesaffer, A.; Dewettinck, K. Edible Oleogels Based on water Soluble Food Polymers: Preparation, Characterization and Potential Application. *Food Funct.* **2014,** *5,* 2833-2841.

Patel, A.; Shah, N.; Verma, D. K. Lactic Acid Bacteria (LAB) Bacteriocins: An Ecological and Sustainable Biopreservative Approach to Improve the Safety and Shelf-life of Foods. In *Microorganisms in Sustainable Agriculture, Food and the Environment.* as Part of Book Series on *Innovations in Agricultural Microbiology;* Verma, D. K., Srivastav, P. P., Eds.; Apple Academic Press: USA, 2017.

Pearson, D. A.; Holt R. R.; Rein D.; Paglieroni T.; Schmitz H. H.; Keen C. L. Flavanols and Platelet Reactivity. *Clin. Dev. Immunol.* **2005,** *12,* 1–9.

Pineiro, M.; Stanton, C. Probiotic Bacteria: Legislative Framework—Requirements to Evidence Basis. *J. Nutr.* **2007,** *137*(3), 850S–853S.

Possemiers, S.; Marzorati, M.; Verstraete, W.; Van de Wiele, T. Bacteria and Chocolate: a Successful Combination for Probiotic Delivery. *Int. J. Food Microbiol.* **2010,** *141*(1–2), 97–103.

Ramiro-Puig, E.; Urpi-Sarda, M.; Perez-Cano, F. J.; Franch, A.; Castellote, C.; Andres-Lacueva, C.; Izquierdo-Pulido, M.; Castell, M. Cocoa-Enriched Diet Enhances Antioxidant Enzyme Activity and Modulates Lymphocyte Composition in Thymus from Young Rats. *J. Agric. Food Chem.* **2007,** *55,* 6431–6438.

Rezende, N. V.; Benassi, M. T.; Vissotto, F. Z.; Augusto, P. P. C.; Grossman, M. V. E. Mixture Design Applied for The Partial Replacement of Fat with Fibre in Sucrose-Free Chocolates. *LWT-Food Sci. Technol.* **2015,** *62,* 598-604.

Ried, K.; Sullivan T.; Fakler P.; Frank O.; Stocks N. Effect of Cocoa on Blood Pressure. Cochrane Database *Syst. Rev.* **2012;** *8,* CD008893.

Rijkers, G. T.; Bengmark, S.; Enck, P.; Haller, D.; Herz, U.; Kalliomaki, M.; Kudo, S.; Lenoir-Wijnkoop, I.; Mercenier, A.; Myllyluoma, E.; Rabot, S.; Rafter, J.; Szajewska, H.; Watzl, B.; Wells, J.; Wolvers, D.; Antoine, J. M. Guidance for Substantiating the Evidence for Beneficial Effects of Probiotics: Current Status and Recommendations for Future Research. *J. Nutr.* **2010,** *140,* 671S–676S.

Robbins, R. J.; Leonczak, J.; Li, J.; Johnson, J. C.; Collins, T.; Kwik-Uribe, C. Determination of Flavanol and Procyanidin (by Degree of Polymerization 1–10) Content of Chocolate, Cocoa Liquors, Powder(s), and Cocoa Flavanol Extracts by Normal Phase High-Performance Liquid Chromatography: Collaborative Study. *J. AOAC Int.* **2012,** *95,* 1153–1160.

Roberfroid, M. B.; Van Loo, J.; Gibson, G. The Bifidogenic Nature of Chicory Inulin and its Hydrolysis Products. *J. Nutr.* **1998,** *128,* 11–19.

Rogers, M. A., Tang, D.; Ahmadi, L.; Marangoni, A. Fat Crystal Network. In *Food Material Science - Principles and Practice;* Aguilera, J. M., Lillford, P. J., Eds.; Springer: New York, 2008; pp 369–414.

Saarela, M.; Virkajarvi, I.; Nohynek, L.; Vaari, A.; Matto, J. Fibres as Carriers for Lactobacillus Rhamnosus During Freeze-Drying and Storage in Apple Juice and Chocolate-Coated Breakfast Cereals. *Int. J. Food Microbiol.* **2006,** *112,* 171–178.

Saka, E. K. The Design of Packaging Graphics for the Expansion of Ghanaian Chocolate Products. Graduate Theses and Dissertations, Lowa State University, U. S. Paper 10,239, 2011, p 111.

Scheid, M. M. A.; Moreno, Y. M. F.; Marostica-Júnior, M. R.; Pastore, G. M. Effect of Prebiotics on the Health of the Elderly. *Food Res. Int.,* **2013,** *53,* 426–432.

Scholey, A.; Owen, L. Effects of Chocolate on Cognitive Function and Mood: a Systematic Review. *Nutr. Rev.* **2013,** *71,* 665–681.

Scholey, A. B.; French, S. J.; Morris, P. J.; Kennedy, D. O.; Milne, A. L.; Haskell, C. F. Consumption of Cocoa Flavanols Results in Acute Improvements in Mood and Cognitive Performance During Sustained Mental Effort. *J. Psychopharmacol.* **2010,** *24,* 1505–1514.

Schumacher, A. B.; Brandelli, A.; Schumacher, E. W.; Macedo, F. C.; Pieta, L.; Klug, T. V.; de Jong, E. V. Development and Evaluation of a Laboratory Scale Conch for Chocolate Production. *Int. J. Food Sci. Technol.* **2009,** *44,* 616–622.

Schwab, U. S.; Maliranta H. M.; Sarkkinen E. S.; Savolainen M. J.; Kesaniemi Y. A.; Uusitupa M. I. Different Effects of Palmitic and Stearic Acid-Enriched Diets on Serum Lipids and Lipoproteins and Plasma Cholesteryl Ester Transfer Protein Activity in Healthy Young Women. *Metabolism* **1996,** *45,* 143–149.

Servais, C.; Ranc, H.; Roberts, I. D. Determination of Chocolate Viscosity. *J. Texture Stud.* **2004,** *34,* 467–497.

Shah, N. P. Functional Cultures and Health Benefits. *Int. Dairy J.* **2007,** *17,* 1262–1277.

Shah, A. S.; Jones, G. P.; Vasiljevic, T. Sucrose-Free Chocolate Sweetened with Stevia Rebaudiana Extract and Containing Different Bulking Agents Effects on Physicochemical and Sensory Properties. *Int. J. Food Sci. Technol.* **2010,** *45,* 1426–1435.

Shourideh, M.; Taslimi, A.; Azizi, M. H.; Mohammadifar, M. A. Effects of D-Tagatose and Inulin of some Physicochemical, Rheological and Sensory Properties of Dark Chocolate. *Int. J. Biosci. Biochem. Bioinf.* **2012,** *2*(5), 314–319.

Smit, H. J.; Gaffan, E. A.; Rogers, P. J. Methylxanthines are the Psycho-Pharmacologically Active Constituents of Chocolate. *Psychopharmacology (Berl)* **2004,** *176,* 412–419.

Sokmen, A.; Gunes, G. Influence of some Bulk Sweeteners on Rheological Properties of Chocolate. *LWT-Food Sci. Technol.* **2006,** *39*(10), 1053–1058.

Sokolov, A. N.; Pavlova, M. A.; Klosterhalfen, S.; Enck, P. Chocolate and the Brain: Neurobiological Impact of Cocoa Flavanols on Cognition and Behavior. *Neurosci. Biobehav. Rev.* **2013,** *37,* 2445-2453.

Spadafranca, A.; Martinez Conesa C.; Sirini, S.; Testolin, G. Effect of Dark Chocolate on Plasma Epicatechin Levels, DNA Resistance to Oxidative Stress and Total Antioxidant Activity in Healthy Subjects. *Br. J. Nutr*. **2010**, *103,* 1008–1014.

Srisuvor, N.; Chinprahast, N.; Prakitchaiwattana, C.; Subhimaros, S. Effects of Inulin and Polydextrose on Physicochemical and Sensory Properties of Low-Fat Set Yoghurt with Probiotic-Cultured Banana Puree. *LWT-Food Sci. Technol*. **2013**, *51,* 30–36.

Statista. Statistics and Facts on the Chocolate Industry. www.statista.com/statistics/238849/global-chocolateconsumption/ (accessed Dec 7, 2015).

Statista. Statistics and Facts on the Chocolate Industry. https://www.statista.com/statistics/ 263779/per-capita-consumption-of-chocolate-in-selected-countries-in-2007/ (accessed Nov 30, 2016).

Statistica. 2015. https://www.statista.com/chart/3668/the-worlds-biggest-chocolate-consumers/

Suter A. The effect of galactooligosaccharide addition to a chocolate system. MSc Desertation, Department of Food Science and Technology, The Ohio State University, Ohio, USA, 2010.

Torres-Moreno, M.; Tarrega, A.; Costell, E.; Blanch, C. Dark Chocolate Acceptability: Influence of Cocoa Origin and Processing Conditions. *J. Sci. Food Agric*. **2012**, *92,* 404–411.

USDA *National Nutrient Database for Standard Reference;* Release 23, 2017. Accessed Feb 5, 2017, URL: https://ndb.nal.usda.gov/ndb/

Verma, S. K.; David, J.; Chandra, R. Synbiotics: Potential Dietary Supplements in Functional Foods. *Indian Dairyman* **2012**, 58–62.

Vicente, F.; Sandra-Ruiz, S.; Rabanal, M.; Rodríguez-Lagunas, M. J.; Pereira, P.; Perez-Cano, F. J.; Castell, M. A New Food Frequency Questionnaire to Assess Chocolate and Cocoa Consumption. *Nutrition* **2016**, *32,* 811–817.

Villegas, B.; Tarrega, A.; Carbonell, I.; Costell, E. Optimising Acceptability of New Prebiotic Low-Fat Milk Beverages. *Food Qual. Preference* **2010**, *21,* 234–242.

Vinson, J. A.; Proch, J.; Bose, P.; Muchler, S.; Taffera, P.; Shuta, D.; Samman, N.; Agbor G. A. Chocolate is a Powerful Ex vivo and in vivo Antioxidant, an Antiatherosclerotic Agent in an Animal Model, and a Significant Contributor to Antioxidants in the European and American Diets. *J. Agric. Food Chem*. **2006**, *54,* 8071–8076.

Vitali, B.; Ndagijimana, M.; Cruciani, F.; Carnevali, P.; Candela, M.; Guerzoni M. E.; Brigidi, P. Impact of a Synbiotic Food on the Gut Microbial Ecology and Metabolic Profiles. *BMC Microbiol*. **2010**, *10*(4), 1-13.

Volpini-Rapina, L. F.; Sokei, F. R.; Conti-Silva, A. C. Sensory Profile and Preference Mapping of Orange Cakes with Addition of Prebiotics Inulin and Oligofructose. *LWT-Food Sci. Technol*. **2012**, *48,* 37-42.

Vriesmanna, L. C.; Amboni, R. D. M.; Lúcia, C.; Petkowicz, O. Cacao Pod Husks (*Theobroma cacao* L.): Composition and Hot-Water-Soluble Pectins. *Ind. Crops Prod.* **2011**, *34,* 1173–1181.

Wang, Y. Prebiotics: Present and Future in Food Science And Technology. *Food Res. Int.* **2009**, *42,* 8–12.

Williams, S.; Tamburic, S.; Lally, C. Eating Chocolate can Significantly Protect the Skin from UV Light. *J. Cosmet. Dermatol*. **2009**, *8,* 169–173.

Wilson, P. K. Centuries of Seeking Chocolate's Medicinal Benefits. *Lancet* **2010,** *376,* 158–159.

Winardi-Liem, M. Rheology of Prebiotic Chocolate. Thesis, The Ohio State University (Advisor; Vodavots, Y.) **2011,** p. 17.

Zaric, D. B.; Bulatovic, M. L.; Rakin, M. B.; Krunic, T. Z.; Loncarevic, I. S.; Pajin, B. S. Functional, Rheological and Sensory Properties of Probiotic Milk Chocolate Produced in a Ball Mill. *RSC Adv.* **2016,** *6,* 13934–13941.

CHAPTER 2

FERMENTED WHEY BEVERAGES: AS FUNCTIONAL FOODS AND THEIR HEALTH EFFECTS

TATIANA COLOMBO PIMENTEL[1,*], MICHELE ROSSET[2], SUELLEN JENSEN KLOSOSKI[1,3], CARLOS EDUARDO BARÃO[1,4], and ADRIANO GOMES DA CRUZ[5]

[1]Federal Institute of Paraná (IFPR), Paranavaí PR 87703-536, Brazil, Tel.: +55 44 34820110

[2]Federal Institute of Paraná (IFPR), Colombo, PR 83403-515, Brazil, Tel.: +55 41 35351835, E-mail: michele.rosset@ifpr.edu.br

[3]E-mail: suellen.jensen@ifpr.edu.br

[4]E-mail: carlos.barao@ifpr.edu.br

[5]Instituto Federal de Educação, Ciência e Tecnologia do Rio de Janeiro (IFRJ), Mestrado Profissional em Ciência e Tecnologia de Alimentos (PGCTA), Maracanã, Rio de Janeiro, PR 20270-021, Brazil, Tel: +55 21 22641146, E-mail: food@globo.com

*Corresponding author. E-mail: tatiana.pimentel@ifpr.edu.br

2.1 INTRODUCTION

Whey is the residual liquid obtained after coagulation and drainage of the curd during the manufacture of cheeses. It is the main by-product of the dairy industry and its disposal in the environment, without prior treatment, represents an important source of pollution due to its high organic load. The treatments of the whey are expensive; therefore, numerous efforts have been made to turn the large whey volumes generated into ingredients suitable for food use.

Whey beverages are the main products in which whey is used as an ingredient. From the technological point of view, the main difference between fermented whey beverages and yogurts is the addition of whey as ingredient in the former. In addition, the yogurts are necessarily fermented by the microorganisms *Streptococcus thermophilus* and *Lactobacillus delbrueckii* ssp. *bulgaricus*, whereas in fermented whey beverages other microorganisms can be used in the fermentation process (Verma et al., 2017a).

The use of whey in whey beverages has some advantages, such as reduction of environmental problems related to its disposal, reduction of production costs, and the possibility of using the existing equipment in the milk processing plant. In addition, whey is a source of proteins of high biological value and is rich in minerals and vitamins, mainly riboflavin. It also has antithrombotic, antihypertensive, antimicrobial and antiviral properties, and hypocholesterolemic effects.

The current consumers prefer foods that promote good health and reduce the risk of disease. Such foods should be practical and convenient to use, have a pleasant taste, aroma, and texture, and have acceptable price. Probiotics are living microorganisms that beneficially affect the health of consumers when administered in adequate amounts. To exert their beneficial effect they must be maintained in quantities greater than 10^6 cfu/ml throughout the shelf life of the products.

In the past, probiotic cultures intended for inclusion in foods were selected for their technological properties. Today, they are selected based on the health benefits desired for the product; therefore, the processing of the products has to be often adapted in order to prevent lethal damage to probiotic cells.

The conversion of fermented whey beverages into probiotic products does not require substantial changes in processing; however, several points should be evaluated to ensure minimum counts of probiotic culture and to result in products with adequate physicochemical and sensory characteristics. This chapter will discuss the problem of whey as an industrial by-product, define probiotic cultures, selection criteria, and the main strains used and, finally, discuss the technological processing of probiotic fermented whey beverages, describing the technological challenges and possible solutions.

2.2 WHEY: AT A GLANCE

The dairy industries are manufacturing plants that process milk, producing the most diverse derivatives, among them, cheese. Cheese manufacturing is a method of transforming milk components into a product that is easy to store, has low volume, high nutritional value, pleasant flavor, and good digestibility. In this process, there is not 100% conversion of raw milk into cheese product. The cheese yield can vary between 8.5 and 20%, depending on the consistency of the cheese, resulting in two products: cheese and whey.

The significant pollutant power of whey, with a biochemical oxygen demand (BOD) about 175 times greater than typical sewage effluents, was in the recent past, prompting governments and other regulators to restrict and/or prohibit the disposal of untreated whey. Although the potential of whey as food ingredient is proven, many industries still consider it as waste and discard it as an effluent.

Whey, when considered as industrial liquid waste and dumped along with other liquid waste from the dairy industry, can overload the treatment system, as it has BOD between 25,000 and 80,000 mg/l. Owing to its high concentration of organic matter and nitrogen deficiency, its stabilization by conventional methods of biological treatment is difficult (Braile, 1971).

Whey may be defined as the liquid part, yellowish green in color, resulting from the coagulation of milk by acids or proteolytic enzymes. According to the type of coagulation suffered by casein in the manufacture of cheeses, it has two types: sweet whey, obtained by the enzymatic coagulation of casein, and acid whey, obtained by coagulation caused by lactic acid bacteria (LAB) or by the direct addition of mineral acid to milk. Their characteristics are differentiated and are presented in Table 2.1.

TABLE 2.1   Chemical Composition of Whey.

Components of whey	Types of whey	
	Sweet whey	Acid whey
pH	6.2–6.4	4.6–5.0
Titratable acidity (°D)	11.50	–
Moisture (g/100 ml)	93–94	94–95
Total solids (g/100 ml)	6–6.5	5–6
Fat (g/100 ml)	0.30	0.10
Protein (g/100 ml)	0.8–1.0	0.8–1.0
Lactose (g/100 ml)	4.5–5.0	3.8–4.3
Ash (g/100 ml)	0.5–0.7	0.5–0.7
Calcium (mg/100 ml)	40–60	60–80
Phosphates (mg/100 ml)	10–30	20–45

Source: Adapted from Dairyforall (2013); Jelen (2001).

According to its average composition, whey is approximately 93–95% water and contains about 50% of total solids present in the milk of which lactose is the main constituent. Whey proteins constitute less than 1% of dry matter.

The protein fraction contains approximately 50% of β-lactoglobulin, 25% of α-lactalbumin, and 25% of other protein fractions, including immunoglobulins. Owing to the high content of essential amino acids (notably lysine, cysteine, and methionine) and cystin, whey proteins are one of the nutritionally valuable proteins. Owing to such amino acid composition, whey proteins have much higher biological value (but also other parameters that determine nutritional value) in comparison with casein or other proteins of animal origin, including egg proteins which have been considered for a long time as referent proteins (Jelicic et al., 2008).

β-lactoglobulin (β-lg) is the protein present in higher quantities in the whey, has a genetic polymorphism, is a typical globular protein, relatively rich in sulfur amino acids and exhibits reversible conformational changes at temperatures lower than 70°C. Therefore, β-lg is a protein of great nutritional value and high solubility and stability.

The α-lactoalbumin (α-la) is tryptophan rich, approximately 6% by weight, suiting very well for human nutrition. One of the main physiological functions of α-la is related to its participation in the synthesis of lactose in the mammary glands. Immunoglobulins represent a special class of high-molecular-weight proteins found in blood serum and other body fluids. In whey, three classes of immunoglobulins (IgG, IgA, and IgM) are found in small amounts, with a higher concentration in colostrum. In bovine milk, IgG is the found in the largest amount, about 75% of the total immunoglobulin content of milk, while IgA is the one with the highest content in human milk.

Whey still contains most of the water-soluble vitamins present in milk. It is particularly rich in vitamin B, supplying an appreciable amount of the human needs of vitamins B_2, B_5, and B_6.

In addition to the whey in its liquid form, whey powder is obtained, which according to the Codex Alimentarius (Codex Alimentarius, 2011), is the product obtained by the dehydration of whey by means of technologically adequate processes, suitable for human consumption. The main unit operations that may be involved in the production of whey powder are membrane separation, vacuum evaporation, crystallization, and spray drying. Reverse osmosis can be used as the initial phase of water withdrawal from the whey and is characterized by low energy consumption.

Several technological problems can occur during the production of whey powder, especially colloidal particles adhesion, equipment clogging, and shortening of the shelf life of the product. The development of operational and technological attributes for the processing of whey powder is necessary for an efficient equipment design in the industry of the concentrated and dehydrated products. Well-designed process equipment and lines will contribute to a greater competitiveness of these industries in the national and international markets, as well as the better use of the whey.

Whey protein concentrate (WPC) is known for its industrial applications because of the variable functionality, aroma, and the ability to modify the texture and desorption properties of the products to which it is applied. WPC can vary its protein composition from 35 to 80%. When it contains 53% of protein its composition is 35% of lactose, 5% of fat, and 7% of ashes. When the protein concentration increases to 80%, the lactose content decreases by an average of 7%, and fat and ashes contents are between 4 and 7%. WPC is obtained by removal of sufficient amount of nonprotein constituents from pasteurized whey and is manufactured by physical separation techniques, such as precipitation, filtration, or dialysis. This concentrate can be used in dairy products, bakery, and confectionery, nutritional products as an economical source of milk solids, as a partial substitute for skim milk powder and as soluble protein in acidic solutions.

2.2.1 INDUSTRIAL APPLICATION AND MARKET OF WHEY

Coming from the dairy industry's need to develop and use functional ingredients, whey proved to be a raw material with great functional potential and commercial advantages. The company Mordor Intelligence released a report revealing that only the whey protein market was valued US$5.4 billion in 2014 and is expected to reach US$8.4 billion by 2020, with increase of 7.6% per year (Mordon Intelligence, 2016). The report indicated that the increased demand for the special benefits of whey is related to its nutritional content and ability to increase shelf life of the products. There is a consumer interest in health and well-being, due to the busy lifestyle and work schedule, with investments in meals that provide all the nutrients they need. According to Research and Markets, North America has a 32% share of the whey market. Much of this may be due to increased interest in sports nutrition, a US$6.7 million industry in the United States, according to Euromonitor.

Some limitations are related to the use of whey as an ingredient in the food industry, such as the lack of industrial technologies, problems related to its physical stability, and limitations on its use in food. The researchers' attention in studies on the use of proteins as functional ingredients is increasing, for example, there are commercial dairy products with functional appeals based on bioactive peptides obtained from milk proteins (caseins and whey proteins) in the international market. In many cases, this is possible by employing proteins as functional agents to develop products with special characteristics and add value to by-products, which in general pose a problem for industries.

The membrane systems allow the separation of milk and whey in several fractions. It is possible to use whey in the form of lactose, whey powder, with high protein concentrates (close to 89%), enriched protein isolates, or in the form of isolated components, such as glycomacropeptides or lactoferrin.

2.3 LACTIC ACID BACTERIA (LAB) AND PROBIOTIC CULTURES

2.3.1 LAB

LABs are gram-positive, facultative anaerobes or microaerophilic, and nonsporulating bacteria that accumulate lactic acid in the environment as a product of their primary metabolism (Patel et al., 2017). This group is distinguished by the ability to carry out fermentation of carbohydrates to form lactic acid and includes cocci and bacilli, representatives of the species belonging to the genus *Lactococcus, Lactobacillus, Streptococcus, Pediococcus, Leuconostoc, Bifidobacterium*, and several others (Wyszyńska et al., 2015; Verma and Srivastav, 2017; Verma et al., 2017a). Many LABs have the "generally regarded as safe" status from the US Food and Drug Administration (FDA), but it is worthwhile mentioning that pathogenic bacteria such as *Streptococcus pyogenes* and *Streptococcus pneumonia* also belong to this group (Daniel et al., 2011; Verma et al., 2017b).

The bacteria in the LAB genera are classified by their cell morphology and the fermentation pathway used to ferment glucose. They are widespread, natural habitants of many plants and are also part of gastrointestinal microflora (Abadias et al., 2008; Espinoza and Navarro, 2010; Verma and Srivastav, 2017).

LABs are among the most important groups of microorganisms used in food fermentations. They are important to the food industry because of their ability to transform fermentable sugars into lactic acid, ethanol, and other metabolites, which changes the characteristics of the product by lowering the pH and creating unfavorable conditions for the growth of potentially pathogenic microorganisms in both food products and the human intestinal microflora (Chiu et al., 2007; Espinoza and Navarro, 2010; Verma and Srivastav, 2017). They also contribute to the flavor and texture of fermented products and inhibit food spoilage bacteria by producing growth-inhibiting substances and large amounts of lactic acid. As agents of fermentation, LABs are involved in making yoghurt, cheese, cultured butter, sour cream, sausage, cucumber pickles, olives, and sauerkraut, but some species may spoil beer, wine, and processed meats (Todar, 2016).

LABs have been widely used in the production of fermented foods, where most do not present pathogenic effects to the host. The most important LAB genera are *Leuconostoc, Lactobacillus, Streptococcus,* and *Pediococcus.*

2.3.2 PROBIOTIC CULTURES

In recent years, both research and consumer were interested in probiotic cultures, because of the increased clinical evidence that support some of the proposed health benefits related to the use of probiotics, particularly in managing certain diarrheal diseases (Williams, 2010). These evidence occur due to the dynamic interaction between the colonic content and the intestinal mucosa by the consumption of these microorganisms in the diet (Burns and Rowland, 2000).

The term probiotic comes from a Greek word meaning "for life." It is considered as a new word, although the beneficial effects of certain foods containing live bacteria have been recognized for centuries (Williams, 2010). According to the United Nations and World Health Organization Expert Panel, probiotics are defined as "live microorganisms which when administered in adequate amounts confer health benefit to the host" (FAO, 2002). They have the ability to resist the digestion process and are mostly able to populate the gut and provide consumer benefits, in addition to improving the balance of the intestinal microflora.

The FDA allows the use of microorganisms in foods, especially probiotics. It is worth noting that in Brazil, 65% of the functional foods sold are products that use probiotic cultures (Cruz et al., 2007). In Europe, the functional product market has Germany, France, the United Kingdom, and the Netherlands as main countries (Makinen-Aakula, 2006). Despite the possibility of using microorganisms as probiotics in food, some criteria should be considered in the selection of probiotic cultures, which are divided into three groups safety, technological, and functional criteria (Fig. 2.1).

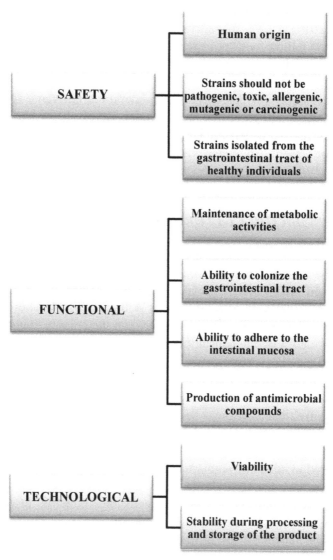

FIGURE 2.1   Criteria for selection of probiotic strains.

Regarding the safety criterion, the microorganisms should preferably be of human origin, as it is believed that probiotic strains present greater functionality in similar environments to which they have been isolated. However, the specificity of action, that is, the beneficial effect provided, has been considered a factor of greater relevance than the origin of the microorganism. In addition, microorganisms considered as probiotic should have a history of nonpathogenicity and should not be associated with diseases such as endocarditis and gastrointestinal disorders. Finally, they cannot negatively affect the immune system, produce toxins, undo the colonocyte function, or present genes that determine antibiotic resistance.

Stability against acid and bile salts, the ability to adhere and colonize the human gastrointestinal tract and the ability to produce antimicrobial compounds are considered as fundamental criteria for the selection of probiotic strains.

In addition, strains of probiotic microorganisms should maintain their stability and viability during food processing and storage. The supplementation of the products with the strains should result in a product with sensory qualities compatible with the conventional products, such as flavor and texture. The dairy products are especially considered as ideal vehicles for delivering probiotic bacteria to the human gastrointestinal tract. The matrices used most frequently are cheese, yoghurt, ice cream, and frozen dairy desserts.

Other matrices, such as fruits, vegetables, cereals, and legumes, have been used in the development of nondairy probiotic products, due to the presence of minerals, vitamins, dietary fibers, and antioxidants. However, the incorporation of probiotics in fruit juices requires the protection against acid conditions. In this case, probiotic microencapsulation should be evaluated.

Among the microorganisms, the bacteria belonging to the genus *Lactobacillus* are considered the safest one. On the other hand, certain bacteria of the genus *Streptococcus* are opportunistic pathogens.

Several microorganisms have probiotic potential (Table 2.2) and are traditionally used in fermented dairy products. The majority of probiotic products available in the marketplace contain species of *Lactobacillus* and *Bifidobacterium*, which are the main genera of Gram-positive bacteria currently characterized as probiotics (FAO, 2002). The most used strains in yoghurt supplementation are *Lactobacillus acidophilus* and *Bifidobacterium bifidum*. However, some other species of bacteria and fungi such as

Bacillus, Enterococcus, and *Saccharomyces* have also been used as probiotics in food and feed. *L. acidophilus* has been considered the predominant lactobacilli in the intestinal tract of healthy humans and therefore is the organism most commonly used in probiotic products. Its optimum conditions are temperature between 35 and 40°C, 0.3–1.9% of titratable acidity, and pH at 5.5–6.0 (Espinoza and Navarro, 2010).

TABLE 2.2 Mains Probiotic Microorganisms.

Lactobacillus	Bifidobacterium	Other lactic acid bacteria	Non-LAB
L. acidophilus	*B. adolescentis*	*Enterococcus faecalis*	*Bacillus cereus var. toyoi*
L. amylovorus	*B. animalis*	*Enterococcus faecium*	
L. casei	*B. bifidum*	*Lactococcus lactis*	*Bacillus subtilis*
L. crispatus	*B. breve*	*Leuconostoc mesenteroides*	*Bacillus clausii*
L. delbrueckii ssp. *bulgaricus*	*B. infantis*		*Bacillus coagulans*
	B. lactis	*Pediococcus acidilactici*	*Bacillus licheniformi*
L. gallinarum	*B. longum*	*Streptococcus thermophilus*	*Saccharomyces cerevisiae*
L. gasseri	*B. thermophilum*		
L. johnsonii	*B. essensis*	*Streptococcus diacetylactis*	*Escherichia coli* ssp. *nissle*
L. paracasei			
L. plantarum		*Streptococcus cremoris*	*Propionibacterium freudenreichii*
L. reuteri			
L. rhamnosus			
L. brevis			
L. cellobiosus			
L. curvatus			
L. fermentum			
L. helveticus			

Source: Adapted from Holzapfel et al. (2001); Saad et al. (2013).

The genus *Enterococcus* is constituted by transient colonizing organisms of the gastrointestinal tract, that is, they are found in external and internal surfaces, during a few hours, days, or even weeks, indispensable in the treatment of diarrhea. Among the beneficial effects is the ability to reduce low-density lipoprotein cholesterol levels. The species of *Enterococcus faecium* stands out for its marked activity. However, *E. faecium* has long been confused with *Enterococcus faecalis*, pathogenic species, and its studies were impaired, with its late onset in relation to other bacteria beneficial to the human organism.

A dairy probiotic product should not only exhibit the minimum number of cells that are capable of providing beneficial health effects, but also be sensory acceptable to consumers, and should be evaluated throughout the shelf life of the product to avoid potential problems during marketing. It is possible to find in the literature studies like that of Khorokhavar and Mortazavian (2010) that developed a probiotic stirred drink added with *L. acidophilus* with low acceptability due to off-flavors.

Among the several benefits that the bacteria of the genus *Bifidobacterium* spp. and *Lactobacillus* have, the following stand out: recolonization of the intestinal microbiota after administration of antibiotics, treatment of individuals with diarrhea, reduction of constipation, treatment of inflammatory diseases in the intestine, reduction of lactose intolerance, reduction of blood cholesterol level, resistance to microbial infections, and potential anticancer action. Several studies demonstrated that the continuous intake of foods containing these microorganisms leads to a reduction in the activity of β-glucuronidase and nitroreductase enzymes, which are present in inflammatory processes.

Frequently the term probiotic is related to LAB belonging to the genera *Lactobacillus* and *Bifidobacterium*. However, it is possible to find other microorganisms that present the potential to be probiotic cultures, especially the species *Bacillus subtilis*, *Bacillus clausii*, *Bacillus cereus*, *Bacillus coagulans*, and *Bacillus licheniformis*, which present heat resistance, storage stability at room temperature, and ability to colonize the small intestine, as long as they can support stomach acid pH.

In relation to probiotic yeasts, *Saccharomyces cerevisiae* stands out, because it presents tolerance to the acidic pH and the bile acids, favoring reduction of the intestinal pro-inflammatory process.

According to Psomas et al. (2001), some technological and probiotic properties of selected yeast isolates from infant feces and feta cheese were investigated. The tested strains exhibited lipolytic and proteolytic activities. They also tolerated low pH and survived satisfactorily in gastric juice in vitro as well as in the presence of bile salts. In general, the isolates from feces were more resistant to low pH and bile than those from feta cheese. The selected strains could be used as starter supplements for industrial fermentations. However, the adhesion properties of these yeasts should be better investigated.

2.3.3 HEALTH BENEFITS

In Brazil, according to the National Sanitary Surveillance Agency (ANVISA), the claim of functional or health properties of a probiotic culture must be proposed by the manufacturer of the product and evaluated, based on the definitions and principles established in resolution number 18 of April 30, 1999, which establishes the basic guidelines for the analysis and verification of functional and/or health properties alleged in food labeling (ANVISA, 2016; Brasil, 1999). The requirements are presented in Table 2.3. These requirements are also used in other countries.

TABLE 2.3 Requirements for Claiming Functional Properties of Probiotic Products.

Requirement	Description
Adverse effects	Provide studies describing the adverse effects observed with the strain.
Antimicrobial resistance	Provide information on the genetic basis of antimicrobial resistance.
Characterization of the microorganism	Identification of the genus, species, and strain. Information on the deposit of the microorganism strain in an internationally recognized culture bank. Origin and form of obtaining, including information if the microorganism is genetically modified. Production of toxins and bacteriocins.
Evidence of effectiveness	Provide robust scientific evidence, constructed through randomized, double-blind, placebo-controlled clinical studies, whose outcomes show the proposed relationship in the claim between the consumption of the product subject to the petition, or product with the equivalent matrix, and the functional effect.
Hemolytic activity	Determination of hemolytic activity for species with hemolytic potential.
Viability	Submit an analysis report that verifies the minimum viable quantity of microorganism to exert the functional property at the end of the shelf life of the product and under the conditions of use, storage, and distribution.

Source: Adapted from ANVISA (2016).

With respect to evidence of efficacy, where the effect cannot be measured directly, the validated biomarkers that are related to the claimed effect should be identified. The identification and measurement of the effect must be clearly defined. Identification of the strain and the amounts tested in the studies used as references are essential. The size of the sample should be duly justified and the participating population must match the size for which the product is intended. Health claims should be based on epidemiological studies. In the case of products with more than one microorganism or products that mix prebiotic fibers with microorganisms, the probiotic effect must be checked for the combination (ANVISA, 2016).

Regarding the nutritional benefits related to probiotic microorganisms (Table 2.4), the most important are the production of vitamins, the availability of minerals used by digestive enzymes, and the integration of the intestinal mucosa. It should be emphasized that the functionality of the probiotic culture will depend on the strain, that is, there is no probiotic microorganism that will perform all beneficial functions for the host. Table 2.4 demonstrates some beneficial effects of some probiotic strains.

TABLE 2.4 Probiotics Health Benefits Effects.

Probiotic strains	Potential health benefit
Bifidobacterium breve	• Immune modulation and stimulation • Reduced symptoms of irritable bowel disease
B. animalis	• Increased IgA secretion
B. lactis Bb12	• Shortening the frequency of rotavirus and traveler's diarrhea Inhibitory effect against *Helicobacter pylori*
B. longum BB536	• Treatment of allergy
Escherichia coli Nissle 1917	• Inhibitory effect of *H. pylori* • Fewer relapses of inflammatory bowel disease • Immune modulation • Recovery of ulcerative colitis; exclusion of pathogenic *E. coli*
Lactobacillus plantarum 299v	• Relief of irritable bowel syndrome • Reduction of low-density lipoprotein cholesterol • Reduction of the recurrence of *Clostridium difficile* diarrhea
Lactobacillus acidophilus La5	• Shortening rotavirus and antibiotic-associated diarrhea
L. acidophilus M92	• Immune system activation in patients with irritable bowel syndrome • Lowering of serum cholesterol
L. casei DN114001	• Immune modulation
L. casei Shirota	• Downregulation of lipopolysaccharide-induced interleukin 6 and interferon-α

TABLE 2.4 *(Continued)*

Probiotic strains	Potential health benefit
L. reuteri ATCC PTA 6475	• Immune modulation
L. reuteri DSM 12246	• Shortening of rotavirus diarrhea
L. rhamnosus GG	• Treatment of acute rotavirus and antibiotic-associated diarrhea
	• Immune modulation
	• Relief of inflammatory bowel disease
	• Treatment and prevention of allergy
	• Postsurgical prevention of pouchitis
L. salivarius UCC118	• Relief symptoms of inflammatory bowel disease and modulation of gut microflora
Saccharomyces boulardii	• Fewer relapses of inflammatory bowel disease
	• Reduction of antibiotic-associated diarrhea
	• Prevention of recurrent Clostridium *difficile diarrhea*
Streptococcus thermophilus	• Enhance lactose intolerance
	• Prevention of rotavirus diarrhea

Sources: Adapted with permission from Saad et al. (2013).

2.4 TECHNOLOGICAL PROCESSING OF PROBIOTIC FERMENTED WHEY BEVERAGES

Fermented whey beverage is a dairy product which contains whey as a mandatory ingredient and is fermented by the action of specific microorganisms and/or added fermented milk. This product cannot be subjected to heat treatment after fermentation. Fermented whey beverage is more liquid and refreshing than liquid yogurts available in the market. As it presents whey in its composition, it has a lower cost and attracts consumers, mainly, of lower purchasing power.

The mandatory ingredients in the manufacture of fermented whey beverage are milk (raw, pasteurized, sterilized, ultrahigh temperature processing [UHT], reconstituted, concentrated, powdered, whole, partly skimmed or skimmed), whey (liquid, concentrated, or powdered), and lactic cultures or fermented milk.

Optional dairy ingredients are cream, dairy solids, butter, butter oil, food caseinates, dairy proteins, buttermilk, and other dairy products. Optional nondairy ingredients are sugars, maltodextrin, nutritive and nonnutritive sweeteners, fruit pieces, pulp or juice, and other preparations based on fruit, honey, cereals, vegetables, vegetable fat, chocolate, dried fruits, coffee, spices and other natural and innocuous flavoring foods, starches or modified starches, gelatin, or other ingredients.

2.4.1 INGREDIENTS FOR DEVELOPMENT OF PROBIOTIC FERMENTED WHEY BEVERAGES

Milk should be fresh, produced in the best possible sanitary conditions, with low counts of spoilage microorganisms and somatic cells, and absence of pathogenic microorganisms and inhibitory substances such as antibiotic, mycotoxin, pesticide, or sanitizing residues (Niamah and Verma, 2017). As probiotic fermented whey beverages require the action of lactic and probiotic cultures, interference of these factors in their development in milk can occur, causing the inhibition of the cultures, compromising product development and its quality parameters, and decreasing the counts of the probiotic culture to lower levels than those recommended to have a beneficial effect on consumer health.

The milk used may have been pre-pasteurized and refrigerated but after the addition of sugars, thickeners, or other ingredients, it has to be thermally treated under suitable conditions to destroy spoilage and pathogenic microorganisms that may be present in these ingredients, making the environment more conducive to the development of the microorganisms from lactic and probiotic cultures.

The whey used in the preparation of probiotic fermented whey beverage may be liquid, concentrated, or powdered. The liquid whey is highly perishable due to its high water activity, making the use of heat treatment necessary. Equipment for the concentration and dehydration of whey are not available in small and medium-sized dairy producers, which causes them to import this raw material in the form of powder. Therefore, the use of whey in the liquid form reduces the costs of obtaining the raw material and, consequently, the costs of the fermented whey beverages.

The liquid whey must have a maximum acidity of 13°D and must not contain particles of cheese mass. Thus, it is subjected to filtration or clarification prior to its use as raw material in fermented whey beverages. In Brazil, the whey may be added to the quantity desired by the industry, provided that the milk base in the final product is at least 51% (m/m). Ideally, the effect of the addition of whey in the physicochemical and sensory characteristics of the products should be studied in order to use it without compromising the quality parameters. Concentrations of 15–35% (v/v) of whey have already been observed in commercial products, but up to 50% have already been studied.

Whey has low total solids content (6–7%), which makes the oral tactile sensation of fermented whey beverages lower than that observed in fermented milk. Therefore, in order to promote the improvement of texture characteristics, hydrocolloids are typically added, which have the property of increasing the viscosity of the products and preventing the settling of particles during refrigerated storage. The selection of hydrocolloids should be judicious since they must be able to fulfill their technological function in acidic environmental conditions (pH 4–4.6) and without masking the natural flavor of the product. Carboxymethyl cellulose, pectin, xanthan gum, gelatin, and alginate are some examples of hydrocolloids for this purpose (Jelicic et al., 2008). The concentration of hydrocolloids varies, as a rule, between 0.1 and 0.4%.

Gelatin is effective for controlling moisture and results in a product of pleasant texture, not interfering with the flavor of the beverage. Low methoxyl pectin is used to promote viscosity enhancement, while high methoxyl pectin has the stabilizing function (Damin et al., 2009).

In recent years, the addition of probiotic cultures capable of producing exopolysaccharides (EPS) has been also studied. EPS have the ability to improve fermented whey beverage texture similarly to hydrocolloids, as they act as texturizing and stabilizing agents, first increasing the viscosity of the final product and then retaining water and interacting with other milk constituents, such as proteins and micelles, increasing the viscosity of the final product. Therefore, the selection of probiotic cultures with EPS production capacity would result in consumer health benefits and important technological properties to the products.

Fermented whey beverages are added sugars or sweeteners to promote sweet flavor in products. Owing to the high availability, the most commonly used sugar is sucrose, in concentrations ranging between 8 and 12% in relation to the volume of milk. However, there are products in the market that contain 16% sugar. In the case of the development of products without the addition of sugars or with partial substitution of them, it is common to use sweeteners such as sucralose, saccharin, cyclamate, or aspartame.

In recent years, with increased consumer demand for natural products and with health benefits, other ingredients have been studied as substitutes for sugars in fermented dairy products. Oligofructose, a nondigestible carbohydrate obtained by partial hydrolysis of inulin extracted from chicory roots, is one of the components studied. Oligofructose have beneficial effects associated with prebiotics. However, its sweetening power is around 30% when compared to that of sucrose.

The high amount of minerals results in whey beverages with a characteristic and undesirable flavor of "whey"; being characterized as bitter and salty flavor. To overcome this problem, fermented whey beverages are conventionally flavored. The use of different flavors in whey beverages also favors whey consumption by a very varied public, from children to the elderly.

Fruit preparations are products widely used in the production of whey beverages. They consist of fruits, components of fruits and sugars, as well as extracts, aromas, dyes, stabilizers, and acids. They are preserved by heat treatment or chemical compounds. The fruit content in these preparations varies with the type of fruit, being usually between 20 and 40% and they have sugar concentrations between 35 and 65°Bx. The pH can vary between 3 and 4.5 and the amount used can reach 5% of the volume of the beverages (Damin et al., 2009).

Flavoring is essential in the flavor composition of fruit-flavored whey beverages since flavor is the main factor that influences the consumer's choice of a product. Natural or artificial flavoring and coloring agents may be used. Permission to use vegetable fats, which can also be vegetable oils, allows the development of new products that can be sources of fatty acids that are important for health.

2.4.2 MANUFACTURING TECHNOLOGY FOR PROBIOTIC FERMENTED WHEY BEVERAGES

The technology of manufacturing fermented whey beverages is quite simple. There are two main processes employed in the industry (Fig. 2.2).

The first process is quite similar to that used in the manufacture of fermented milk and yogurts and, therefore, is the most used by the industry. The ingredients (milk, whey, sugar, thickener, and stabilizer) are homogenized, heat treated, cooled and lactic acid culture is added. The base medium is incubated at the optimum temperature of the micro-organisms of the lactic acid culture until the desired acidity is reached. Then, cooling, the addition of the fruit preparations, aromas and colorings, beating, packaging, and storage take place.

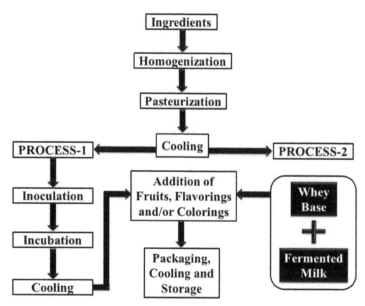

FIGURE 2.2   Manufacturing processes of fermented whey beverages.

In the second process, a whey base is prepared by homogenizing the whey and other ingredients (sugar, thickener, and stabilizer). The whey base is heat treated and then cooled to 20°C. Then, it is mixed with fermented milk, fruit preparations, flavorings, and colorings and subjected to beaten, packaging, cooling, and storage. The production of the fermented milk follows the steps for this type of product, similar to those used in the previous process (addition of lactic acid culture, fermentation, cooling, and storage).

Therefore, the difference between the processes resides in the fact that in the first process all the base medium is submitted to the fermentation process and, in the second process, the whey and other ingredients do not participate in the fermentation, being mixed to the fermented milk. The selection of the best process depends on the industrial conditions.

Fermented whey beverages are widely available in the market. The use of probiotic cultures in their manufacture does not require substantial changes in processing, which, from an industrial point of view, is important, as it does not result in an increase in operational costs. However, minor changes may be necessary in order to obtain a product with adequate probiotic culture viability and appropriate physicochemical and sensory characteristics.

Table 2.5 presents the main technological challenges in the development of probiotic fermented whey beverages.

Homogenization affects the lipid phase of milk, and therefore, is important in whey beverages made with whole milk, as it promotes the homogeneous dispersion of the constituents, improving the flavor, consistency, presentation, and digestibility of the products.

The objective of the heat treatment is to eliminate pathogenic microorganisms; to destroy or reduce the spoilage microorganisms to an acceptable number; and to reduce the total microbial population, so that it does not interfere in the development of lactic bacteria and probiotic cultures. Furthermore, the heat treatment hydrates the stabilizers which dissolve under heat.

Different time versus temperature binomials can be used for the heat treatment of the base medium of fermented whey beverages. Milk pasteurization or sterilization standards such as 62–65°C/30 min (slow pasteurization), 72–75°C/15–20 s (rapid pasteurization), or 130–150°C/2–4 s (UHT) may be applied. However, to produce products with better texture characteristics, 85°C/30 min or 90–95°C/5 min are commonly used. In batch processes, the lower temperatures (85°C) are used, while in continuous processes, with heat exchangers, intermediate temperatures (90–95°C) are proposed.

After the heat treatment, the base medium is cooled to the appropriate temperature to perform the inoculation with the lactic and probiotic cultures, which is usually 42–45°C. Very high temperatures inhibit or destroy the inoculated microorganisms, while very low temperatures cause delay in the fermentation process, resulting in longer times to obtain the product and higher operating costs.

The optimum temperature for the development of probiotic cultures is usually 37°C. Therefore, if the selected strain has low milk growth capacity and/or low stability, it is recommended to evaluate the decrease in fermentation temperature from 42°C to 37–40°C. The evaluation should

TABLE 2.5 Technological Challenges in the Development of Probiotic Fermented Whey Beverages.

Stage	Challenge	Solution
Ingredients	Sugar, sweeteners, flavorings, preservatives, and fruits can inhibit probiotic culture	• Test the compatibility of the ingredients with the probiotic culture • Use less acid fruits • Avoid adding preservatives
Lactic and probiotic cultures	Possible antagonism between lactic and probiotic cultures	• Select the lactic culture based on preliminary compatibility tests • Reduce the size of the lactic acid culture inoculum
	Low growth of probiotic culture in milk	• Selection of strains capable of growing in milk • Addition of nutrients to milk • Addition of growth factors (prebiotics)
	Moment of the addition of probiotic culture influences its survival	• Select the moment of addition by preliminary tests. Suggested addition before fermentation process
	Size of probiotic inoculum has influence on the probiotic viability	• Tests with concentrations suggested by manufacturers and/or preliminary tests
Fermentation	Optimal temperature of probiotics is different from the temperature conventionally used in the fermentation of dairy products	• Change fermentation temperature to 37–40°C, if necessary
	Probiotic cultures are sensitive to low pH	• Stop the fermentation process at higher pH (4.8–4.9) • Selection of strains resistant to acidity • Reduction or elimination of *L. bulgaricus* from lactic culture • Microencapsulation of probiotic culture

TABLE 2.5 *(Continued)*

Stage	Challenge	Solution
Beating	Incorporated oxygen is toxic to probiotic cultures	• Addition of antioxidants to the medium • Use of lactic cultures with antioxidant capacity • Avoid strains that produce peroxide • Selection of probiotic cultures more resistant to oxygen
Packaging	Oxygen permeating the packaging is toxic to probiotics	• Use of multilayer packaging • Use of selectively permeable packaging • Addition of antioxidants to the packaging material • Store in glass packaging

consider the viability of the probiotic culture, but also the technological characteristics of the product, such as fermentation time, and the final sensory characteristics. This is because the use of temperatures below 40°C results in prolonged fermentation times, with a consequent economic impact, as well as, on product flavor (lower acetaldehyde production) and post-acidification. For the fermentation of probiotic fermented whey beverages, the traditional cultures of yogurts (*S. thermophilus* and *L. bulgaricus*) and probiotic cultures, such *as L. acidophilus, L. casei, B. bifidum, B. longum,* and *B. adolescentis* are used.

The number of total viable bacteria must comply with a minimum of 10^6 cfu/ml of product throughout the shelf life. In the case of using claims due to the presence of probiotic cultures, the minimum number of viable microorganisms depends on the legislation of each country. However, minimum counts of 10^6–10^7 cfu/ml of the product have been suggested (Donkor et al., 2007). It is worth mentioning that when working with probiotic cultures, the count should be performed for the specific microorganism and not the total counts of LAB.

Lactic and probiotic cultures are usually marketed in lyophilized form, for direct use in milk and presenting high concentrations of bacteria. The choice for lyophilized cultures lies in the fact that they are in standard-ized form; there is no need to activate them, dispense propagation, reduce labor, and inhibit phage contamination. Other options of lactic cultures are in liquid or ultra-frozen forms.

The main function of the lactic culture is to produce lactic acid from the fermentation of lactose, resulting in a fermented product with pH 4.5–4.6. Probiotic microorganisms are generally neutrophils, that is, they have an optimal pH growth between 5 and 9. It is therefore recommended that the pH of the product in which the probiotic culture is added is maintained at levels above 4.6. In this case, it is possible to stop the fermentation process at higher pH (4.7–4.8) so that during the refrigerated storage, the pH will not become too low and compromises the survival of the probiotic culture in the product. In general, lactobacilli are more acid-resistant than *Bifidobacteria.*

Other alternatives would be: selection of probiotic strains more resistant to the acidity of the medium and reduction or elimination of *L. bulgaricus* from the lactic acid culture in order to reduce the post-acidification of the product and the production of hydrogen peroxide. However, *L. bulgaricus* contributes significantly to the fermentation process because of the proteolytic capacity, as well as, to the production of acetaldehyde, and the removal of this strain can affect the rate of acidification and the flavor of the product. Fermented whey beverages obtained by process 2 (Fig. 2.2), commonly have a higher pH, since the fermented milk is added to the mixture containing whey. In this way, this type of process can benefit the stability of probiotic cultures in products.

Concomitant to the development of acidity, other characteristics of the products are improved by the action of lactic cultures, such as flavor, by the production of flavor compounds (acetaldehyde and diacetyl), nutritional value by liberation of free amino acids and synthesis of vitamins of the B-complex, and the texture by the increases in the viscosity. In addition, although some LAB microorganisms of the lactic culture are not considered probiotics in many countries, they can have beneficial effects on consumer health.

The concentration of each of the microorganism component of the lactic culture may vary with 1:1, 2:1, or 3:2 is commonly found for *S. thermophilus* to *L. bulgaricus*. In some cases, the concentration of *L. bulgaricus* can be much lower than that of *S. thermophilus*, due to its high post-acidification capacity, reducing the pH of the products during refrigerated storage and, consequently, decreasing the shelf life of the products. This is because, in general, consumers do not like very acid beverages.

The importance of the use of these two specific microorganisms as components of the lactic culture resides in the fact that they act in symbiosis in the products. *S. thermophilus* releases compounds (formic acid and sulfur dioxide) that benefit the growth of *L. bulgaricus* and lower the pH to optimum values for its development. *L. bulgaricus*, in return, releases amino acids and peptides from milk protein, compounds that stimulate the growth of *S. thermophilus*. In general, the addition of the lactic acid culture takes place at a concentration of 2–3%.

As the purpose is to produce fermented whey beverages with functionality, due to the action of probiotic cultures, it is important to evaluate the compatibility of strains constituting the lactic culture with the probiotic strain that will be added. This is because there may be antagonistic or synergistic effects between them. Antagonist effects would be related to the production of acids, peroxides, bacteriocins, or other metabolites by the lactic cultures with consequent inhibition of the probiotic culture. In cases, where the cultures have little compatibility, but alterations in the composition of the lactic culture are not possible, the alternative is to decrease the lactic acid culture concentration, reducing the observed antagonistic effects. This is because the pH decline would be slower, keeping the medium more conducive to the development of the probiotic culture for longer. In addition, the formation of toxic compounds would also be lower.

The addition of the probiotic culture can be done in four different ways: at the beginning of the fermentation process (as a component of the lactic culture or by direct addition of a concentrated probiotic culture); after the fermentation of the milk; by double fermentation (probiotic culture fermented separately for 2 h at 40–42°C and then added to the milk with the lactic acid culture. The fermentation proceeds until the desired pH); or in separate tanks (probiotic culture and lactic culture are propagated separately and then put together).

The addition of the probiotic culture before fermentation has the advantage that the probiotic culture becomes more adapted to the medium because it participates in all the changes that occur in the process, including the gradual decrease of the pH. In addition, probiotic culture can take advantage of the peptides that are released by lactic microorganisms during milk fermentation. From the manufacturing point of view, it would be the most practical and compatible with the industrial routine.

In the case of process 1, the probiotic culture would be added together with the lactic culture after the heat treatment and cooling of the product and, in the case of process 2, in the same step, but in the manufacture of the fermented milk (Fig. 2.2).

The amount probiotic culture inoculum is a key factor in ensuring sufficient viable cells in the final product, and, in general, higher levels of inoculation result in higher counts. Care must be taken, however, because an excessive inoculation level can affect the quality of the final product, in addition to resulting in higher costs. Therefore, the moment of the inoculation of the probiotic culture and the inoculum size should be evaluated to ensure the preparation of products with quality and adequate quantities ($>10^6$ cfu/ml) of probiotic culture throughout the shelf life. The recommended concentration of probiotic culture is provided by the manufacturer, with 0.1 g/l of milk being common when added before the fermentation process. When the cultures are added after the fermentation process, higher concentrations are suggested because the microorganism will be inserted into an already acid product and with a large amount of metabolites that can be harmful.

It is possible to prepare fermented whey beverages only with the use of probiotic cultures, that is, without the addition of lactic culture. However, several changes must be made to obtain products with adequate physicochemical and sensory characteristics. Probiotic cultures usually have limited growth in milk due to the low concentration of free amino acids and low-molecular-weight peptides in this substrate. Therefore, if the probiotic culture is used in order to promote fermentation, the addition of amino acids, peptides, and other micronutrients to the milk should be evaluated, as well as, an increase in its solids content. Another alternative would be the selection of probiotic cultures with adequate growth in milk.

The curd is subjected to cooling from the time the desired acidity for the product is reached, the pH generally being between 4.5 and 4.6, with fermentation times between 4 and 5 h. After this, the product is beaten, with breakage of the curd and submitted to a homogenization step, eliminating the presence of pieces or large flakes, and reducing the viscosity.

During the beverage beating stage, there is the incorporation of oxygen, a critical factor for maintaining the viability of probiotic cultures. Probiotic cultures have anaerobic or microaerophilic metabolism, and oxygen is directly toxic to the cells and, in its presence, some cultures such as *L. bulgaricus* produce peroxide, which inhibits the growth of probiotics. Therefore, the survival of probiotic cultures to the beat process should be carefully evaluated. As this step cannot be suppressed from the process, alternatives should be considered, such as the addition of compounds with antioxidant activity (vitamin C), use of lactic cultures with antioxidant capacity, or selection of probiotic strains resistant to oxygen.

Flavoring is done after fermentation, characterizing a critical point to be controlled. For this reason it is fundamental to choose raw materials with guaranteed quality, otherwise the risks of contamination are very high. In addition, the compatibility of fruit preparations with probiotic culture should be checked, because there may be too much reduction of pH due to the acidity of the fruits, with the compromised viability of the probiotic culture. Inhibitory effects of components present in some fruits have been reported.

After this, the products are packed in plastic containers commonly made of polyethylene. Plastic packaging allows the permeation of oxygen during the storage of the product, which is not the case with glass containers. As mentioned earlier, this oxygen can cause cell death of probiotic cultures. In this case, the development of multilayer packages with selective permeability and the inclusion of antioxidants in the packaging material are potential applications in probiotic fermented whey beverages. In addition, there are probiotic strains with greater resistance to oxygen.

Probiotic fermented whey beverages should be kept at a temperature between 5 and 10°C throughout their shelf life, which is around 35 days. Probiotic fermented whey beverages added with fruit pulps can gain an additional shelf life due to the presence of sorbate, a preservative allowed in this type of product.

2.4.3   PHYSICOCHEMICAL AND SENSORY CHARACTERISTICS OF PROBIOTIC FERMENTED WHEY BEVERAGES

The chemical composition of probiotic fermented whey beverages is similar to that of yogurts and other commonly marketed fermented milk, with the possibility of lower protein and calcium concentration due to the addition of liquid whey. However, whey proteins have interesting characteristics, such as favoring the immune system of the individual, in addition to having compounds that aid in digestion. In addition, whey beverages, in some cases, have lower caloric intake.

Probiotic cultures generally have no negative effect on the sensory characteristics of probiotic fermented whey beverages, provided the inoculum size is adequate. However, *Bifidobacteria* produce 3:2 acetic and lactic acids, which can develop aroma and flavor of acetic acid in products. As probiotic fermented whey beverages are commonly flavored with fruit preparations, these ingredients may mask or minimize this effect in this type of product. In cases of non-flavoring products, it is possible to perform the microencapsulation of the probiotic, which decreases the negative effects.

Dairy products containing *L. acidophilus* are characterized as being poor in flavor since this bacterium has the enzyme alcohol dehydrogenase, which converts acetaldehyde to ethanol, and acetaldehyde is considered the most important compound for the typical aroma of yoghurt. In addition, highly proteolytic probiotic cultures can produce peptides, which impart cheese flavor to fermented dairy drinks.

2.5 SUMMARY AND CONCLUSIONS

Fermented whey beverage is a dairy product resulting from the mixing of milk and whey added or not with food products or substances, vegetable fat, fermented milk, selected milk lactic culture, and other dairy products. The milk base represents at least 51% (m/m) of the total ingredients of the product. The use of whey enables greater profit for dairy products, reduction of environmental pollution problems, and the availability of a nutritious product to consumers.

For the manufacture of fermented whey beverage, the same LAB used in the manufacture of yoghurt (*S. thermophilus* and *L. bulgaricus*) is traditionally used. These microorganisms are symbiotic cultures because they produce more lactic acid in the form of mixed culture than in the isolated form, thus reducing fermentation time.

Probiotic fermented whey beverage present great economic and innovative potential for dairy products, attracting differentiated consumers, increasingly concerned with health and well-being. The conversion of fermented whey beverages into probiotic products does not require substantial changes in processing; however, several points should be evaluated to ensure minimum counts of probiotic culture and to result in products with adequate physicochemical and sensory characteristics. Among them are the compatibility between the ingredients or the lactic culture and the probiotic culture and the process parameters (final pH, fermentation temperature, oxygen incorporation, and type of packaging).

KEYWORDS

- α-lactalbumin
- acid tolerance
- acid whey
- adhesion properties
- antagonism
- antibiotic resistance
- antioxidants
- beating
- *Bifidobacterium*
- bile salts tolerance
- by-products
- cheese
- constipation
- *Enterococcus faecium*
- fermentation
- gastrointestinal tract

- gut microflora
- health effects
- hydrocolloids
- immune system
- lactic acid bacteria
- lactic culture
- *Lactobacillus*
- *Lactobacillus bulgaicus*
- oligofructose
- optimal temperature
- oxygen toxicity
- package permeability
- pH

- prebiotic fiber
- probiotic
- probiotic inoculum
- probiotic viability
- *Saccharomyces cerevisiae*
- *Streptococcusthermophiles*
- sweet whey
- symbiosis
- whey
- whey powder
- whey protein concentrate
- β-lactoglobulin

REFERENCES

Abadias, M.; Usall, J.; Anguera, M.; Solsona, C.; Viñas, I. Microbiological Quality of Fresh, Minimally-Processed Fruit and Vegetables, and Sprouts from Retail Establishments. *Int. J. Food Microbiol.* **2008,** *123,* 121–129.

ANVISA. Brazilian Agency of Sanitary Surveillance. *Foods With Claims of Functional and/or Health Properties,* 2016. http://portal.anvisa.gov.br/alimentos/alegacoes (accessed March 12, 2017).

Braile, P. M. *Tratamento de despejos de laticínios.* In: Manual de tratamento de águas residuárias industriais. CETESB, FESB: São Paulo, 1971.

Brasil. Resolution No. 18 of April 30, 1999. Approves the Technical Regulation that Establishes the Basic Guidelines for the Analysis and Verification of Functional and/or Health Properties Alleged in Food Labeling, Included in the Annex to this Ordinance. Official Diary of the Union; Executive Power of May 3, 1999.

Burns, A. J.; Rowland, I. R. Anti-Carcinogenicity of Probiotics and Prebiotics. *Curr. Issues Intest. Microbiol.* **2000,** *1,* 13–24.

Chiu, H. H.; Tsai, C. C.; Hsih, H. Y.; Tsen, H. Y. Screening from Pickled Vegetables the Potential Probiotic Strains of Lactic Acid Bacteria Able to Inhibit the Salmonella Invasion in Mice. *J. Appl. Microbiol.* **2007,** *104,* 605–612.

Codex Alimentarius. *Milk and Milk Products.* World Health Organization and Food and Agriculture Organization of the United Nations, 2011. http://www.fao.org/3/a-i2085e.pdf (accessed Dec 12, 2016).

Cruz, A. G.; Faria, J. A. F.; Van Dender, A. G. F. Packaging System and Probiotic Dairy Foods. *Food Res. Int.* **2007**, *40*, 951–956.

Dairyforall. *Whey.* 2013. http://www.dairyforall.com/whey.php (accessed Dec 12, 2016).

Damin, M. R.; Sivieri, K.; Lannes, S. C. S. Bebidas Lácteas Fermentadas e Não Fermentadas e Seu Potencial Funcional. In *Tecnologia De Produtos Lácteos Funcionais;* Oliveira, M. N., Ed.; Atheneu: São Paulo, 2009; p 384

Daniel, C.; Roussel, Y.; Kleerebezem, M.; Pot, B. Recombinant Lactic Acid Bacteria as Mucosal Biotherapeutic Agents.*Trends Biotechnol.* **2011**, *29*, 499–508.

Donkor, O. N.; Nilmini, S. L. I.; Stolic, P.; Vasiljevic, T.; Shah, N. P. Survival and Activity of Selected Probiotic Organisms in Set-Type Yoghurt During Cold Storage. *Int. Dairy J.* **2007**, *17*, 657–665.

Espinoza, Y.; Navarro, Y. G. Non-Dairy Probiotic Products. *Food Microbiol.* **2010**, *27*, 1–11.

FAO. Food and Agriculture Organization of the United Nations, World Health Organization. *Guidelines for the Evaluation of Probiotics in Food.* Ontario: Joint FAO/WHO Working Group Report on Drafting London Canada, 2002.

Holzapfel, W. H.; Haberer, P.; Geisen, R.; Björkroth, J.; Schillinger, U. Taxonomy and Important Features of Probiotic Microorganisms in Food and Nutrition. *Am. J. Clin. Nutr.* **2001**, *73*, 365S–373S.

Jelen, P. Utilization and Products. Whey Processing. In *Encyclopedia of Diary Sciences*, 2nd Ed.; Fuquay, J. F., Ed.; Academic Press: Cambridge, USA, 2001; pp 731–738.

Jelicic, I.; Bozanic, R.; Tratinik, L. Whey-Based Beverages—A New Generation of Dairy Products. *Mljekarstvo* **2008**, *58*, 257–274.

Khorokhavar, R.; Mortazavian, A. M. Effects of Probiotic-Containing Microcapsules on Viscosity, Phase Separation and Sensory Attributes of Drink-Based on Fermented Milk. *Milchwissenschaft* **2010**, *65*, 177–179.

Makinen-Aakula, M. In *Trends in Functional Foods Dairy Market*. Proceedings of The Third Functional Food Net Meeting, 2006.

Mordon Intelligence. Global Whey Protein Market—Growth, Trends and Forecast (2015–2020). 2016. https://www.mordorintelligence.com/industry-reports/global-whey-protein-market-industry (accessed Dec 12, 2016).

Niamah, A. K.; Verma, D. K. Microbial Intoxication in Dairy Food Product. In *Microorganisms in Sustainable Agriculture, Food and the Environment;* Verma, D. K., Srivastav, P. P., Eds.; (as part of book series on *Innovations in Agricultural Microbiology*); Apple Academic Press: USA, 2017; pp 143–170.

Patel, A.; Shah, N.; Verma, D. K. Lactic Acid Bacteria (LAB) Bacteriocins: An Ecological and Sustainable Biopreservative Approach to Improve the Safety and Shelf-life of Foods. In *Microorganisms in Sustainable Agriculture, Food and the Environment;* Verma, D. K., Srivastav, P. P., Eds.; (as part of book series on *Innovations in Agricultural Microbiology);* Apple Academic Press: USA, 2017;pp 197–258.

Psomas, E.; Andrighetto, C.; Litopoulou-Tzanetaki, E.; Lombardi, A.; Tzanetakis, N. Some Probiotic Properties of Yeast Isolates from Infant Faeces and Feta Cheese. *Int. J. Food Microbiol.* **2001**, *69*, 125–133.

Saad, N.; Delattre, C.; Urdaci, M.; Schmitter, J. M.; Bressollier, P. An Overview of the Last Advances in Probiotic and Prebiotic Field. *LWT—Food Sci. Technol.* **2013**, *50*, 1–16.

Todar, K. *Todar's Online Textbook of Bacteriology;* University of Wisconsin: Department of Bacteriology: Madison, Wisconsin, 2016. http://textbookofbacteriology.net/lactics. html (accessed Dec 12, 2016).

Verma, D. K.; Srivastav, P. P. *Microorganisms in Sustainable Agriculture, Food and the Environment;* (as part of book series on *Innovations in Agricultural Microbiology)*; Apple Academic Press: USA, 2017.

Verma, D. K.; Mahato, D. K.; Billoria, S.; Kapri, M.; Prabhakar, P. K.; Ajesh Kumar, V.; Srivastav, P. P. Microbial Approaches in Fermentations for Production and Preservation of Different Food. In *Microorganisms in Sustainable Agriculture, Food and the Environment;* Verma, D. K. et al., Eds.; (as part of book series on *Innovations in Agricultural Microbiology*); Apple Academic Press: USA, 2017a. pp 105–142.

Verma, D. K.; Mahato, D. K.; Billoria, S.; Kapri, M.; Prabhakar, P. K.; Ajesh Kumar, V. and Srivastav, P. P. Microbial Spoilage in Milk Products, Potential Solution, Food Safety and Health Issues. In *Microorganisms in Sustainable Agriculture, Food and the Environment;* Verma, D. K. et al., Eds.; (as part of book series on *Innovations in Agricultural Microbiology*); Apple Academic Press: USA, 2017b. pp 171-196.

Williams, N. T. Probiotics. *Am. J. Health-Syst. Pharm.* **2010,** *67,* 449–458.

Wyszyńska, A.; Kobierecka, P.; Bardowski, J.; Katarzyna Jagusztyn-Krynicka, E. Lactic Acid Bacteria—20 Years Exploring their Potential as Live Vectors for Mucosal Vaccination. *Appl. Microbiol. Biotechnol.* **2015,** *99,* 2967–2977.

PART II
Emerging Applications of
Microorganisms in Safe Food Production

CHAPTER 3

GENETICALLY MODIFIED ORGANISMS PRODUCED ENZYMES: MULTIFARIOUS APPLICATIONS IN FOOD MANUFACTURING INDUSTRIES

DEEPAK KUMAR VERMA[1,*], KIMMY G.[2], PRATIBHA SINGHAL[3], RISHIKA VIJ[4], DIPENDRA KUMAR MAHATO[1,5], AMI PATEL[6], and PREM PRAKASH SRIVASTAV[1,7]

[1]*Agricultural and Food Engineering Department, Indian Institute of Technology Kharagpur, Kharagpur, West Bengal 721302, India, Mob.: +00–91–7407170260, Tel.: +91–3222–281673, Fax: +91–3222–282224*

[2]*Department of Food Engineering and Technology, Sant Longowal Institute of Engineering and Technology, Sangrur, Longowal, Punjab 148106, India, Mob.: +00–91–8699271602, E-mail: kishorigoyal09@gmail.com*

[3]*Department of Bioscience and Biotechnology, Banasthali University, Rajasthan 304022, India, Mob.: +00–91–9643850855, E-mail: pratibhasinghal89@gmail.com*

[4]*Department of Veterinary Physiology and Biochemistry, COVAS, Chaudhary Sarwan Kumar Himachal Pradesh Krishi Vishvavidyalaya, Palampur, Himachal Pradesh 176062, India, Mob.: +00–91–9466493086, E-mail: rishikavij@gmail.com*

[5]*Mob.: +91–9911891494, E-mail: kumar.dipendra2@gmail.com*

[6]*Division of Dairy and Food Microbiology, Mansinhbhai Institute of Dairy and Food Technology (MIDFT), Dudhsagar Dairy Campus, Mehsana, Gujarat 384002, India, Mob.: +00–91–9825067311, Tel.: +00–91–2762243777 (O), Fax: +91–02762–253422, E-mail: amiamipatel@yahoo.co.in*

[7]*Tel: +91 3222281673, Fax: +91 3222282224, E-mail: pps@agfe.iitkgp.ernet.in*

Corresponding author. E-mail: deepak.verma@agfe.iitkgp.ernet.in; rajadkv@rediffmail.com

3.1   INTRODUCTION

The application of genetically modified organisms (GMOs) or microorganism, such as fungus, bacteria, yeast, has contributed to a great extent in the food industry. Microorganisms have been involved in the production of various products such as in the production of alcoholic beverage, ethyl alcohol, organic products, dairy products, and drugs including antibiotics through fermentation (Verma and Srivastav, 2017; Verma et al., 2017a). Since ancient times, a variety of enzymes found in nature, has been widely used in the production of numerous food products such as cheese, beer, wine, vinegar along with lots of applications in the field of biotechnology and food technology. Enzymes which break down complex molecules into smaller units, such as carbohydrates into sugars, are basically natural substances involved in all biochemical processes. Owing to enzymes specificities, each substance has a corresponding enzyme. Basically, it has the function indispensable to maintenance and activity of life. It has been found that there are various sources of enzymes such as plants, fungi, bacteria, and yeasts but enzymes from microbial sources are much more important and advantageous than their equivalents from animal or vegetable sources (Soares et al., 2012). Nowadays, microorganisms are the major enzyme-producing sources along (Lambert et al., 1983) with animals and plant-derived enzymes, which contribute to approximately 10 and 5% of the total enzyme market, respectively (Illanes, 2008). This chapter is trying to influence GMOs used in various enzymes production in the food industry and other applications.

3.2   GENETICALLY MODIFIED ORGANISMS (GMOS): SOURCE OF ENZYME

Genetic modifications can potentially improve quality and productivity. Genetic engineering and recombinant DNA (rDNA) technology are continuously progressing with reference of microorganism. In accession to increasing the fundamental understanding of microbiology, transgenic microbes are being introduced commercially at an increasing value (Klein and Geary, 1997). The GMOs are most commonly used to refer to microbe created for human consumption. They are defined as the microbes that have specific changes introduced into genetic material by using genetic engineering techniques by a scientist or an expert, in order to produce desired biological products. This process is known as genetic engineering or modification. Direct manipulation of genes within a species may also involve the transfer of genes, thus the characteristics ordered from one species to another. These microbes have been modified in the laboratory to enhance desired traits.

The food industry consisting baking, beverages, and dairy products is widely using microorganisms for amelioration. Researchers are on the way to cross the limitation for improving the commercialization of enzymes and their products are formulated based on the enzymes produced by microorganisms with increasing effect spectrum. Desired alterations of food by microorganisms and their product formulations are referred to as fermentation, irrespective of the type of metabolism. According to the definition of fermentation, it is an anaerobic breakdown of an organic substance through enzymatic system, in which the final hydrogen acceptor is an organic compound. Hence, since the enzyme that produced by microorganisms catalyze the changes in foods, for some reactions, it is advantageous, and for some, it is not. In this chapter, the production of enzymes through few genetically modified microorganisms (GMMs) is detailed.

3.3   RECOMBINANT MICROORGANISMS TO PRODUCED ENZYMES

Nowadays, many researchers have endeavored to improve food quality and quantity by using GMOs. Microorganisms are among the first tools used for the discovery of biologically active compounds (Klein and Geary, 1997). Recombinant microorganisms have been used to synthesize enzymes for further application in food. The rDNA techniques have been available since the 1970s. At present, various microbes are genetically modified to produce enzymes as desire trait, scientific classification (Table 3.1), and microscopic image (Fig. 3.1), of them, some are given and described as below:

3.3.1   FUNGAL GROUP

3.3.1.1   ASPERGILLUS NIGER

Aspergillus niger belongs to a member of genus *Aspergillus* which includes a set of fungi that are generally conceived asexually although

TABLE 3.1 Scientific Classification of Different Species of Microorganisms.

Scientific classification	Fungal group			
	Aspergillus niger	Aspergillus oryzae	Trichoderma longibrachiatum	Trichoderma reesei
Domain	Eukarya	Eukarya	Eukarya	Eukarya
Kingdom	Fungi	Fungi	Fungi	Fungi
Phylum	Ascomycota	Ascomycota	Ascomycota	Ascomycota
Subphylum	Pezizomycotina	Pezizomycotina	Pezizomycotina	Pezizomycotina
Class	Eurotiomycetes	Eurotiomycetes	Sordariomycetes	Sordariomycetes
Order	Eurotiales	Eurotiales	Hypocreales	Hypocreales
Family	Trichocomaceae	Trichocomaceae	Trichocomaceae	Trichocomaceae
Genus	*Aspergillus*	*Aspergillus*	*Trichoderma*	*Trichoderma*
Species	*A. niger*	*A. oryzae*	*T. longibrachiatum*	*T. reesei*
Scientific	Bacterial group			
Classification	*Bacillus amyloliquefaciens*	*Bacillus licheniformis*	*Bacillus subtilis*	*Streptomyces lividans*
Domain	Bacteria	Bacteria	Bacteria	Bacteria
Kingdom	Bacteria	Bacteria	Bacteria	Bacteria
Phylum	Firmicutes	Firmicutes	Firmicutes	Actinobacteria
Subphylum	Pezizomycotina	Pezizomycotina	Pezizomycotina	Actinobacteria
Class	Bacilli	Bacilli	Bacilli	Actinobacteridae
Order	Bacillales	Bacillales	Bacillales	Actinomycetales
Family	Bacillaceae	Bacillaceae	Bacillaceae	Streptomycetaceae
Genus	*Bacillus*	*Bacillus*	*Bacillus*	*Streptomyces*
Species	*B. amyloliquefaciens*	*B. licheniformis*	*B. subtilis*	*S. lividans*

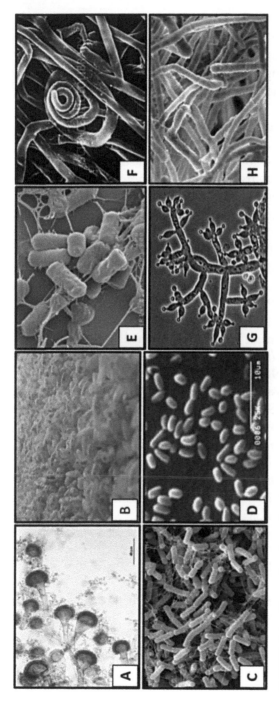

FIGURE 3.1 (See color insert.) Microscopic image of microorganisms. (a) *Aspergillus niger*, (b) *Aspergillus oryzae*, (c) *Bacillus amyloliquefaciens*, (d) *Bacillus licheniformis*, (e) *Bacillus subtilis*, (f) *Streptomyces lividans*, (g) *Trichoderma harzianum*, and, (h) *Trichoderma reesei*.

Source: Adapted from a: http://medicinembbs.blogspot.in/2013/02/microscopic-morphology-of-aspergillus.html; b: https://en.wikipedia.org/wiki/Aspergillus oryzae; c: http://www.gettyimages.in/detail/photo/bacillus-amyloliquefaciens-is-a-gram-high-res-stock-photography/128628136; d: http://www.sciencephoto.com/media/13170/view; e: Reprinted from Zweers et al. (2008). ©Zweers et al; licensee BioMed Central Ltd. 2008. https://creativecommons.org/licenses/by/2.0/; f: (to come); g: http://www.mycolog.com/CHAP4a.htm; h: Public domain.

perfect forms have been found. Aspergilla are ubiquitous in nature and geographically widely distributed worldwide. This fungus is able to grow between the wide ranges of temperature from 6 to 47°C with a relative at 35–37°C optimum high temperature and extremely wide range of pH, that is, 1.4–9.8. Due to their small size of 2–3.5 µ, about 70% of fumigatus spores are able to penetrate inside the host (Rippel-Baldes, 1955). Fungal spores are widespread, and often associated with organic material and soil.

A. *niger* has been subjected to research and industry used in food for many decades. In 1919, for the first time, this microorganism industrially tapped due to its ability to produce citric acid (Roukas, 2000). It is a prolific citrate producer and a remarkably acid-tolerant fungus. A. *niger* is used for the production of lactic acid (LA) (Dave and Punekar, 2015), itaconic acid (IA) (Hossain et al., 2016), phytase (Shivanna and Venkateswaran, 2014), fumonisin (FUM), and ochratoxin (OTA) synthesis (Susca et al., 2016). A. *niger* increased the production of amyloglucosidase (AMG) and exo-polygalacturonase (Exo-PG) (Colla et al., 2017).

Using genetic engineering, the researchers have increased the production processes. The development of rDNA-mediated transformation of *Aspergilli*, initially in A. *nidulans* (Balance et al., 1983; Tilburn et al., 1983; Balance and Turner, 1985; Campbell et al., 1989) and subsequently in A. *niger* has been successfully done (Buxton et al., 1985; Kelly and Hynes, 1985; Van Hartingsveldt et al., 1987; Ward et al., 1988). Kluyver and Perquin (1932) have shown the cultivation of submerged culture of filamentous fungi such as A. *niger*. The US Food and Drug Administration (FDA) has accepted a variety of enzymes for food making and processing: in the early-1960s, the FDA also addressed the potential of A. *niger* in food production which help to improve industrialization under the nonpathogenic and nontoxigenic conditions.

3.3.1.2   ASPERGILLUS ORYZAE

Aspergilus oryzae is a filamentous asexual ascomycetous fungus which mainly occurs in soil. It has been used for hundreds of years to make certain fermented foods and chemical. A. *oryaze* fungus is widely used in the synthesis of adenylate deaminase—it is an aminohydrolase that is widely used in the food and medicine industries (Li et al., 2016a). A.

oryzae is also used in the production of adenylic acid deaminase, also known as adenosine monophosphate deaminase (AMP deaminase). It is one of the most common and important enzymes for the food industry, which catalyzes an irreversible hydrolysis of adenosine monophosphate (AMP) to inosine monophosphate (IMP) and ammonia (Li et al., 2017a). Moreover, it has the ability to produce γ-aminobutyric acid (GABA) (Ab Kadir et al., 2016) and succinic acid by wheat-based bioprocess (Dorado et al., 2009). It is also used in the production of asparaginase, an enzyme that hydrolyzes asparagine to aspartic acid (Hendriksen et al., 2009). There are various low levels of secondary metabolites with low to moderate toxicity, such as 3-β-nitropropionic acid, kojic acid, and cyclopiazonic acid, which are produced by different strains of *A. oryzae* (Barbesgaard et al., 1992; Blumenthal, 2004). Kojic acid, which is one of the inhibitors of tyrosinase, inhibits its activity by chelating copper ions requisite for the active site of the enzyme. It is widely utilized as a skin-lightening agent, and thus in cosmetics industry (Bentley, 2006). During the production of products, always follow conditions that are optimized for industrial production of enzymes.

3.3.1.3 *TRICHODERMA LONGIBRACHIATUM*

Trichoderma spp. are among the most predicting biocontrol agents and have an activity against a wide range of plant-pathogenic fungi (Chet, 1987*).* *Trichoderma longibrachiatum* helps in the production of cellulase, an enzyme that degrades cellulolytic materials (Leghlimi et al., 2013). They are considered plant saprophytes but some documented studies show that they have the potential as an emerging pathogen in immune-compromised patients (Alanio et al., 2008). *Trichoderma* spp. are opportunistic fungi found in various ecosystems. Some strains have tremendous potential to reduce the severity of plant diseases. These strains normally inhibit the plant pathogens in the soil or on plant roots through their high antagonistic and mycoparasitic property, thus helpful for plants to fight against the biotic stress (Viterbo and Horwitz, 2010). Lignocellulosic materials, mostly obtained from agricultural and forestry residues, provide a potential and one of the cheap renewable resources for sustainable biorefineries. Lignocellulose can be usually produced after the processing of *T. longibrachiatum.*

3.3.1.4   TRICHODERMA REESEI

During the Second World War, the strain was isolated and initially identified as *Trichoderma viride* and named QM6a. It was later recognized as being distinct from *T. viride* and was given the new species name of *Trichoderma reesei* (Simmons, 1977). *T. reesei* has a capacity to secrete big amounts of cellulolytic enzymes, thus produced cellulose by using sugar beet pulp (Moosavi-Nasab and Majdi-Nasab, 2007). *T. reesei* has become a potentially major source of the enzyme industry. It used in the production of industrial enzymes such as β-mannanase, which helps in the degradation of mannan (Ohmiya et al., 1997); α-L-arabinofuranosidase, used in cleavage of side groups in xylan (Roche et al., 1995); α-galactosidase (Golubev et al., 2004); pectin methylesterases (Haltmeier et al., 1983); acetylxylan esterases (Hakulinen et al., 2000); and laccases, used in the oxidation of wide variety of compounds (Kiiskinen et al., 2004). The vast majority of industrial enzymes are mainly hydrolytic enzymes which are secreted by microbes. These enzymes are usually produced with highly developed microbial hosts with the help of genetic modifications (Paloheimo et al., 2016).

3.3.2   BACTERIAL GROUP

3.3.2.1   BACILLUS AMYLOLIQUEFACIENS

Bacillus amyloliquefaciens is an important organism. Protection provided against phytopathogens relies on its potential to produce multiple anti-microbial metabolites (Cawoy et al., 2015). *B. amyloliquefaciens* spp. such as N2–4 and N3–8 were obtained from soil. Metabolites produced from these species could kill *Burkholderia pseudomallei*, a Gram-negative pathogenic bacterium usually found in soil in its endemic areas (Boottanun et al., 2017). In some *B. amyloliquefaciens* strains, protection against phytopathogens relies on their potential to produce multiple antimicrobial metabolites (Cawoy et al., 2015) and on its ability to induce systemic resis-tance in plants (Ongena et al., 2005). The strains, FZB42 and SQR9, have been used as biofertilizers and biocontrol agents in agricultural industry (Xu et al., 2013; Chowdhury et al., 2015). The use of genetic engineering and rDNA technology can reveal some physiological behavior and help to improve the strains by metabolic engineering for better productivity.

3.3.2.2   BACILLUS LICHENIFORMIS

A common bacterium, named as *Bacillus licheniformis* is mesophilic, Gram-positive, rod-shaped in nature. Its survival mainly occurs at much higher temperatures. This bacterium normally requires an optimal temperature of 37°C for enzyme secretion. *B. lichesniformis* is used for the production of wide range of enzymes and small metabolites such as, trehalose synthase (*Tre*S), which carry out transformation of maltose into trehalose by the process of isomerization. It is one of the crucial enzymes for enzymatic transformation of trehalose (Li et al., 2017b), glutamic acid, and independent production of poly-γ-glutamic acid (γ-PGA) (Kongklom et al., 2016), along with induced calcium carbonate precipitation. It is one of the emerging and popular processes for the production of self-healing concrete (Seifan et al., 2016). This process can be useful in the production of α-1,4-glucosidase, which can be used to fulfill the accelerating demand of food and pharmaceutical industries (Nawaz et al., 2017). Some strains of *B. licheniformis* are used in the production of biosurfactant so as to enhancing oil recovery (Joshi et al., 2016). Overall review suggested that *B. licheniformis* is one of the most industrially used microorganisms.

3.3.2.3   BACILLUS SUBTILIS

A rod-shaped, Gram-positive bacterium, *Bacillus subtilis* is able to grow in the mesophilic temperature range and the optimum temperature is 25–35°C. This bacterium has a weak capacity for occurrence of disease in humans unless the load of bacteria will be high. *B. subtilis* also has probiotic potential and other applications. Genetically engineered strains are also used in various enzymes and polymers production. They are also used in the production of medically important enzymes and vitamins such as vitamin B_6, designation for the six vitamins pyridoxal, pyridoxine, pyridoxamine, and pyridoxal 5'-phosphate (PLP) (Rosenberg et al., 2016), *B. subtilis* strain, MUR1 has a potential to produce LA, which can be used in industrial production of LA due to its high productivity (Gao et al., 2012). Researchers have working on the improvement of the production of extracellular proteases, which can improve the degrading efficiency of feathers by random mutation in *B. subtilis* (Wang et al., 2016b). *B. subtilis* has the ability to produce new oligosaccharides in a maltose-containing medium and improve biosynthesis of D-ribose without acid formation (Shin, 2016; Cheng et al., 2017).

3.3.2.4   STREPTOMYCES LIVIDANS

Streptomyces lividans is a Gram-positive bacteria belonging to the *Actinomycetes,* having high guanine–cytosine content. They has been studied as a heterologous host for more than 20 years (Anné and Mellaert, 1993; Brawner, 1994), and during that time a number of important developments have been made, including an efficient expression system, characterization of a number of promoters and secretion–signal peptides, and molecular characterization of the export systems (Sec and Tat), creation of a wide range of cloning vectors, and establishment of protocols for genetic and molecular work. They are extremely well proposed or capable for the expression of rDNA or foreign DNA. Furthermore, along with the expression of foreign DNA and its high innate extracellular secretion capacity, *Streptomyces* has found to be a better, useful and most advantageous than Escherichia *coli* (Rebets et al., 2017). Therefore, *S. lividans* can be a prototypical model system to study and industrially used, such as, produce highly thermostable esterase protein by *S. lividans* TK64 strain (Wang et al., 2016a).

3.4   ENZYMES DERIVED FROM GMOs

The regulation of enzymes including those produced by genetic engineering is major concern in the food industry. Enzymes, the biological molecules act as catalysts, altering both the rate and specificity of chemical reactions. It occurs about 10^6 times faster than in the case of lack of an enzyme. Enzymes lower the activation energy and the reactions proceed toward equilibrium more rapidly than the uncatalyzed reaction. Enzymes are also involved in the metabolic activity of living organism. Enzymes are classified according to their nomenclature, nature, and mode of action. Today, there is no exaggeration to say, "Enzymes are functional catalytic proteins." This is due to growing interest in research and development of enzymes used in food industries.

There are about 3000 known enzymes produced by all animals, green plants, fungi, and bacteria that catalyze about 4000 biochemical reactions (Bairoch, 2000). Microorganisms are known to secrete a number of enzymes, namely α-acetolactate decarboxylase, α-amylase, glucoamylase, β-glucanase, catalase, chymosin, hemicellulose, lipase, phospholipase, polygalacturonase (PG), protease, pullulanase, xylanase, pectin-associated enzymes, and so forth. All these have been known to play an important role in the food manufacturing industries (Underkolfer et al., 1958). In addition of this organism, there are many other microorganisms that have been genetically modified to produce a variety of enzymes, which are said to contribute to their role in industries such as baking, beverages, cheese, dairy, egg-based products, and so forth.

3.4.1 α-AMYLASE

The α-amylase (Fig. 3.2) is a protein enzyme, *systematic name:* 1,4-α-D-glucan glucohydrolase (IUBMB, 1992). It catalyzes the end hydrolysis of 1,4-α-D-glucosidic linkages of large polysaccharides that leads the production of mainly α-maltose from amylose. α-amylase is routinely used for the production of bread, cakes, and pastries; but, conversely, some studies also identified it as an inhalative allergen for occupational diseases (bakers' asthma) (Baur et al., 1994). The α-amylase is secreted by many microorganisms and also present in seeds as reserve food in the form of starch.

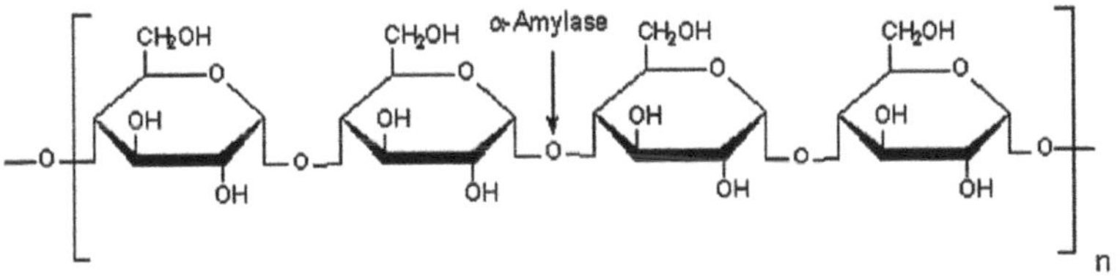

FIGURE 3.2 Structure of α-amylase.

Olempska-Beer (2004) presented a chemical and technical assessment entitled "α-amylase from *B. licheniformis*, having a genetically engineered α-amylase gene isolated from *B. licheniformis* (a thermostable bacterium)" at 61st Joint FAO/WHO Expert Committee on Food Additives (JECFA). He suggested that α-amylase is thermostable genetically engineered enzyme. Its activation mainly occurs at a relatively lower pH and lower calcium (Ca) concentration. In starch hydrolysis, this enzyme was observed particularly suitable for use due to said characteristics. Starch liquefaction can be taken as an example for the production of nutritive sweeteners from starch. The α-amylase enzyme is genetically engineered through introducing DNA into the production strain of *Bacillus,* that is, *B. licheniformis,* which should be nontoxigenic and nonpathogenic by using pure culture fermentation. Appropriate purification, concentration, and formulation of enzymes with appropriate substances are crucial for further use in food industries (Olempska-Beer et al., 2006).

3.4.2 α-ACETOLACTATE DECARBOXYLASE

The α-acetolactate decarboxylase (ALDC), *systematic name:* (S)-2-hydroxy-2-methyl-oxobutanoate carboxylase (IUBMB, 1992) has the unique ability to catalyze the chemical reaction involving 2-hydroxy-2-methyl-3-oxobutanoate into 2-acetoin along with CO_2. ALDC enzyme is produced from a submerged culture of *B. subtilis,* which causes a direct decarboxylation of α-acetolactate to acetoin. The enzyme α-acetolactate decarboxylase has a great significance as it leads to alter the acceleration phenomenon of beer fermentation/maturation. This is because it shunts diacetyl formation, whose removal or elimination from the process is the rate-limiting step, thus reducing maturation time (Godtfredsen and Ottesen, 1982). In other words, we can say ALDC is used to avoid rate-limiting step during fermentation.

3.4.3   AMINOPEPTIDASE

Aminopeptidases (APs) were some of the proteases discovered exopeptidases enzymes which constitute a large group of enzymes sub-subclass EC 3.4.11 (IUBMB, 1992) with closely related activities that exist in microorganisms as well as in animal and plant tissues (McDonald, 1986; Sanderink et al., 1988; Gonzalez and Boudouy, 1996; Strater and Hipscomb, 1998; Ito et al., 2006). These enzymes are capable of hydrolyzing the peptide bond of *N*-terminal amino acids in peptides and proteins (Taylor, 1993; Lin et al., 2008). These enzymes also have the ability to split amino acid into 2-naphthylamides, and this process of hydrolysis is generally being named as arylamidase activity. Arylaminopeptidases are obtained from numerous sources and show their activity upon *N*-terminal of various amino acids such as alanine, arginine (or lysine), aspartic acid (or glutamic acid), and leucine (McDonald and Schwabe, 1977; Sanderink et al., 1988). These enzymes are mainly involved in the catabolism of peptides and proteins, which is one of their major roles and suggested usually, but some evidences for their physiological role in cells have been unequivocally shown. One of the promising exceptions is the inactivation of oxytocin (Tuppy and Nesvadba, 1957).

These enzymes participate in a variety of biological processes such as protein maturation, metabolism of biologically active peptides of food origin, antigen presentation on immune cells, and regulation of hormone activity, neurotransmitters, and so forth. (Ragheb et al., 2009). Besides, APs are very important in food processing enzymes and have many commercial applications in food industries (Sriram et al., 2012) due to its ability to change the flavor and aroma of proteins. The food industries are widely using APs enzyme to modify the proteins by debituminize and improve the functional properties for the production of a variety of food products such as cheese, beverages, flavorings, meat, milk, and so forth (Izawa et al., 1997; Raksakulthai and Haard, 2003; Lin et al., 2008; Sriram et al., 2012; Smily et al., 2014). Lots of commercial APs, particularly those from microbial sources have been used to produce protein hydrolysates (PHs) for food applications because they present an additional advantage over other antioxidants and also confab functional and nutritional properties (Moure et al., 2006; Monod et al., 2008). In 2012, a study was conducted by Rahulan et al. (2012) to determine the efficiency of AP enzyme from *Streptomyces gedanensis*. These are used as a useful tool and one of the most effective enzymes to produce PHs with improved antioxidant, functional, and nutritional properties.

3.4.4 ARABINOFURANOSIDASE

α-*N*-arabinofuranosidase (EC 3.2.1.55), also known as α-L-arabinosidase, α-arabinofuranosidases; systematic name is α-L-arabinofuranoside arabinofuranohydrolase (IUBMB, 1992). Several natural polymers, including hemicellulose and pectic polysaccharides, contain arabinose and galactose residues either as main or side-chain components. These side groups may alter the form and functional properties of the polymers and they can be specifically hit by α-L-arabinofuranosidases and galactosidases, respectively. In additions, L-arabinose has been the best monomeric sugar and complex polysaccharide inducer of α-L-arabinofuranosidase, for example, in *A. niger* (Van der Veen et al., 1991), *A. nidulans* (Fernández-Espinar et al., 1994), and *Streptomyces olivochromogenes* (Higashi et al., 1983).

3.4.5 β-GLUCANASE

The glucanase enzymes help in the breakdown of glucan, a polysaccharide made of several glucose subunits. As they perform hydrolysis of the glucosidic bond, they are hydrolases. Debituminized glucanases can be produced by Neocallimastigomycota, a phylum of anaerobic fungi. They are usually found in the digestive tracts of herbivores. The β-glucanase is capable of causing lysis of cell walls (Dake et al., 2004). They are active on insoluble substrates, such as beta β-glucan components of fungal cell wall, laminarin and pachyman. They act both as endo- or exo-hydrolases. β-glucanase is further classified into various glucanases such as β-1,3-glucanase, β-1,6-glucanase, β-1,4-glucanase on the basis of the types of glucosidic linkage cleaved by them (Cabib et al., 1982).

3.4.6 CATALASE

Catalase (EC 1.11.1.6) (IUBMB, 1992) is responsible to catalyze the decomposition of hydrogen peroxide (H_2O_2) to water (H_2O) and oxygen (O_2). This enzyme has antioxidant property as it helps to protect the cell from oxidative damage by reactive oxygen species (ROS). Catalase may also play a role in the graying process of human hair at low levels (Wood et al., 2009). It is one of the common enzymes, which is ubiquitous in nearly all living organisms.

$$H_2O_2 \xrightarrow{\text{Catalase}} H_2O + O_2$$

The catalase enzyme can be produced by genetic modifications in microorganisms. There is a wide range of organisms which have an ability to produce catalase enzyme at industrial level, such as, with *A. niger*, it is now possible to increase catalase production by alteration of molecular genes using recombinant techniques (Berka et al., 1994a, b). *Kat*B and *Kat*G are two distinct catalases produced by *Vibrio cholera*, which is known to maintain ROS level, thus making an important and relevant contribution in ROS homeostasis (Goulart et al., 2016). *Geobacillus* spp. have a unique combination of several industrially important extremophilic properties. They lead to hyperproduction of catalase (Kauldhar and Sooch, 2016). *B. subtilis* also enhances the production of catalase after self-cloning (Xu et al., 2014).

3.4.7 CHYMOSIN

Chymosin (rennin) (EC 3.4.23.4) (IUBMB, 1992) is an aspartyl proteinase. Its occurrence is in the fourth stomach of the unweaned calf. It leads to limited or partial proteolysis of K-casein in milk (Foltmann, 1970), resulting in clotting (Kaye and Jolles, 1978). Chymosin is a major factor to bring about the extensive precipitation and curd formation in cheese-making (Emtage et al., 1983). The enzyme is also used in the food industry as a processing aid for the manufacture of cheese and curd and other milk coagulants (Kumar et al., 2010). The GMOs comply with the organization for large-scale production. It has been described by some studies that cloning and expression of yak chymosin in gene-recombinant *Pichia pastoris* is one of the great and novel sources. This strain is generally used for rennet production (Luo et al., 2016). Some *Lactococcus* spp. also help in the production of chymosin at industrial scale (Luerce et al., 2014). Some other common microorganisms are also involved in the production of chymosin, directly or indirectly such as, *E. coli* (Kumar et al., 2007), *Kluyveromyces lactis* (Vega-Hernández et al., 2004), *P. pastoris* (Vallejo et al., 2008), *Aspergillus awamori* (Cardoza et al., 2003).

3.4.8 CYCLODEXTRIN GLUCOSYLTRANSFERASE

Cyclodextrin glycosyltransferase (EC 2.4.1.19) (IUBMB, 1992) is a bacterial enzyme. It belongs to the same family of that of α-amylase, specifically known as glycosyl hydrolase family. Glucosylation of steroidal saponins takes place with help of cyclodextrin glucanotransferase. It is well known that the sugar chains of steroidal saponins are very important as they perform great functions both in biological and pharmacological activities. The synthesis of steroidal saponins with novel sugar chains in one-step reaction (Wang et al., 2010). *Bacillus cereus* and alkalophilic *Bacillus* spp.

have potential to synthesize cyclodextrin glucosyltransferase (Jamuna et al., 1993; Kometani et al., 1996).

3.4.9 GLUCOAMYLASE

Glucoamylase (α-1, 4-glucan glucohydrolase, amyloglucosidase, EC 3.2.1.3) is another enzyme (Fig. 3.3), which is greatly utilized by food industry, in order to obtain high glucose syrup. It is also known to be used in the production of beverages such as beer and ethanol by fermentation process (Pavezzi et al., 2008). Various microorganisms like *A. niger*, *A. awamori*, and *Rhizopus oryzae* are important sources of this enzyme, which also has enormous industrial applications (Coutinho and Reilly, 1997). Researchers recently improved production of glucoamylase using raw-starch digestion by *A. niger* F-01 (Sun and Peng, 2017). Some other organisms also have the ability to produce glucoamylase, such as the novel gene-encoding glucoamylase GlucaM from *Corallococcus* spp. strain EGB (Li et al., 2017c). *A. niger* produces amyloglucosidase, which is widely used in glucose syrup and alcohol industries. There is a possibility to manipulate genes and increase productivity by using molecular tools, that is, recombinant gene cloning (Mullaney et al., 1991; Van Gorcom et al., 1991; Piddington et al., 1993; Van Hartingsveldt et al., 1993).

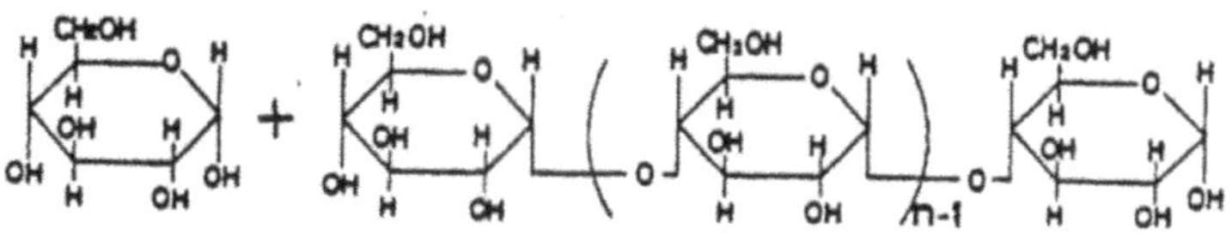

FIGURE 3.3   Structure of glucoamylase.

3.4.10 GLUCOSE ISOMERASE

Glucose isomerase (GI) (also known as D-xylose ketol-isomerase; EC 5.3.1.5) (IUBMB, 1992) carries out reversible isomerization of D-glucose and D-xylose resulting in D-fructose and D-xylulose, respectively. This enzyme known to have biggest market value in the food industry leads to its utilization for ethanol production and development of high-fructose corn syrup (Bhosale et al., 1996). Evaluation of its suitability for industrial application and cost of production of enzyme are crucial factors. Thus, extracellular GI has been reported to be produced by *E. coli* BL21 (Yaman and Çalık, 2016), *Streptomyces glaucescens* (Weber, 1976), and *Streptomyces flavogriseus* (Chen et al., 1979). Thus, in order to release the enzyme extracellularly, there is a great need, and major priority is, to change the permeability of cells and carry out partial lysis. The extracellular xylose isomerases from *Chainia* spp. (Srinivasan et al., 1983; Vartak et al., 1984) and an alkalothermophilic *Bacillus* spp. (Chauthaiwale and Rao, 1994) have been purified to homogeneity by various conventional purification techniques such as gel filtration, ion-exchange chromatography, poly-acrylamide gel electrophoresis, and so forth. Several species of genus *Streptomyces* and *Bacillus* are good producers of GI. The occurrence of GI in a few yeasts such as *Candida utilis* (Wang et al., 1980) and *Candida boidinii* (Vongsuvanlert and Tani, 1988) has also been reported. The only fungus, which is documented to possess GI activity, is *A. oryzae*.

3.4.11 GLUCOSE OXIDASE

A very important non-hydrolytic enzyme, glucose oxidase (GO) (EC 1.1.3.4) (IUBMB, 1992) belongs to the oxidoreductase family. This is also called as glucose aerodehydrogenase (Witteveen et al., 1990; Fiedurek and Gromada, 1997). GO has successful applications in pharmaceutical and food industries. It is used to remove glucose and oxygen to enhance the color, taste, and consistency of different products (Field et al., 1986; Hanft and Koehler, 2006). It also enhances the shelf life of food products, thus saving the food from rottening. It is also used for the preservation of the stored foods and their products. GO has also emerged its use in the biofuel cells (Zhu et al., 2006).

A majority of microorganisms have great capability to produce GO. Different species of fungi such as *Penicillium* and *Aspergillus* have prominent source among all microbes. *Penicillium adametzii* is an important fungus for the production of extracellular GO (Eremin et al., 2001). *A. niger* (Berka et al., 1992) and some other species from the genera *Penicillium, Gliocladium, Scopulariopsis,* and *Gonatobotyrs* also produce gluconic acid along with the GO. GO is also produced by many bacterial species. Among these, *Acetobacter methanolicus, Enterobacter, Gluconobacter oxydans, Micrococcus, Pullularia, Scopulariopsis,* and *Zymomonas mobilis* are the most dominant. For the large-scale production of this enzyme, at industrial scale, most of the fungal species are considered to be best and more beneficial as compared to bacterial one (Ramachandran et al., 2006).

3.4.12　HEMICELLULASE

Hemicellulase (EC 3.2.1.8) (xylanase), (galactanase) (EC 3.2.1.89) (IUBMB, 1992) is a collective term for enzymes that carry out hydrolysis of hemicellulose. These enzymes usually work on hemicellulose (also called pentosan), which is a polymer of pentose sugars. These enzymes are known to have a major role, both in food and beverages industry. In bakery products, these are used to develop the quality of dough, softness of the crumb, and volume (Schuster et al., 2002). Hemicellulose is also utilized in wine production (Montedoro, 1976). There is a wide range of microorganisms which have an eminent ability to produce hemicellulose at the industrial level, such as *Penicillium oxalicum* (Li et al., 2016b) and *A. nidulans, Chaetomium globosum* (Wanmolee et al., 2016). *Trichoderma asperellum* KIF125 has prominent and quiet important ability to produce hemicellulase among other fungal species (Inoue et al., 2016). In the same way, some other species are also used in enzyme production, like *Streptomyces* sp., but selection of type of carbon source is an important factor in the production of hemicellulases (Brito-Cunha et al., 2013). *T. reesei* is today a paradigm for the commercial-scale production of different plant cell-wall-degrading enzymes. These two important enzymes are cellulases and hemicellulases (Seiboth et al., 2012).

3.4.13　TRIACYLGLYCEROL LIPASE

Triacylglycerol lipase (EC 3.1.1.3) (IUBMB, 1992) is an enzyme with systematic name *triacylglycerol acylhydrolase*. Lipases are found in nature in an adequate amount but only a few microbial lipases have commercial significance. These enzymes have various applications which include specific organic syntheses, hydrolysis of fats and oils, modification of numerous fats, and so forth. Flavor enhancement during food processing, resolution of racemic mixtures, and chemical analyses are also important applications (Sharma et al., 2001).

Lipases are important ubiquitous enzymes having desirable physiological significance and various industrial potential. Lipases help in catalyzing the hydrolysis of triacylglycerols that leads to the production of diacylglycerol and carboxylate.

$$\text{Triacylglycerol} + H_2O \rightleftharpoons \text{Diacylglycerol} + \text{a carboxylate}$$

Lipases are generally found in microorganisms for commercial purposes. Microbes usually produce a wide range of extracellular lipases. The main emphasis is on microbial lipases, particularly produced by fungal and bacterial species. In 1994, Novo Nordisk introduced the first commercial recombinant lipase named as "Lipolase" which originated from the fungus *Thermomyces lanuginosus* and was also expressed in *A. oryzae*. Two bacterial lipases were introduced which is "Lumafast" from *Pseudomonas mendocina* and other is "Lipomax" from *Pseudomonas alcaligenes* by Genencor International (Jaeger and Reetz, 1998). Other lipase producing organisms are *Bacillus megaterium* (Godtfredsen, 1990), *B. subtilis* (Kennedy and Rennarz, 1979), *Staphylococcus carnosus* (Tahoun et al., 1985), *Staphylococcus aureus* (Lee and Yandolo, 1986), *Streptococcus lactis* (Sztajer et al., 1988), *Pseudomonas aeruginosa* (Aoyama et al., 1988), and some fungus species also used in lipase production such as, *Rhizopus delemar* (Klein et al., 1997), *A. flavus* (Long et al., 1996), *A. oryzae* (Ohnishi et al., 1994a, b), *Penicillium cyclopium* (Chahinian et al., 2000), and some yeast like *Candida rugosa* (Frense et al., 1996), *Saccharomyces lipolytica* (Tahoun et al., 1985). Some lipases vary according to the degrees of glycosylation. Fungal lipases usually exist as monomers. Their molecular mass ranges from about 30 to 60 kDa. These lipases vary in specificity, specific activity, temperature stability, and other properties; however, some dimeric lipases have been reported in Table 3.2 (Mozaffar and Weete, 1993; Mase et al., 1995).

3.4.14 MALTOGENIC AMYLASE

These amylases catalyze the hydrolysis of starchy material, thus plays central role in carbohydrate metabolism. Maltogenic amylases exhibit catalytic versatility: carry out hydrolysis of α-D-1,4 and α-D-1,6-glycosidic bonds and transglycosylation of oligosaccharides to C_3-, C_4-, or C_6-hydroxyl groups of various acceptor mono- or disaccharides. This

TABLE 3.2 Lipases Enzymes Produce from Legion Microbial Species.

Microbial species	Studies for microbial lipase production	References
Neurospora sp. TT-241	Purification and characterization of a lipase from Neurospora sp. TT-241	Lin et al. (1996)
Pseudomonas fluorescens AK102	Purification and characterization of an alkaline lipase from Pseudomonas fluorescens	Kojima et al. (1994)
Rhizopus niveus 1f04759	Purification, characterization, and crystallization of two types of lipase from Rhizopus niveus	Kohno et al. (1994)
Candida parapsilosis	Peculiar properties of lipase from Candida parapsilosis (Ashland) Langeron and Talice	Riaublanc et al. (1993)
Propionibacterium acidipropionici	Purification and properties of lipase from the anaerobe Propionibacterium acidipropionici	Sasada and Joseph (1992)
Neurospora crassa	Purification and characterization of an extracellular lipase from the conidia of Neurospora crassa	Kundu et al. (1987)
Pythium ultimum #144	Purification of properties of an extracellular lipase from Pythium ultimum	Mozaffar and Weete (1993)
Rhizopus delemar ATCC 34612	Purification and characterization of an extracellular lipase from the fungus Rhizopus delemar	Haas et al. (1992)
Fusarium heterosporium	Purification and characterization of a novel solvent-tolerant lipase from Fusarium heterosporum	Shimada et al. (1993)
Penicillium roqueforti	Purification and characterization of Penicillium roqueforti 1AM 7268 lipase	Mase et al. (1995)
Penicillium sp. uzim-4	Purification and properties of a Penicillium spp. which discriminates against diglycerides	Gulomova et al. (1996)

activity is different from other amylases that are able to hydrolyze only α-D-1,4-glycosidic bonds. It has been speculated that the catalytic property of the enzymes is linked to the additional ~130 residues at the N-terminus that are usually absent in other typical α-amylases. The maltogenic amylase from *Bacillus stearothermophilus* are capable to catalyze the synthesis of 2-(α-maltosyloxy)-ethyl acrylate (Glc–Glc–EA) transglycosylation product (Wouter et al., 2014).

3.4.15   PECTIN-ASSOCIATED ENZYMES

Pectin, a heteropolysaccharide, is one of the principal components of commercially important fruits and vegetables. Various enzymes, including pectinesterases, endo- and exo-polygalacturonidases and pectin lyases, produced from *A. niger.* These enzymes degrade pectin. Thus, they are used in wine and fruit juice production. These enzymes are used to reduce juice viscosity before pressing and also improve clarification (Grassin and Fauguenbergue, 1999). Pectinase production occupies about 10% of the overall manufacturing of enzyme production. Pectinolytic enzymes are widely used in the food industry mainly for juice and wine production (Semenova et al., 2006).

3.4.15.1   PECTINLYASE

Pectinases are important enzyme group that catalyzes pectic substance degradation through the process of depolymerization (hydrolases and lyases) and de-esterification (esterases) reactions. It is determined to contain two types of pectinase, endo-PG (EC 3.2.1.15), endo-pectin lyase (EC 4.2.2.10) along with maceration stimulating factor (IUBMB, 1992). In enzymology, a pectin lyase (EC 4.2.2.10), also known as pectolyase, is a naturally occurring pectinase that degrades pectin. This enzyme belongs to the family of lyases, specifically those carbon–oxygen lyases acting on polysaccharides. It is produced commercially for the food industry mostly from fungi where it is used to destroy residual fruit starch (pectin), in wine and cider (Godfrey and Reichelt, 1983). In plant cell culture, it is used in combination with the enzyme cellulase to generate protoplasts by degrading the plant cell walls. Thus, it has great importance in plant biotechnology (Yadav et al., 2009). Pectin lyase is an enzyme that also catalyzes the chemical reaction: eliminative cleavage of 1-4-α-D-galacturonan methyl ester to give oligosaccharides with 4-deoxy-6-O-methyl-alpha-D-galact-4-enuronosyl groups at their nonreducing ends. Pectinylase is one of the valuable enzymes that aim to increase the productivity of food products.

3.4.15.2   PECTINESTERASE (PE)

Pectin methylesterase or pectinesterase (PE) (EC 3.1.1.11) (IUBMB, 1992) is one of the ubiquitous cell-wall-associated enzymes. It is present in several isoforms that facilitate plant cell wall modification and its subsequent breakdown. This enzyme is usually found in all higher plants as well as in some bacteria and fungi. Pectin is one of the important components of plant cell wall. PE functions primarily by altering the localized pH of the cell wall, thus resulting in alterations in cell wall integrity. PE catalyzes the de-esterification of pectin into pectate and methanol. This enzyme plays an important role in cell wall metabolism and fruit ripening in plants (Micheli, 2001). In plant bacterial pathogens interactions, such as *Erwinia carotovora* and in fungal pathogens such as *A. niger*, PE is involved in maceration and soft rotting of plant tissue. Plant PEs are mainly regulated by PE inhibitors, which are quite ineffective against microbial enzymes.

3.4.15.3   POLYGALACTURONASE

PG (EC 3.2.1.15) (IUBMB, 1992) is an enzyme which is involved in the ripening process in plants. It is also produced by some bacteria and fungi and helps in the rotting process. PGs degrade polygalacturonan, which is a component of cell walls of plants by hydrolysis of the glycosidic bonds. These bonds link galacturonic acid. Polygalacturonan is a significant carbohydrate component of the pectin network of plant cell wall.

3.4.16   PHOSPHOLIPASE

Phospholipase enzyme family members are those enzymes which are categorized on the basis stereospecifically numbered sites within phospholipids where they promote cleavage process. These are ubiquitous enzymes involved in several diverse processes such as membrane homeostasis, nutrient acquisition, and generation of numerous bioactive molecules. Some phospholipases play a vital role in microbial pathogenesis and virulence; other phospholipases are also found in venoms (Djordjevic, 2010; Ghannoum, 2000). Within the phospholipase families, there are isoforms, which have differential distribution patterns and specific functions within specific cell types. There are four major classes, termed A, B, C, and D, each of which is distinguished by the type of reaction they catalyze. Phospholipase C enzymes are phosphodiesterases that cleave the glycerophosphate bond. Phospholipase D enzymes help in the removal of the base group of phospholipids, and phospholipases A and B have a potential role, which is described in the below sections.

3.4.16.1   PHOSPHOLIPASE A

This category of enzymes hydrolyzes phospholipids into fatty acids and other lipophilic substances. Phospholipase A1 cleaves the SN-1 acyl chain, while A2 cleaves the SN-2 acyl chain, thus releasing arachidonic acid. Phospholipase A2 was the first phospholipase to be reported and recognized. It is ubiquitous in nature and occurs naturally both in animal and plant cells. Its isolation is done from a number of food sources (including wheat flour) as well as it is a natural constituent of the digestive pancreatic juice of mammals, including humans (De Haas et al., 1968; Rossiter, 1968). This enzyme also found as a component of many animal- and plant-derived foods. Phospholipase enzyme is also produced by few fungi for example; *Aspergillus* spp. and *T. reesei.*

3.4.16.2   PHOSPHOLIPASE B

This enzyme is also known as lysophospholipase. In combination of both PLA1 and PLA2 activities, it can cleave acyl chains from both the sn-1 and sn-2 positions of a phospholipid. Many fungal species appear to produce phospholipase B enzymes that hydrolyze both acyl groups resulting in a very minimal accumulation of lysophospholipid product. Hence, these enzymes often have lysophospholipase activity as well (Ghannoum, 2000). Some fungal phospholipases B have also been shown to promote transacylase activity. They are capable enough to convert lysopholipids and free fatty acids into phospholipids.

3.4.17 PHYTASE

Phytase (Fig. 3.4) is one of the most widely used diet feed enzymes in the world. This feed phytase was first introduced in the late 1980s that control phosphorus pollution and also improves nutrient uptake. This enzyme specifically acts on phytate, breaking it down to release phosphorus, thus making this form available to the animal. This overall process greatly reduces the need for supplemental inorganic phosphorus and also promotes the nutritional value of feed stuffs. Phytase is non-starchy polysaccharide-degrading exogenous enzyme that has been consistently shown to be highly beneficial to pigs (Selle and Ravindran, 2008). The proven efficacy of phytase has resulted its worldwide acceptance and vast use in pig production. rDNA technology leads to 1000-fold improvement in the expression level for *A. niger* phytase (Van Gorcom et al., 1991; Van Hartingsveldt et al., 1993; Selten, 1994).

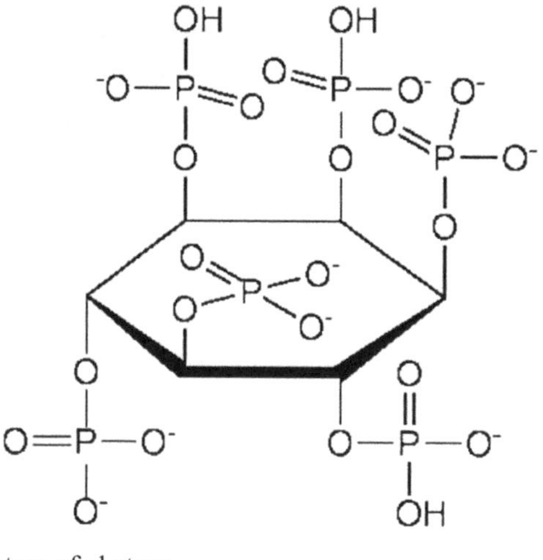

FIGURE 3.4   Structure of phytase.

3.4.18 PROTEASE

A protease is a type of enzyme that mainly performs proteolysis. Proteolytic enzymes are those that break down the peptide bonds in the protein foods to liberate the amino acids needed by the body; thus, these are very important in the digestion process. These enzymes also lead to the beginning of protein catabolism in which large polypeptide chain are broken down by hydrolysis of peptide bonds. Over a long time, microbial proteases had vital roles in the production of traditional fermented foods. Nowadays, a large and vibrant enzyme industry, dominated by microbial protease products, supplies the world with large biocatalysts and their use in food processing, for detergents, for use as therapeutics, and for organic chemical synthesis (Ward et al., 2009). Proteases have evolved in multiple ways. Different classes of protease carry out the same reaction by completely different catalytic mechanisms. There are various types of proteolytic enzymes whose classification is based on the cleavage of proteins at a particular site, catalyzed by these enzymes. Two major groups are the exopeptidases, which usually target the terminal ends of proteins, while other is the endopeptidases whose target sites are within proteins. Endopeptidases perform various catalytic mechanisms. This group includes aspartic endopeptidases, cysteine endopeptidases, glutamic endopeptidases, metalloendopeptidases, serine endopeptidases, and threonine endopeptidases. The term oligopeptidase is mainly reserved for those enzymes that act specifically on peptides. Proteases are ubiquitous, found in animals, plants, bacteria, archaea, fungi, and viruses (Rodarte et al., 2011).

3.4.19   PULLULANASE

Pullulanase (EC 3.2.1.41) is a specific kind of glucanase (Fig. 3.5). It is an amylolytic exoenzyme that usually degrades pullulan (Bender and Wallenfels, 1966; Lee and Whelan, 1972; Jensen, and Norman, 1984; Manners, 1997). This enzyme is produced as an extracellular, cell-surface-anchored lipoprotein by Gram-negative bacteria that belongs to the genus *Klebsiella*. Type I pullulanases specifically attack α-1,6 linkages, while type II pullulanases are able to hydrolyze α-1,4 linkages. Pullulnanase is produced by various other bacteria and archaea. Pullulanase is used as a processing agent in grain processing biotechnology (Jensen and Norman, 1984). Thus, it is mainly used in the production of ethanol and sweeteners. The α-dextrin endo-1,6-α-glucosidase (EC 3.2.1.41) (also known as pullulanase, pullulan 6-glucanohydrolase limit dextrinase, debranching enzyme; amylopectin 6-glucanohydrolase, the starch-debranching enzyme) hydrolyzes 1-6-α-D-glucosidic linkages in pullulan and starch to form maltotriose. An example is from *Klebsiella pneumonia*.

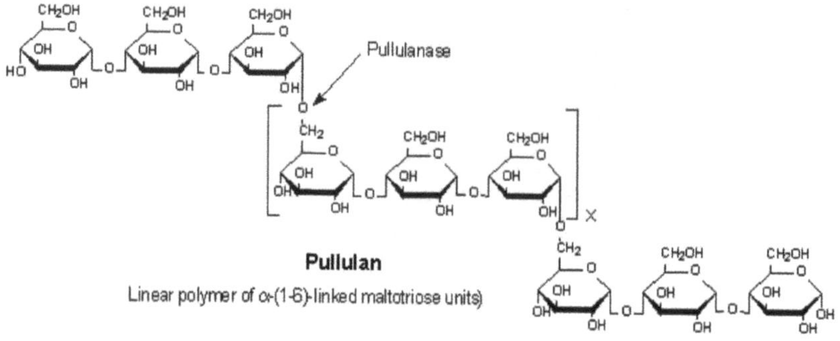

FIGURE 3.5   Structure of pullulanase.

Source: Adapted from http://www.sigmaaldrich.com/life-science/metabolomics/enzyme-explorer/learning-center/carbohydrate-analysis/carbohydrate-analysis-iii.html.

3.4.20   XYLANASE

Xylanase (EC 3.2.1.8) is the name usually given to a class of enzymes (IUBMB, 1992) which degrade the linear polysaccharide β-1,4-xylan into xylose, that break down hemicelluloses as shown in Figure 3.6. Hemicellulose is one of the important components of the cell wall of plants. It plays a vital role in microorganisms thriving on plant sources, which leads to the degradation of plant matter into useful or vital nutrients. Xylanases are produced by various microorganisms such as fungi (Suprabha et al., 2008), bacteria, yeast, marine algae, protozoans, snails, crustaceans, insect, seeds, and so forth.

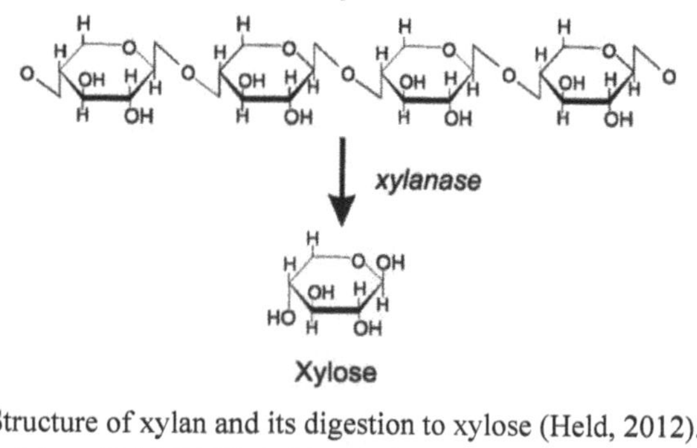

FIGURE 3.6   Structure of xylan and its digestion to xylose (Held, 2012).
Source: Reprinted from Held (2012) (with permission from Biotek).

3.5 APPLICATION OF GMO ENZYMES IN FOOD

Nowadays, GMOs are used at a very high level of production in industries. They give arm in biotechnology and food industry to enhance productivity as well as betterment of the products. GMOs used in a production of various food products are mentioned below.

3.5.1 BAKING

Various enzymes such as malt and fungal α-amylases have been widely used in bread-making process. Rapid advances and developments in the field of biotechnology have made a number of exciting or novel enzymes and are available for the baking industry. The raw material in the baking industry is flour: a mixture of starch, protein (gluten), lipids, glucan, and some wheat (naturally occurring) enzymes. The most important enzymes contained in flour for the baking process are the amylases and proteases.

3.5.2 BEVERAGES

In the food industry, several enzymes are regarded as nuisance substances that must be deactivated or destroyed to develop an acceptable product. However, for others, particularly in the beverage industry and wine and alcohol production, enzymes and the reactions they catalyze are indispensable and crucial.

In beverages, as well as in several food products, enzymes may occur naturally or their presence may be due to intentional formulation. Enzymes perform several functions in beverage production (Urlaub, 1999). Enzymes lead to increased yields, form nutrients for the fermentation process, facilitate processing, and also affect the color, aroma flavor, and finally clarity of the finished product. Several enzyme preparations containing PE in pure form help to retain the original shape and structure of individual fruit pieces during processing as it carries out complete degradation of pectin and thus increases juice extraction and also enhances clarification of juice (Han, 2004).

In the beer industries, ALDC enzyme allows the acceleration or alteration of the total time for beer fermentation/maturation as it shunts diacetyl formation whose elimination is important and it is a rate-limiting step of the whole process (Dulieu et al., 2000). These enzymes are also used to make it possible to shorten beer primary fermentation (Aschengreen and Jepsen, 1992; Jepsen, 1993; Kabaktschieva et al., 1994) until no more maturation is needed in regard to diacetyl (Linko and Kronlof, 1991). The encapsulation of ALDC enzyme in small spheres of polyelectrolyte complexes was evaluated by Dulieu et al. (1997). The encapsulated ALDC enzyme shows more advantage because it leads to acceleration of beer fermentation as efficiently as free ALDC enzyme suggested by Dulieu et al. (2000) and reported the great use of immobilized ALDC as compared to free ALDC enzyme, which is recoverable and reusable, thus reducing the cost of production of the process.

3.5.3   CHEESE

Dairy industry also utilized vast majority of enzymes for the production of a variety of food and dairy products. Some enzymes are required for the production of cheeses, yoghurt, and other dairy products, while others are used in a more specialized fashion or way to improve texture or flavor of the products. The gas-forming culture of *Propionibacterium shermanii* is essential for giving Swiss cheese its eye, or holes and flavor (Smily et al., 2014).

3.5.4   DAIRY

The dairy enzymes market forms a part of the food and beverage industry. The dairy enzymes market consists of products that are used for processing/producing dairy products such as cheese, yoghurt, or milk. Within the overall dairy industry, the utilization of enzymes extends across applications such as the production of milk, cheese, yoghurt, and other dairy products. Dairy enzymes are widely used for refining the texture and quality of various products. Moreover, these enzymes are often used as flavor enhancers. The dairy enzymes market comprises applications dependent on both nonmicrobial and microbial enzymes. Lactase is used commercially to prepare lactose-free products, particularly milk, for lactose-intolerant individuals. It is also used in the preparation of ice cream, to make a creamier and sweeter-tasting product. Lactase is usually obtained from *Kluyveromyces* spp. of yeast and *Aspergillus* spp. of fungi. Another important GMO-derived enzyme used in dairy industry is chymosin (microbially produced rennet-like enzyme) employed for cheese-making. In 1990, this chymosin obtained from GMO was granted Generally Regarded as Safe (GRAS) status by the FDA in the United States (Verma et al., 2017b). About 90% of the cheese produced in the United States is made using fermentation-produced chymosin (FPC) making the use of *K. lactis*, *Aspergillus* spp., *Mucor* spp., as well as bacterial cell like *Bacillus* spp. (Entine and Lim, 2015).

3.5.5   EGG-BASED PRODUCTS

In the food and beverage industry, enzymes play an indispensable or crucial role in the production, processing, and preservation of dairy and egg-based products, chocolates, syrups, infant formula and baby food, sweeteners, food colors and flavors, cheese, milk, and others.

Various industrially produced cream products normally use dried egg powder instead of fresh eggs. Two enzymes, lipase and GO, are often produced with the help of GMMs that are normally added in order to preserve and maintain egg powder along with its color. Few microorganisms which produced lipase is *Rhizopus* and for GO *Aspergillus* (Chaplin, 2014). A majority of microbial enzymes come from a very limited number of genera, like *Aspergillus* spp., *Bacillus* spp., and *Kluyveromyces* spp. (also called *Saccharomyces*).

3.5.6   STARCH

Starch is basically a long chain of sugar molecules which is normally linked together like a chain; thus, it is a polysaccharide. Its single sugar molecule is called monosaccharide. Basic function of starch in plant cells is to store energy in the form of glucose. In animals, cells have a different way of storing energy usually in the form of glycogen, which is similar to the plant's starch form of amylopectin.

Starch is found in several dietary foods such as in potatoes, wheat, rice, and other foods. It varies greatly in appearance, depending on its source. Starches have limited use in the food industry in unmodified form. Generally, native starches produce weak-bodied, cohesive, rubbery pastes when heated, and undesirable gels are formed when pastes are allowed to cool (Adzahan, 2002). This is a reason, that majority of food manufacturers generally prefer starches with better or improvised behavioral characteristic than those usually provided by native starches. Researchers have been developed various methods to modify starch, which is done with the help of chemicals and enzymes.

Chemical modifications aim to improve several chemical properties such as viscosity, resistant to shear, low pH, and high temperature (Abbas, 2010). Enzymatic modification of starch generally results in the hydrolysis of some part of starch into a low molecular weight of starch called maltdextrin or dextrin with the help of amylolytic enzymes (Guzmán-Maldonado et al., 1995). They are widely used for food and pharmaceutical industries. Physical modification involves various processes such as pre-gelatinization, heat treatment of starch, and so forth (Miyazaki et al., 2006). Starches are α-glucan polymers which are joined by α-1,4 linkages with the additional branches of α-1,6 linkages. Various organisms such as bacteria, fungi, plants, and animals that generally contain gene encoding enzymes, which serve their demand for α-glucan digestion along with its biosynthesis (Ball et al., 2011; Cenci et al., 2013). Some of these genes can also be used to generate various GMMs, which can produce such enzymes on an industrial scale, having a wide role in starch modifications.

3.6   PROSPECTIVE FUTURES AND RESEARCH OPPORTUNITIES

Production levels of metabolites of viable value, such as enzymes, amino acids, vitamins, bacteriocins, and antibiotics, are quite low by intact natural producing microorganisms (Patel et al., 2017). Improving metabolite yield is therefore requisite for meeting the product demands and for upholding an economically possible process. By future research, we can overcome several obstructions and guide the usage of right approaches for increasing the production of desired products through manipulating the producing microorganisms (Parekh, 2004), for example, manipulating regulatory genes, overcoming rate-limiting steps, eliminating feedback regulation, distressing essential breakdown, removing challenging trails, increasing product transportation, and providing economic strength.

3.7 CONCLUSION

Enzymes manufactured by GMOs have been used in the food industry for more than 15 years. The well-known examples include the usage of chymosin for cheese production and pectinases for fruit and beverage processing. Conventionally, chymosin containing rennet from calf stomach is required for cheese production to provide the essential proteolytic activity for coagulation of milk proteins. The production of enzymes by the process of genetic engineering emphasizes on economic production, enhanced enzyme purity, and naturally responsive production processes.

With the development of rDNA technology and genetic engineering, the potential of microorganisms (metabolic application) is being explored and harnessed in different ways. Currently, GMMs have found applications in anthropoid healthiness, agronomy, and bioremediation and in certain industries, namely paper, food, and textiles. Increasing molecular diversity and improved chemical selectivity are the advantages of genetic engineering over traditional approaches. In addition, it also offers sufficient supplies of desired and cheaper production of products and safe handling of dangerous agents (if any). This chapter defines several molecular tools, strategies to modify microorganisms, to produce useful enzymes and various other applications.

3.8 SUMMARY

GMOs are concerned generally with organisms which are produced when genes of a selected individual are transferred from a given donor organism to another target system. Genetic modifications can potentially improve quality and productivity. The usage of microorganisms (mainly bacteria, yeasts, and fungi) by the food industry has led to a highly broadened industry with relevant economical assets/values. Fermentation technology with special reference to the production of alcoholic products, enzymes, dairy products, organic acids, and drugs (including antibiotics also) are the most important examples of microbiological processes. In microorganisms, genetic engineering continues to make rapid progress. In summation to increasing the fundamental understanding of microbiology, transgenic microbes are being introduced commercially with an increasing value. This chapter makes an attempt to show how GMOs can be a useful potential to improve food quality and productivity. This may arm to improve industrialization as well as economical empower.

KEYWORDS

- **aminopeptidases**
- **antimicrobial metabolite**
- ***aspergillus***
- **biosurfactant**
- **esterases**
- **galactosidases**
- ***Gliocladium***
- **glucanase**
- **metalloendopeptidases**
- **ochratoxin**
- **oligosaccharides**
- **phosphodiesterases**
- **phospholipase**
- **poly-γ-glutamic acid**
- **serine endopeptidases**
- **xylanase**
- **γ-aminobutyric acid**

REFERENCES

Ab Kadir, S.; Wan-Mohtar, W. A.; Mohammad, R.; Abdul, H. L. S.; Sabo Mohammed, A.; Saari, N. Evaluation of Commercial Soy Sauce *Koji* Strains of *Aspergillus oryzae* for γ-Aminobutyric Acid (GABA) Production. *J. Ind. Microbiol. Biotechnol.* **2016,** *43,* 1387–1395.

Abbas, K. A. Modified Starches and Their usages in Selected Food Products: a Review Study. *J. Agric. Sci.* **2010,** *2,* 90–100.

Adzahan, N. M. Modification on Wheat, Sago and Tapioca Starches by Irradiation and its Effect on the Physical Properties of Fish Cracker (Keropok). MSc Thesis, Food Science and Biotechnology, University of Putra Malaysia, Malaysia, 2002.

Alanio, A.; Brethon, B.; de Chauvin, M. F.; de Kerviler, E.; Leblanc, T.; Lacroix, C.; Baruchel, A.; Menotti, J. Invasive Pulmonary Infection Due to *Trichoderma longibrachiatum* Mimicking Invasive *Aspergillosis* in a Neutropenic Patient Successfully Treated with Voriconazole Combined with Caspofungin. *Clin. Infect. Dis.* **2008,** *46*(10), e116–e118.

Anné, J.; van Mellaert, L. *Streptomyces lividans* as Host for Heterologous Protein Production. *FEMS Microbiol. Lett.* **1993,** *4*(2), 121–128.

Aoyama, S.; Yoshida, N.; dan Inouya, S. Cloning, Sequencing and Expression of Lipase Gene from *Pseudomonas fragi* IFD-12049 in *Escherichia coli. FEBS Lett.* **1988,** *242,* 36–40.

Aschengreen, N. H.; Jepsen, S. Use of Acetolactate Decarboxylase in Brewing Fermentations. *Proc. Conv.-Inst. Brew. (Aust. N. Z. Sect.)*, **1992,** *22*, 80–83.

Bairoch, A. The ENZYME database. *Nucleic Acids Res.* **2000,** *28*(1), 304–305.

Balance, D. J.; Buxton, F. P.; Turner, G. Transformation of *Aspergillus nidulans* by the Orotidine-5′-Phosphate Decarboxylase Gene of *Neurospora crassa. Biochem. Biophys. Res. Commun.* **1983,** *112,* 284–289.

Balance, D. J.; Turner, G. Development of a High-Frequency Transforming Vector for *Aspergillus nidulans. Gene* **1985,** *36*, 321–331.

Ball, S.; Colleoni, C.; Cenci, U.; Raj, J. N.; Tirtiaux, C. The Evolution of Glycogen and Starch Metabolism in Eukaryotes Gives Molecular Clues to Understand the Establishment of Plastid Endosymbiosis. J. Exp. Bot. **2011,** *62*, 1775–1801.

Barbesgaard, P.; Heldt-Hansen, H. P.; Diderichsen, B. On the Safety of *Aspergillus oryzae*: A Review. *Appl. Microbiol. Biotechnol.* **1992,** *36*, 569–572.

Baur, X.; Sander, I.; Jansen, A.; Czuppon, A. B. Are Amylases in Bakery Products and Flour Potential Food Allergens? *Schweiz. Med. Wochenschr.* **1994,** *124*(20), 846–51.

Bender, H.; Wallenfels, K. Pullulanase (An Amylopectin and Glycogen Debranching Enzyme) from *Aerobacter aerogenes. Methods Enzymol.* **1966,** *8,* 555–559.

Bentley, R. From Miso, Sake and Shoyu to Cosmetics: A Century of Science for Kojic Acid. *Nat. Prod. Rep.* **2006,** *23,* 1046–1062.

Berka, R. M.; Dunn-Coleman, N. S.; Ward, M. Industrial Enzymes from *Aspergilli*. In *Aspergillus, Biology and Industrial Applications;* Bennett, J. W., Klich, M. A., Eds.; Butterworth Heinemann: London, 1992; pp 155–202.

Berka, R. M.; Fowler, T.; Rey, M. W. Gene Sequence Encoding *Aspergillus niger* Catalase-R. U.S. Patent 5,360,901, 1994a.

Berka, R. M.; Fowler, T.; Rey, M. W. Production of *Aspergillus niger* Catalase-R. U.S. Patent 5,360,732, 1994b.

Bhosale, S. H.; Rao, M. B.; Deshpande, V. V. Molecular and Industrial Aspects of Glucose Isomerase. *Microbiol. Rev.* **1996,** *60*(2), 280–300.

Blumenthal, C. Z. Production of Toxic Metabolites in *Aspergillus niger, Aspergillus oryzae,* and *Trichoderma reesei*: Justification of Mycotoxin Testing in Food Grade Enzyme Preparations Derived from the Three Fungi. *Regul. Toxicol. Pharmacol.* **2004,** *39,* 214–228.

Boottanun, P.; Potisap, C.; Hurdle, J. G.; Sermswan, R. W. Secondary Metabolites from *Bacillus amyloliquefaciens* Isolated from Soil Can Kill *Burkholderia pseudomallei. AMB Express* **2017,** *7,* 16. DOI: 10.1186/s13568–016–0302–0.

Brawner, M. E. Advances in Heterologous Gene Expression by *Streptomyces. Curr. Opin. Biotechnol.* **1994,** *5,* 475–481.

Brito-Cunha, C. C.; de Campos, I. T.; de Faria, F. P.; Bataus, L. A. Screening and Xylanase Production by *Streptomyces* sp. Grown on Lignocellulosic Wastes. *Appl. Biochem. Biotechnol.* **2013,** *170,* 598–608.

Buxton, F. P.; Gwynne, D. I.; Davies, R. W. Transformation of *Aspergillus niger* using the *arg*B Gene of *Aspergillus nidulans. Gene* **1985**, *37*, 207–214.

Cabib, E.; Roberts, R.; Bowers, B. Synthesis of the Yeast Cell Wall and its Regulation. *Annu. Rev. Biochem.* **1982**, *51*, 763–793.

Campbell, E. I.; Unkles, S. E.; Macro, J. A.; Van den Hondel, C.; Contreras, R.; Kinghorn, J. R. Improved Transformation Efficiency of *Aspergillus niger* Based on the *pyr*G Gene. *Mol. Gen. Genet.* **1989**, *206*, 71–75.

Cardoza, R. E.; Gutiérrez, S.; Ortega, N.; Colina, A.; Casqueiro, J.; Martín, J. F. Expression of a Synthetic Copy of the Bovine Chymosin Gene in *Aspergillus awamori* from Constitutive and pH-Regulated Promoters and Secretion using Two Different Pre-Pro Sequences. *Biotechnol. Bioeng.* **2003**, *83*, 249–259.

Cawoy, H.; Debois, D.; Franzil, L.; De Pauw, E.; Thonart, P.; Ongena, M. Lipopeptides as Main Ingredients for Inhibition of Fungal Phytopathogens by *Bacillus subtilis/amyloliquefaciens. Microb. Biotechnol.* **2015**, *8*, 281–295.

Cenci, U.; Chabi, M.; Ducatez, M.; Tirtiaux, C.; Nirmal-Raj, J.; Utsumi, Y. Convergent Evolution of Polysaccharide Debranching Defines a Common Mechanism for Starch Accumulation in Cyanobacteria and Plants. Plant Cell **2013**, *25*, 3961–3975.

Chahinian, H.; Vanot, G.; Ibrik, A.; Rugani, N.; Sarda, L.; Comeau, L. C. Production of Extracellular Lipases by *Penicillium cyclopium* Purification and Characterization of Partial Acylglycerol Lipase. *Biosci. Biotechnol. Biochem.* **2000**, *64*, 215–222.

Chaplin, M. Enzyme Technology: Sources of Enzymes, 2014. http://www1.lsbu.ac.uk/water/enztech/sources.html (accessed April 26, 2017).

Chauthaiwale, J. V.; Rao, M. B. Production and Purification of Extracellular D-Xylose Isomerase from an Alkaliphilic, Thermophilic *Bacillus* sp. *Appl. Environ. Microbiol.* **1994**, *60*, 4495–4499.

Chen, W. P.; Anderson, A. W.; Han, Y. W. Production of Glucose Isomerase by *Streptomyces flavogriseus. Appl. Environ. Microbiol.* **1979**, *37*, 324–331.

Cheng, J.; Zhuang, W.; Li, N. N.; Tang, C. L.; Ying, H. J. Efficient Biosynthesis of D-Ribose using a Novel Co-Feeding Strategy in *Bacillus subtilis* without Acid Formation. *Lett. Appl. Microbiol.* **2017**, *64*(1), 73–78.

Chet, I. *Trichoderma* Application, Mode of Action, and Potential as Biocontrol Agent of Soilborne Plant Pathogenic Fungi. In *Innovative Approaches to Plant Disease Control;* Chet, I., Ed.; John Wiley and Sons.; 1987; pp 137–160.

Chowdhury, S. P.; Hartmann, A.; Gao, X. W.; Borriss, R. Bio-Control Mechanism by Root Associated *Bacillus amyloliquefaciens* FZB42-A Review. *Front. Microbiol.* **2015**, *6*, 780–790.

Colla, E.; Santos, L. O.; Deamici, K.; Magagnin, G.; Vendruscolo, M.; Costa, J. A. Simultaneous Production of Amyloglucosidase and Exo-Polygalacturonase by *Aspergillus niger* in a Rotating Drum Reactor. *Appl. Biochem. Biotechnol.* **2017**, *181*(2), 627–637.

Coutinho, P. M.; Reilly, P. J. Glucoamylase Structural, Functional and Evolutionary Relationships. *Protein Eng.* **1997**, *29*, 334–347.

Dake, M. S.; Jadhav, J. P.; Patil, N. B. Induction and Properties of (1→3)-β-D-Glucanase from *Aureobasidium pullulans. Indian J. Biotechnol.* **2004**, *3*, 58–64.

Dave, K. K.; Punekar, N. S. Expression of Lactate Dehydrogenase in *Aspergillus niger* for L-Lactic Acid Production. *PLoS ONE* **2015**, *10*, e0145–459. DOI: 10.1371/journal.pone.0145459.

De Haas, G. H.; Postema, N. M.; Nieuwenhuizen, W.; van Deenen, L. L. Purification and Properties of Phospholipase a from Porcine Pancreas. *Biochim. Biophys. Acta.* **1968,** *159*(1), 103–117.

Djordjevic, J. T. Role of Phospholipases in Fungal Fitness, Pathogenicity, and Drug Development—Lessons from *Cryptococcus neoformans. Front. Microbiol.* **2010,** *1*(125), 1–13. DOI: 10.3389/fmicb.2010.00125.

Dorado, M. P.; Lin, S. K.; Koutinas, A.; Du, C.; Wang, R.; Webb, C. Cereal-Based Biorefinery Development: Utilisation of Wheat Milling by-Products for the Production of Succinic Acid. *J. Biotechnol.* **2009,** *143,* 51–59.

Dulieu, C.; Dautzenberg, H.; Poncelet, D. *Immobilized Enzyme System as a Tool for Beer Maturation;* International Workshop Bioencapsulation VI: Barcelona, 1997; pp 5–4.

Dulieu, C.; Moll, M.; Boudrant, J.; Poncelet, D. Improved Performances and Control of Beer Fermentation using Encapsulated Alpha-Acetolactate Decarboxylase and Modeling. *Biotechnol. Prog.* **2000,** *16*(6), 958–965.

Emtage, J. S.; Angal, S.; Doel, M. T.; Harris, T. J.; Jenking, B.; Lilley, G.; Lowe, P. A. Synthesis of Calf Prochymosin (Prorennin) in *Escherichia coli. Proc. Nat. Acad. Sci. USA* **1983,** *80,* 3671–3675.

Entine; J.; Lim, X.; Genetic Literacy Project. Cheese: The GMO food die-hard GMO Opponents Love (and Oppose a Label for), 2015. https://www.geneticliteracyproject. org/2015/05/15/cheese-gmo-food-die-hard-gmo-opponents-love-and-oppose-a-label-for/. (accessed April 26, 2017.

Eremin, A. N.; Metelitsa, D. I.; Shishko, Z. F.; Mikhaĭlova, R. V.; Iasenko, M. I.; Lobanok, A. G. Thermal Stability of Penicillium Adametzii Glucose Oxidase. *Prikl. Biokhim. Mikrobiol.* **2001,** *37,* 678–686.

Fernández-Espinar, M. T.; Peña, J. L.; Piñaga, F.; Vallés, S. Alpha-L-Arabinofuranosidase Production by *Aspergillus nidulans. FEMS Microbiol. Lett.* **1994,** *115*(1), 107–112.

Fiedurek, J.; Gromada, A. Screening and Mutagenesis of Molds for Improvement of the Simultaneous Production of Catalase and Glucose Oxidase. *Enzyme Microbial. Technol.* **1997,** *20,* 344–347.

Field, C. E.; Pivarnik, L. F.; Barnett, L. S. M.; Rand, Jr., A. G. Utilizing of Glucose Oxidase for Extending the Shelf-Life of Fish. *J. Food Sci.* **1986,** *51,* 66–70.

Foltmann, B. Prochymosin and Chymosin (Prorennin and Rennin). *Methods Enzymol.* **1970,** *19,* 421–436.

Frense, D.; Lange, U.; Hartmeier, U. Immobilization of *Candida rugosalipase* in Lyotropic Liquid Crystals and Some Properties of Immobilized Enzymes. *Biotechnol. Lett.* **1996,** *18,* 293–298.

Gao, T.; Wong, Y.; Ng, C.; Ho, K. L-Lactic Acid Production by *Bacillus subtilis* MUR1. *Bioresour. Technol.* **2012,** *121,* 105–110.

Ghannoum, M. A. Potential Role of Phospholipases in Virulence and Fungal Pathogenesis. *Clin. Microbiol. Rev.* **2000,** *13*(1), 122–143.

Godfrey, T.; Reichelt, J. *Industrial Enzymology: The Application of Enzymes in Industry;* Nature Press: New York, 1983.

Godtfredsen, S. E.; Ottesen, M. Maturation of Beer with R-Acetolactate Decarboxylase. *Carslberg Res. Commun.* **1982,** *47,* 93–102.

Godtfredsen, S. E. Microbial Lipases. In *Microbial Enzymes and Biotechnology;* Fogartty, W. M., Kelly, C. T., Eds.; Springer: Netherlands, 1990; pp 255–274.

Golubev, A. M.; Nagem, R. A.; Brandao Neto, J. R.; Neustroev, K. N.; Eneyskaya, E. V.; Kulminskaya, A. A.; Shabalin, K. A.; Savel'ev, A. N.; Polikarpov, I. Crystal Structure of Alpha-Galactosidase from *Trichoderma reesei* and its Complex with Galactose: Implications for Catalytic Mechanism. *J. Mol. Biol.* **2004,** *339,* 413–422.

Gonzalez, T.; Baudouy, J. R. Bacterial Aminopeptidases: Properties and Functions. *FEMS Microbiol. Rev.* **1996,** *18,* 319–344.

Goulart, C. L.; Barbosa, L. C.; Bisch, P. M.; von Krüger, W. M. Catalases and PhoB/PhoR System Independently Contribute to Oxidative Stress Resistance in *Vibrio cholerae* O1. *Microbiology* **2016,** *162*(11), 1955–1962.

Grassin, C.; Fauguenbergue, P. Enzymes, Fruit Juice Processing. In *Encyclopedia of Bioprocess Technology: Fermentation, Biocatalysis, and Bioseparation;* Flickinger, M. C., Drew, S. W., Eds.; Wiley and Sons, Inc.: New York, 1999.

Gulomova, K.; Ziomek, E.; Schrag, J. D.; Davranov, K.; Cygler, M. Purification and Properties of a *Penicillium* sp. which Discriminates against Diglycerides. *Lipids* **1996,** *31*(4), 379–284.

Guzmán-Maldonado, H.; Paredes-López, O.; Biliaderisc, C. G. Amylolytic Enzymes and Products Derived from Starch: A Review. Crit. Rev. Food Sci. Nutr. 1995, *35*(5), 373–403.

Haas, M. J.; Cichowicz, D. J.; Bailey, D. G. Purification and Characterization of an Extracellular Lipase from the Fungus *Rhizopus delemar. Lipids* **1992,** *27*(8), 571–576.

Hakulinen, N.; Tenkanen, M.; Rouvinen, J. Three-Dimensional Structure of the Catalytic Core of Acetyl Xylan Esterase from *Trichoderma reesei*: Insights into the Deacetylation Mechanism. *J. Struct. Biol.* **2000,** *132,* 180–190.

Haltmeier, T.; Leisola, M.; Ulmer, D.; Waldner, R.; Fiecher, A. Pectinase from *Trichoderma reesei* QM9414. *Biotechnol. Bioeng.* **1983,** *25,* 1685–1690.

Han, L. Genetically Modified Microorganisms-Development and Applications. In *The GMO Handbook: Genetically Modified Animals, Microbes, and Plants in Biotechnology;* Parekh, S. R., Ed.; Humana Press Inc.: Totowa, NJ, 2004; pp 29–51.

Hanft, F.; Koehler, P. Studies on the Effect of Glucose Oxidase in Bread Making. *J. Sci. Food Agric.* **2006,** *86,* 1699–1704.

Held, P. Part II: Optimization of Polymer Digestion and Glucose Production in Microplates. In Enzymatic Digestion of Polysaccharides, Application Notes, Biofuel Research, 2012. https://www.biotek.com/resources/single.html?newsid=11087. (Accessed May 5, 2017)

Hendriksen, H. V.; Kornbrust, B. A.; Stergaard, P. R.; Stringer, M. A. Evaluating the Potential for Enzymatic Acrylamide Mitigation in a Range of Food Products using an Asparaginase from *Aspergillus oryzae. J. Agric. Food Chem.* **2009,** *57,* 4168–4176.

Higashi, K.; Kusakabe, I.; Yasui, T.; Ishiyama, T.; Okimoto. Y. Arabinan-Degrading Enzymes from *Streptomyces diastatochromogenes* 065. *Agric. Biol. Chem.* **1983,** *47,* 2903–2905.

Hossain, A. H.; Li, A.; Brickwedde, A.; Wilms, L.; Caspers, M.; Overkamp, K.; Punt, P. J. Rewiring a Secondary Metabolite Pathway Towards Itaconic Acid Production in *Aspergillus niger. Microb. Cell Fact.* **2016,** *15*(130), 1–15. DOI: 10.1186/s12934–016–0527-2

Illanes, A. *Enzyme Biocatalysis Principles and Applications;* Springer: Netherlands, 2008; p 379.

Inoue, H.; Kitao, C.; Yano, S.; Sawayama, S. Production of β-Xylosidase from Trichoderma Asperellum KIF125 and its Application in Efficient Hydrolysis of Pretreated Rice Straw with Fungal Cellulase. *World J. Microbiol. Biotechnol.* **2016,** *32*(186), 1–10. DOI: 10.1007/s11274–016–2145-x

Ito, K.; Nakajima, Y.; Onohara, Y.; Takeo, M.; Nakashima, K. Crystal Structure of Aminopeptidase N (Proteobacteria Alanyl Aminopeptidase) from *Escherichia coli* and Conformational Change of Methionine 260 Involved in Substrate Recognition. *J. Biol. Chem.* **2006,** *281,* 33664–33676.

IUBMB. Enzyme Nomenclature 1992. Recommendations of the Nomenclature Committee of the International Union of Biochemistry and Molecular Biology on the Nomenclature and Classification of Enzymes. Academic Press, Inc. 1992.

Izawa, N.; Tokuyasu, K.; Hayashi, K. Debittering of Protein Hydrolysates using *Aeromonas caviae* Aminopeptidase. *J. Agric. Food Chem.* **1997,** *45*(3), 543–545.

Jaeger, K. E.; Reetz, T. M. Microbial Lipases from Versatile Tools for Biotechnology. *Trends Biotechnol.* **1998,** *16,* 396–403.

Jamuna, R.; Saswathi, N.; Sheela, R.; Ramakrishna, S. V. Synthesis of Cyclodextrin Glucosyl Transferase by *Bacillus cereus* for the Production of Cyclodextrins. *Appl. Biochem. Biotechnol.* **1993,** *43,* 163–176.

Jensen, B. F.; Norman, B. E. *Bacillus acidopullulyticus* pullulanase: Application and Regulatory Aspects for the Food Industry. *Process Biochem.* **1984,** *19,* 129–134.

Jepsen, S. Using ALDC to Speed Up Fermentation. *Brew. Guardian* **1993,** *122,* 55–56.

Joshi, S. J.; A-Wahaibi, Y. M.; Al-Bahry, S. N.; Elshafie, A. E.; Al-Bemani, A. S.; Al-Bahri, A.; Al-Mandhari, M. S. Production, Characterization, and Application of *Bacillus licheniformis* W16 Biosurfactant in Enhancing Oil Recovery. *Front. Microbiol.* **2016,** *7,* 1853. DOI: 10.3389/fmicb.2016.01853

Kabaktschieva, G.; Ginova-Stojanova, T.; Dimitrova, T. The use of an Enzyme Solution with R-Acetolactate Decarboxylase Activity. *Brew. Bever. Ind. Int.* **1994,** *2,* 22–24.

Kauldhar, B. S.; Sooch, B. S. Tailoring Nutritional and Process Variables for Hyperproduction of Catalase from a Novel Isolated Bacterium *Geobacillus sp.* BSS-7. *Microb. Cell Fact.* **2016,** *15*(7), 1–16. DOI: 10.1186/s12934–016–0410–1

Kaye, N. M. C.; Jolles, P. The Involvement of One of the Three Histidine Residues of Cow K-Casein in the Chymosin-Initiated Milk Clotting Process. *Biochim. Biophys. Acta* **1978,** *536,* 329–340.

Kelly, J. M.; Hynes, M. J. Transformation of *Aspergillus niger* by the amdS Gene of *Aspergillus nidulans. EMBO J.* **1985,** *4*(2), 475–479.

Kennedy, M. B.; Rennarz, W. J. Characterization of the Extracellular Lipase of *Bacillus subtilis* and its Relationship to a Membrane Bound Lipase found in a Mutant Strain. *J. Biol. Chem.* **1979,** *254,* 1080–1089.

Kiiskinen, L. L.; Kruus, K.; Bailey, M.; Ylosmaki, E.; Siika-aho, M.; Saloheimo, M. Expression of Melano Carpus Albomyceslaccase in *Trichoderma reesei* and Characterization of the Purified Enzyme. *Microbiology* **2004,** *150,* 3065–3074.

Klein, R. D.; Geary, T. G. Recombinant Microorganisms as Tools for High Throughput Screening for Nonantibiotic Compounds. *J. Biomol. Screen Spring* **1997,** *2,* 41–49.

Klein, R. R.; King, G.; Moreau, R. A.; Haas, M. J. Altered Acyl Chain Length Specificity of *Rhizopus delemar* Lipase through Mutagenesis and Molecular Modeling. *Lipids* **1997,** *32,* 123–130.

Kluyver, A. J.; Perquin, L. H. C. Zur Methodik der Schimmelstoffwechseluntersuchung. *Biochem. Z* **1932,** *266,* 68–81.

Kohno, M.; Kugimiya, W.; Hashemoto, Y.; Moreta, Y. Purification, Characterization, and Crystallization of Two Types of Lipase from *Rhizopus niveus*. *Biosci. Biotechnol. Biochem.* **1994,** *58*(6), 1007–1012.

Kojima, Y.; Yokoe, M.; Mase, T. Purification and Characterization of an Alkaline Lipase from *Pseudomonas fluorescens*. *Biosci. Biotechnol. Biochem.* **1994,** *58*(9), 1564–1568.

Kometani, T.; Terada, Y.; Nishimura, T.; Nakae, T.; Takii, H.; Okada, S. Acceptor Specificity of Cyclodextrin Glucanotransferase from an Alkalophilic Bacillus Species and Synthesis of Glucosyl Rhamnose. *Biosci. Biotechnol. Biochem.* **1996,** *60*, 1176–1178.

Kongklom, N.; Shi, Z.; Chisti, Y.; Sirisansaneeyakul, S. Enhanced Production of Poly-γ-Glutamic Acid by *Bacillus licheniformis* TISTR 1010 with Environmental Controls. *Appl. Biochem. Biotechnol.* **2016,** *24*. DOI: 10.1007/s12010–016–2376–1.

Kumar, A.; Grover, S.; Sharma, J.; Batish, V. K. Chymosin and Other Milk Coagulants: Sources and Biotechnological Interventions. *Crit. Rev. Biotechnol.* **2010,** *30*(4), 243–58.

Kumar, A.; Sharma, J.; Grover, S.; Mohanty, A. K.; Batish, V. K. Molecular Cloning and Expression of Goat (*Capra hircus*) Prochymosin in *E. coli*. *Food Biotechnol.* **2007,** *21*, 57–69.

Kundu, N.; Basu, J.; Guchhait, M.; Chakrabarti, P. Isolation and Characterization of an Extracellular Lipase from the Conidia of *Neurospora crassa*. *J. Gen. Microbiol.* **1987,** *133*, 149–153.

Lambert, P. W.; Meers, J. L.; Best, D. J. The Production of Industrial Enzymes. *Biol. Sci.* **1983,** *300*, 263–282.

Lee, E. Y. C.; Whelan, W. J. Glycogen and Starch Debranching Enzymes. In *The Enzymes*, 3rd Ed.; Boyer, P. D., Ed.; Academic Press: New York, 1972; pp 191–234.

Lee, C. Y.; Yandolo, J. J. Lysogenic Conversion of Staphylococcal Lipase is Caused by Insertion of the Bacteriophage L54a into the Lipase Structural Gene. *J. Bacteriol.* **1986,** *166*, 385–391.

Leghlimi, H.; Meraihi, Z.; Boukhalfa-Lezzar, H.; Copinet, E.; Duchiro, F. Production and Characterization of Cellulolytic Activities Produced by Trichoderma Longibrachiatum (GHL). *Afr. J. Biotechnol.* **2013,** *12*, 465–475.

Li, S.; Chen, L.; Hu, Y.; Fang, G.; Zhao, M.; Guo, Y.; Pang, Z. Enzymatic Production of 5'-Inosinic Acid by AMP Deaminase from a Newly Isolated *Aspergillus oryzae*. *Food Chem.* **2017a,** *216*, 275–281.

Li, S.; Qian, Y.; Liang, Y.; Chen, X.; Zhao, M.; Guo, Y.; Pang, Z. Overproduction, Purification and Characterization of Adenylate Deaminase from *Aspergillus oryzae*. *Appl. Biochem. Biotechnol.* **2016a,** *180*, 1635–1643.

Li, Y.; Gu, Z.; Zhang, L.; Ding, Z.; Shi, G. Inducible Expression of Trehalose Synthase in *Bacillus licheniformis*. *Protein Expr. Purif.* **2017b,** *130*, 115–122.

Li, Y.; Zheng, X.; Zhang, X.; Bao, L.; Zhu, Y.; Qu, Y.; Zhao, J.; Qin, Y. The Different Roles of *Penicillium oxalicum* LaeA in the Production of Extracellular Cellulase and β-xylosidase. *Front. Microbiol.* **2016b,** *7*(2091), 1–14. DOI: 10.3389/fmicb.2016.02091

Li, Z.; Ji, K.; Dong, W.; Ye, X.; Wu, J.; Zhou, J.; Wang, F.; Chen, Q.; Fu, L.; Li, S.; Huang, Y.; Cui, Z. Cloning, Heterologous Expression, and Enzymatic Characterization of a Novel Glucoamylase GlucaM from *Corallococcus* sp. Strain EGB. *Protein Expression Purif.* **2017c,** *129*, 122–127.

Lin, S.; Lee, J. C.; Chion, C. M. Purification and Characterization of a Lipase from *Neurospora* sp. T T-241. *J. Am. Oil Chem. Soc.* **1996,** *73*, 739. DOI: 10.1007/BF02517950

Lin, S.-J.; Chen, Y.-H.; Chen, L.-L.; Feng, H.-H.; Chen, C.-C.; Chu, W.-S. Large-Scale Production and Application of Leucine Aminopeptidase Produced by *Aspergillus Oryzae* Ll1 for Hydrolysis of Chicken Breast Meat. *Eur. Food Res. Technol.* **2008,** *227*(1), 159–165.

Linko, M.; Kronlof, J. *Main Fermentation with Immobilized Yeast*; Eur. Brew. Conv., Oxford; IRL Press: Lisbon, 1991; pp 353–360.

Long, K.; Ghazali, H. M.; Ariff, A.; Ampon, K.; Bucke, C. In-Situ Crosslinking of *Aspergillus flavuslipase*: Improvement of Activity, Stability and Properties. *Biotechnol. Lett.* **1996,** *18,* 1169–1174.

Luerce, T. D.; Azevedo, M. S.; LeBlanc, J. G.; Azevedo, V.; Miyoshi, A.; Pontes, D. S. Recombinant *Lactococcus lactis* Fails to Secrete Bovine Chymosine. *Bioengineered* **2014,** *5*(6), 363–370.

Luo, F.; Jiang, W. H.; Yang, Y. X.; Li, J.; Jiang, M. F. Cloning and Expression of Yak Active Chymosin in *Pichia pastoris*. *Asian-Australas. J. Anim. Sci.* **2016,** *29*(9), 1363–1370.

Manners, D. J. Observations On the Specificity and Nomenclature of Starch Debranching Enzymes. *J. Appl. Glycosci.* **1997,** *44,* 83–85.

Mase, T.; Matsumiya, Y.; Matsuura, A. Purification and Characterization of *Penicillium roqueforti* 1AM 7268 Lipase. *Biosci. Biotech. Biochem.* **1995,** *59,* 329–330.

McDonald, J. K. A Brief History of the Study of Mammalian Exopeptidases. In *Mammalian Piroteases;* McDonald, J. K., Baret, A. J., Eds.; Academic Press: New York, 1986; Vol. 2, pp 7–19.

McDonald, J. K.; Schwabe, C. Proteinases in Mammalian Cells and Tissues. (In *Research Monographs in Cell and Tissue Physiology;* Barrett, A. J., Ed.; Elsevier Science Publisher B.V.: Amsterdam, 1977, Vol. 2, pp 311–391.

Micheli, F. Pectin Methylesterases: Cell Wall Enzymes with Important Roles in Plant Physiology. *TRENDS Plant Sci.* **2001,** *6*(9), 414–419.

Miyazaki, M. R.; Hung, P. V.; Maeda, T.; Morita, N. Recent Advances in Application of Modified Starches for Breadmaking. *Trends Food Sci. Technol.* **2006,** *17,* 591–599.

Monod, M.; Stocklin, R.; Grouzmann, E. Novel Fungal Proteins and Nucleic Acids Encoding Same, U.S. Patent no. 7,468,267, 2008.

Montedoro, G. Use of Enzyme Preparations in Wine Production (Author's Transl). *S TA NU* **1976,** *6*(3), 133–144.

Moosavi-Nasab, M.; Majdi-Nasab, M. Cellulase Production by *Trichoderma reesei* using Sugar Beet Pulp. *Iran Agric. Res.* **2007,** *25–26*(1–2), 107–116.

Moure, A.; Domınguez, H.; Parajo, J. C. Antioxidant Properties of Ultrafiltration-Recovered Soy Protein Fractions from Industrial Effluents and their Hydrolysates. *Process Biochem.* **2006,** *41,* 447–456.

Mozaffar, Z.; Weete, J. D. Purification and Properties of an Extracellular Lipase from *Pythium ultimum*. *Lipids* **1993,** *28*(5), 377–382.

Mullaney, E. J.; Gibson, D. M.; Ullah, A. H. J. Positive Identification of a Lambda Gt11 Clone Containing a Region of Fungal Phytase Gene by Immunoprobe and Sequence Verification. *Appl. Microbiol. Biotechnol.* **1991,** *35,* 611–614.

Nawaz, M. A.; Bibi, Z.; Karim, A.; Rehman, H. U.; Jamal, M.; Jan, T.; Aman, A.; Qader. S. A. Production of α-1,4-Glucosidase from *Bacillus licheniformis* KIBGE-IB4 by utilizing Sweet Potato Peel. *Environ. Sci. Pollut. Res.* **2017,** *24*(4), 4058–4066.

Ohmiya, K.; Sakka, K.; Karita, S.; Kimura, T. Structure of Cellulases and their Applicants. *Biotechnol. Genet. Eng.* **1997,** *14,* 365–414.

Ohnishi, K.; Yoshida, Y.; Sekiguchi, J. Lipase Production of *Aspergillus oryzae. J. Ferment. Bioeng.* **1994a**, *77,* 490–495.

Ohnishi, K.; Yoshida, Y.; Toita, J.; Sekiguchi, J. Purification and Characterization of a Novel Lipolytic Enzyme from *Aspergillus oryzae. J. Ferment. Bioeng.* **1994b**, *78,* 413–419.

Olempska-Beer, Z. Alpha-Amylase from *Bacillus licheniformis* Containing A Genetically Engineered Alpha-Amylase Gene from *B. Licheniformis* (Thermostable) Chemical and Technical Assessment, *61st JECFA*, FAO. pp 1–6. http://www.fao.org/fileadmin/ templates/agns/pdf/jecfa/cta/61/alphaamylase.pdf

Olempska-Beer, Z. S.; Merker, R. I.; Ditto, M. D.; Di Novi, M. J. Food-Processing Enzymes from Recombinant Microorganisms: A Review. *Regul. Toxicol. Pharmacol.* **2006**, *45*(2), 144–158.

Ongena, M.; Duby, F.; Jourdan, E.; Beaudry, T.; Jadin, V.; Dommes, J.; Thonart, P. *Bacillus subtilis* M4 Decreases Plant Susceptibility Towards Fungal Pathogens by Increasing Host Resistance Associated with Differential Gene Expression. *Appl. Microbiol. Biotechnol.* **2005**, *67,* 692–698.

Paloheimo, M.; Haarmann, T.; Makinen, S.; Vehmaanpera, J. Production of Industrial Enzymes in *Trichoderma reesei*. In *Gene Expression Systems in Fungi: Advancements and Applications, Fungal Biology Series,* Schmoll, M., Dattenböck, C., Eds.; Springer International Publishing: Switzerland, 2016; pp 23–57.

Parekh, S. R. *GMO Handbook: Genetically Modified Animals, Microbes, and Plants in Biotechnology;* Humana Press Inc.: Totowa, NJ, 2004; p 374.

Patel, A.; Shah, N.; Verma, D. K. Lactic Acid Bacteria (LAB) Bacteriocins: an Ecological and Sustainable Biopreservative Approach to Improve the Safety and Shelf-life of Foods. In *Microorganisms in Sustainable Agriculture, Food and the Environment.* as Part of Book Series on *Innovations in AGRICULTURAL MICROBIOLOGY,* Verma, D. K., Srivastav, P. P., Eds.; Apple Academic Press: USA, 2017.

Pavezzi, F. C.; Gomes, E.; da Silva, R. Production and Characterization of Glucoamylase from Fungus *Aspergillus awamori* Expressed in Yeast *Saccharomyces cerevisiae* using Different Carbon Sources. *Braz. J. Microbiol.* **2008**, *39,* 108–114.

Piddington, C. S.; Houston, C. S.; Paloheimo, M.; Cantrell, M.; Miettinen-Oinonen, A.; Nevalainen, H.; Rambosek, J. The Cloning and Sequencing of the Genes Encoding Phytase (phy) and pH 2.5 Optimum Acid Phosphatase (aph) from *Aspergillus niger* var. *Awamori. Gene* **1993**, *133,* 55–62.

Ragheb, D.; Bompiani, K.; Dalal, S.; Klemba, M. Evidence for Catalytic Roles for *Plasmodium falciparum* Aminopeptidase P in the Food Vacuole and Cytosol. *J. Biol. Chem.* **2009**, *284*(37), 24806–24815.

Rahulan, R.; Dhar, K. S.; Nampoothiri, K. M.; Pandey, A. Aminopeptidase from Streptomyces Gedanensis as a useful Tool for Protein Hydrolysate Preparations with Improved Functional Properties. *J. Food Sci.* **2012**, *77*(7), C791–797.

Raksakulthai, R.; Haard, N. F. Exopeptidases and their Application to Reduce Bitterness in Food: A Review. *Crit. Rev. Food Sci. Nutr.* **2003**, *43*(4), 401–445.

Ramachandran, S.; Fontanille, P.; Pandey, A.; Larroche, C. Gluconic Acid: Properties, Applications and Microbial Production. *Food Technol. Biotechnol.* **2006**, *44*(2), 185–195.

Rebets, Y.; Kormanec, J.; Lutzhetskyy, A.; Bernaerts, K.; Anné, J. Cloning, Expression of Metagenomic DNA in *Streptomyces lividans* and Subsequent Fermentation for Optimized Production. *Methods Mol. Biol.* **2017**, *1539,* 99–144.

Riaublanc, A.; Ratomahenina, R.; Galzy, P.; Nicolas, M. Peculiar Properties of Lipase from *Candida Parapsilosis* (Ashland) Langeron and Talice. *J. Am. Oil Chem. Soc.* **1993,** *70*(5), 497–500.

Rippel-Baldes, A. *Grundzüge der Mikrobiologie,* 3rd Ed.; Springer, Berlin Heidelberg: New York, 1955.

Roche, N.; Berna, P.; Desgranges, C.; Durand, A. Substrate use and Production of A-L-Arabinofuranosidase During Solid-State Culture of *Trichoderma reesei* on Sugar Beet Pulp. *Enzyme Microb. Technol.* **1995,** *17,* 935–941.

Rodarte, M. P., Dias, D. R., Vilela, D. M.; Schwan, R. F. Proteolytic Activities of Bacteria, Yeasts and Filamentous Fungi Isolated from Coffee Fruit (*Coffea arabica* L.) *Acta Sci. Agron.* **2011,** *33,* 457–464.

Rosenberg, J.; Ischebeck, T.; Commichau, F. M. Vitamin B6 Metabolism in Microbes and Approaches for Fermentative Production. *Biotechnol. Adv.* **2016,** *35*(1), 31–40.

Rossiter, R. J. Metabolism of Phosphatides. In *Metabolic Pathways;* Greenberg, D. M., Ed.; Vol. II, 1968; Academic Press: New York, pp 69–115.

Roukas, T. Citric and Gluconic Acid Production from Fig by *Aspergillus niger* using Solid State Fermentation. *J. Ind. Microbiol. Biotechnol.* **2000,** *25,* 298–304.

Sanderink, G. J.; Artur, Y.; Siest, G. Human Aminopeptidases: A Review of the Literature. *J. Clin. Chem. Clin. Biochem.,* **1988,** *26,* 795–807.

Sasada, R.; Joseph, R. Purification and Properties of Lipase from the Anaerobe *Propionibactrium Acidi-Propionici. J. Am. Oil Chem. Soc.* **1992,** *69*(10), 974–977.

Schuster, E.; Dunn-Coleman, N.; Frisvad, J.; van Dijck, C. P. W. M. On the Safety of *Aspergillus niger*—A Review. *Appl. Microbiol. Biotechnol.* **2002,** *59,* 426–435.

Seiboth, B.; Herold, S.; Kubicek, C. P. Metabolic Engineering of Inducer Formation for Cellulase and Hemicellulase Gene Expression in *Trichoderma reesei. Subcell. Biochem.* **2012,** *64,* 367–390.

Seifan, M.; Samani, A. K.; Berenjian, A. Induced Calcium Carbonate Precipitation using *Bacillus* Species. *Appl. Microbiol. Biotechnol.* **2016,** *100*(23), 9895–9906.

Selle, P. H.; Ravindran, V. Phytate Degrading Enzymes in Pig Nutrition. *Livest. Sci.* **2008,** *113,* 99–122.

Selten, G. The Versatile *Aspergillus niger. Gist* **1994,** *60,* 5–7.

Semenova, M.; Sinitsyna, O.; Morozova, V. Use of a Preparation from Fungal Pectin Lyase in the Food Industry. *Appl. Biochem. Microbiol.* **2006,** *42,* 598–602.

Sharma, R.; Chisti, Y.; Banerjee, U. C. Production, Purification, Characterization, and Applications of Lipases. *Biotechnol. Adv.* **2001,** *19,* 627–662.

Shimada, Y.; Koga, C.; Sugihara, A.; Nagao, T.; Takada, N.; Tsunasawa, S.; Tominaga, Y. Purification and Characterization of a Novel Solvent-Tolerant Lipase from *Fusarium heterosporum. J. Ferment. Bioeng.* **1993,** *75*(5), 349–352.

Shin, K. S. Isolation and Structural Characterization of an Oligosaccharide Produced by *Bacillus subtilis* in a Maltose-Containing Medium. *Prev. Nutr. Food Sci.* **2016,** *21*(2), 124–131.

Shivanna, G. B.; Venkateswaran, G. Phytase Production by *Aspergillus niger* CFR 335 and *Aspergillus ficuum* SGA 01 through Submerged and Solid-State Fermentation. *Scientific World J.* **2014,** *2014,* 6. DOI: 10.1155/2014/392615

Simmons, E. G. Classification of Some Cellulose Producing *Trichoderma* species (Abstract). In *Proceedings of the Second International Mycological Congress.*

University of South Florida, Tampa, Florida, U.S.A., 27th August–3rd September 1977. p 618.

Smily, J. M. B.; Sivakami, R.; Kishore, G. P.; Sumithra, P. A Comparison on the Production and Characterisation of Aminopeptidase from Two Species of *Aspergillus. Int. J. Curr. Microbiol. Appl. Sci.* **2014,** *3,* 829–836.

Soares, I.; Távora, Z.; Barcelos, R. P.; Baroni, S. Microorganism-Produced Enzymes in the Food Industry. In *Scientific, Health and Social Aspects of the Food Industry.* Benjamin, V., Ed.; InTech.: Croatia, 2012; pp 83–94. DOI: 10.5772/31256

Srinivasan, M. C.; Vartak, H. G.; Powar, V. K.; Khire, J. M. High Activity Extracellular Glucose/ (Xylose) Isomerase from a *Chainia* species. *Biotechnol. Lett.* **1983,** *5,* 611–614.

Sriram, N.; Priyadharshini, M.; Sivasakthi, S. Production and Characterization of Amino Peptidase from Marine Aspergillus flavus. *Int. J. Microbiol. Res.* **2012,** *3,* 221–226.

Strater, N.; Lipscomb, W. N. Leucyl Aminopeptidase (Animal and Plant), Chapter 473. In *Handbook of Proteolytic Enzymes CD-ROM.* Barret, A., Rawlings, N. D., Woessner, J. F., Eds.; Academic Press: New York, 1998; pp 1384–1389.

Sun, H.; Peng, M. Improvement of Glucoamylase Production for Raw-Starch Digestion in *Aspergillus niger* F-01 by Maltose Stearic Acid Ester. *Biotechnol. Lett.* **2017,** *39*(4), 561–566.

Suprabha, G. N.; Sindhu, R.; Shashidhar, S. Fungal Xylanase Production under Solid State and Submerged Fermentation Conditions. *Afr. J. Microbiol. Res.* **2008,** *2,* 082–086.

Susca, A.; Proctor, R. H.; Morelli, M.; Haidukowski, M.; Gallo, A.; Logrieco, A. F.; Moretti, A. Variation in Fumonisin and Ochratoxin Production Associated with Differences in Biosynthetic Gene Content in *Aspergillus niger* and *A. welwitschiae* Isolates from Multiple Crop and Geographic Origins. *Front. Microbiol.* **2016,** *7*(1412), 1–15. DOI: 10.3389/fmicb.2016.01412

Sztajer, H.; Maliszewska, I.; Wieczorek, J. Production of Exogenous Lipase by Bacteria, Fungi and Actinomycetes. *Enzyme Microb. Technol.* **1988,** *10,* 492–497.

Tahoun, M. K.; E-Kady, I.; Wahba, A. Production of Lipases from Microorganisms. *Microbiol. Lett.* **1985,** *28,* 133–139.

Taylor, A. Aminopeptidases: Structure and Function. *FASEB J.* **1993,** *7*(2), 290–298.

Tilburn, J.; Scazzocchio, C.; Taylor, G. G.; Zabicky-Zissman, J. H.; Lockington, R. A.; Davis, R. W. Transformation by Integration in *Aspergillus nidulans. Gene* **1983,** *26,* 205–221.

Tuppy, H. C.; Nesvadba, H. Über die Aminopeptidaseaktivität des Schwangerenserums und ihre Beziehung zu dessen Vermögen, Oxytocin zu inaktivieren. *Monatshefte für Chemie,* **1957,** *88,* 977–988.

Underkolfer, L. A.; Barton, R. R.; Rennert, S. S. Production of Microbial Enzymes and their Application. *Appl. Microbiol.* **1958,** *6,* 212–221.

Urlaub, R. Enzymes from Genetically Modified Microorganisms and their use in the Beverage Industry. *Fruit Proc.* **1999,** *9:* 158–163.

Vallejo, J. A.; Ageitos, J. M.; Poza, M.; Villa, T. G. Cloning and Expression of Buffalo Active Chymosin in *Pichia pastoris. J. Agric. Food Chem.* **2008,** *56,* 10606–10610.

Van der Veen, P.; Flipphi, M. J. A.; Voragen, A. G. J.; Visser, J. Induction, Purification and Characterization of Endo-1,4-B-D-Galactanases from *Aspergillus niger. Biosci. Biotech. Biochem.* **1991,** *56,* 1608–1615.

Van Gorcom, R. F. M.; van Hartingsveldt, W.; van Paridon, P. A.; Veenstra, A. E.; Luiten, R. G. M.; Selten, G. C. M. Cloning and Expression of Microbial Phytase. European Patent Application 1991, 0.420.358.

Van Hartingsveldt, W.; Mattern, I. E.; Van Zeijl, C. M. J.; Pouwels, P. H.; van den Hondel, C. A. M. J. J. Development of a Homologous Transformation System for *Aspergillus niger* Based on the *pyr*G Gene. *Mol. Gen. Genet.* **1987**, *206*, 1–75.

Van Hartingsveldt, W.; van Zeijl, C. M. J.; Harteveld, G. M.; Gouka, R. J.; Suykerbuyk, M. E. G.; Luiten, R. G. M.; van Paridon, P. A.; Selten, G. C. M.; Veenstra, A. E.; van Gorcom, R. F. M.; van den Hondel, C. A. M. J. J. Cloning, Characterization and Overexpression of the Phytase-Encoding Gene (*phy*A) of *Aspergillus niger. Gene* **1993**, *127*, 87–94.

Vartak, H. G.; Srinivasan, M. C.; Powar, V. K.; Rele, M. V.; Khire, J. M. Characterisation of Extracellular Substrate Specific Glucose and Xylose Isomerases of *Chainia. Biotechnol. Lett.* **1984**, *6*, 493–494.

Vega-Hernández, M. C.; Gómez-Coello, A.; Villar, J.; Claverie-Martín, F. Molecular Cloning and Expression in Yeast of *Caprine prochymosin. J. Biotechnol.* **2004**, *114*, 69–79.

Verma, D. K.; Srivastav, P. P. *Microorganisms in Sustainable Agriculture, Food and the Environment,* as part of book series on *Innovations in AGRICULTURAL MICROBIOLOGY,* Verma, D. K., Srivastav, P. P., Eds.; Apple Academic Press: USA, 2017.

Verma, D. K.; Mahato, D. K.; Billoria, S.; Kapri, M.; Prabhakar, P. K.; Ajesh Kumar, V.; Srivastav, P. P. Microbial Approaches in Fermentations for Production and Preservation of Different Food. In *Microorganisms in Sustainable Agriculture, Food and the Environment.* as part of book series on *Innovations in AGRICULTURAL MICROBIOLOGY,* Verma, D. K., Srivastav, P. P., Eds.; Apple Academic Press: USA, 2017a.

Verma, D. K.; Mahato, D. K.; Billoria, S.; Kapri, M.; Prabhakar, P. K.; Ajesh Kumar, V.; Srivastav, P. P. Microbial Spoilage in Milk Products, Potential Solution, Food Safety and Health Issues. In *Microorganisms in Sustainable Agriculture, Food and the Environment.* as part of book series on *Innovations in AGRICULTURAL MICROBIOLOGY,* Verma, D. K., Srivastav, P. P., Eds.; Apple Academic Press: USA, 2017b.

Viterbo, A.; Horwitz, B. Mycoparasitism. In *Cellular and Molecular Biology of Filamentous Fungi.* Borkovich, K. A., Ebbole, D. J., Momany, M., Eds.; ASM Press: Washington, DC, 2010; pp 676–693.

Vongsuvanlert, V.; Tani. Y. Purification and Characterization of Xylose Isomerase of a Methanol Yeast, *Candida Boidinii,* which is Involved in Sorbitol Production from Glucose. *Agric. Biol. Chem.* **1988**, *52*, 1817–1824.

Wang, B.; Wang, A.; Cao, Z.; Zhu, G. Characterization of a Novel Highly Thermostable Esterase from the Gram-Positive Soil Bacterium *Streptomyces lividans* TK64. *Biotechnol. Appl. Biochem.* **2016a**, *63*(3), 334–343.

Wang, X. C.; Zhao, H. Y.; Liu, G.; Cheng, X. J.; Feng, H. Improving Production of Extracellular Proteases by Random Mutagenesis and Biochemical Characterization of a Serine Protease in *Bacillus subtilis* S1–4. *Genet. Mol. Res.* **2016b**, *15*(2), 1–11.

Wang, Y. Z.; Feng, B.; Huang, H. Z.; Kang, L. P.; Cong, Y.; Zhou, W. B.; Zou, P.; Cong, Y. W.; Song, X. B.; Ma, B. P. Glucosylation of Steroidal Saponins by Cyclodextrin Glucanotransferase. *Planta Med.* **2010**, *76*(15), 1724–1731.

Wang, P. Y.; Johnson, B. F.; Scneider, H. Fermentation of D-Xylose by Yeasts using Glucose Isomerase in the Medium to Convert D-Xylose to D-Xylulose. *Biotechnol. Lett.* **1980**, *2*, 273–278.

Wanmolee, W.; Sornlake, W.; Rattanaphan, N.; Suwannarangsee, S.; Laosiripojana, N.; Champreda, V. Biochemical Characterization and Synergism of Cellulolytic Enzyme System from *Chaetomium globosum* on Rice Straw Saccharification. *BMC Biotechnol.* **2016**, *16*(82), 1–12. DOI: 10.1186/s12896–016–0312–7

Ward, M.; Wilson, L. J.; Carmonia, C. L.; Turner, G. The *oliC*3 Gene of *Aspergillus niger* isolation, Sequence and use as a Selective Marker for Transformation. *Curr. Genet.* **1988**, *14*, 37–42.

Ward, O. P.; Rao, M. B.; Kulkarni, A. Proteases Production. In *Encyclopedia of Enzymes;* Schaechter, M., Ed.; Elsevier Inc.: New York, 2009; pp 495–511.

Weber, P. Fructose by Isomerisation of Glucose. U.K. *Patent* 1,496,309, 1976.

Witteveen, C. F. B.; Van de Vondervoort, P.; Swart, K.; Visser, J. Glucose Oxidase Over Producing and Negative Mutants of *Aspergillus nidulans. Appl. Environ. Microbiol.* **1990**, *33*, 683–86.

Wood, J. M.; Decker, H.; Hartmann, H.; Chavan, B.; Rokos, H.; Spencer, J. D.; Hasse, S.; Thornton, M. J.; Shalbaf, M.; Paus, R.; Schallreuter, K. Senile Hair Graying: H_2O_2 Mediated Oxidative Stress Affects Human Hair Color by Blunting Methionine Sulfoxide Repair. *FASEB J.* **2009**, *23*, DOI: 10.1096/fj.08–125435.

Wouter, M. J. K.; Jovanovic, D.; Brouwera, S. G. M.; Loos, K. Amylase Catalyzed Synthesis of Glycosyl Acrylates and their Polymerization. *Green Chem.* **2014**, *16*, 203–210.

Xu, S.; Guo, Y.; Du, G.; Zhou, J.; Chen, J. Self-Cloning Significantly Enhances the Production of Catalase in *Bacillus subtilis* WSHDZ-01. *Appl. Biochem. Biotechnol.* **2014**, *173*(8), 2152–2162.

Xu, Z.; Shao, J.; Li, B.; Yan, X.; Shen, Q.; Zhang, R. Contribution of Bacillomycin D in *Bacillus amyloliquefaciens* SQR9 to Antifungal Activity and Biofilm Formation. *App. Environ. Microbiol.* **2013**, *79*, 808–815.

Yadav, S.; Yadav, P. K.; Yadav, D.; Yadav, K. D. S. Pectin Lyase: A Review. *Process Biochem.* **2009**, *44*, 1–10.

Yaman, S.; Çalık, P. Beet Molasses-Based Feeding Strategy Enhances Recombinant Thermostable Glucose Isomerase Production by *Escherichia coli* BL21 (DE3). *Biotechnol. Appl. Biochem.* **2016**, *13*. DOI: 10.1002/bab.1549.

Zhu, Z.; Momeu, C.; Zakhartsev, M.; Schwaneberg, U. Making Glucose Oxidase Fit for Biofuel Cell Applications by Directed Protein Evolution. *Biosens. Bioelectron.* **2006**, *21*, 2046–2051.

Zweers, J. C.; Barák, I.; Becher, D.; Driessen, A. J. M.; Hecker, M.; Kontinen, V. P.; Saller, M. J.; Vavrová, L.; van Dijl, J. M. Towards the Development of *Bacillus subtilis* as a Cell Factory for Membrane Proteins and Protein Complexes. *Microb. Cell. Fact.* **2008**, *7*(10), 1–20. DOI: 10.1186/1475–2859–7–10.

CHAPTER 4

POTENTIAL ROLE OF FOOD MICROBIAL COMMUNITIES IN DIFFERENT FOOD PRODUCTION AND HUMAN HEALTH

DEEPAK KUMAR VERMA[1,*], KIMMY G.[2], ALAA KAREEM NIAMAH[3], AMI PATEL[4], RISHIKA VIJ[5], and PREM PRAKASH SRIVASTAV[1,6]

[1]*Agricultural and Food Engineering Department, Indian Institute of Technology Kharagpur, West Bengal 721302, India, Tel.: +91 3222281673, Mob.: +91 7407170259, Fax: +91 3222282224*

[2]*Department of Food Engineering and Technology, Sant Longowal Institute of Engineering and Technology Longowal, Sangrur, Punjab 148106, India, Mob.: +00–91–8699271602, E-mail: kishorigoyal09@gmail.com*

[3]*Department of Food Science, College of Agriculture, University of Basrah, Basra City, Iraq, Mob.: +00–96–47709042069, E-mail: alaakareem2002@hotmail.com*

[4]*Division of Dairy and Food Microbiology, Mansinhbhai Institute of Dairy and Food Technology (MIDFT), Dudhsagar Dairy Campus, Mehsana, Gujarat 384002, India, Mob.: +00–91–9825067311, Tel.: +00–91–2762243777 (O), Fax: +91–02762–253422, E-mail: amiamipatel@yahoo.co.in*

[5]*Department of Veterinary Physiology and Biochemistry, COVAS, Chaudhary Sarwan Kumar Himachal Pradesh Krishi Vishvavidyalaya, Palampur, Himachal Pradesh 176062, India, Mob.: +00–91–9466493086, E-mail: rishikavij@gmail.com*

[6]*E-mail: pps@agfe.iitkgp.ernet.in*

*Corresponding author. E-mail: deepak.verma@agfe.iitkgp.ernet.in; rajadkv@rediffmail.com

4.1 INTRODUCTION

The 21st century is considered as one of the major challenges era for sustainable food production and environmental security in which microorganisms are biological entities and well-known key players due to their critical role may be a benefit or a potential hazard to humans and to the food manufacturing industry. Typically, they are considered as the first organisms which react to physical, chemical, and biological changes because they are nearer to the food chain bottom. Owing to different changes in the food, microbes are often subjected as a precursor to changes in the health (Itsaranuwat et al., 2003) and viability of the environment as a whole (Shobharani and Agrawal, 2009; Karimi et al., 2011). Thus, food microbial community (FMC) may be defined as the group of microorganism which holds a crucial position in the development of different qualitative aspects of food products namely health benefits, physicochemical and sensory properties (Fig. 4.1) and also provides the information about the different type of changes of food products. In addition, of their different roles, product safety is an issue in food industries

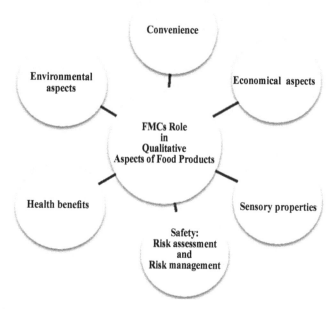

FIGURE 4.1 Food microbial communities (FMCs) in the development of different qualitative aspects of food products.

Source: Adapted from Mortazavian et al. (2012).

is contributed by this FMCs through limiting the growth of spoilage and pathogenic microflora (Caplice and Fitzgerald, 1999; Verma et al., 2017b). Food-grade microorganism develops diverse bioactive components including organic acids, flavor compounds, bacteriocins, enzymes, exopolysaccharides (EPSs), conjugated linoleic acids, and bioactive peptides within the food matrix or different substrates (Patel and Shah, 2016; Verma and Srivastav, 2017). Today, there is a sharp boost in the potential use of FMCs in different food production namely dairy and nondairy-based products. Recently, a number of food products have been developed; the manufacture of these food products involving a microorganism or the microorganisms in the production of food products is a key issue at industrial and commercial level and in the world market, many food products of this kind are available.

4.4.2 FOOD MICROBIAL COMMUNITY IN DIFFERENT FOOD PRODUCTION

The FMCs are dominated by Earth nature and virtually present in the vast majority. They contain prokaryotic cells of about $4–6 \times 10^{30}$ (Whitman et al., 1998) with highly diverse group among the organisms as well as plants and animals which constitute near about 60% of the total biomass of the Earth (Singh et al., 2009). They can be harmful due to their bill to inhabit, create, or contaminate food causing food spoilage (Fratamico and Bayles, 2005) and/or can be beneficial due to their crucial role in the production of different food namely cheese, yoghurt, bread, beer, wine, and other fermented foods, and so forth depicted in Table 4.1 and Figure 4.2 (Caplice and Fitzgerald, 1999; Vinderola and Reinheimer, 2003; Rivera-Espinoza and Gallardo-Navarro, 2010). FMCs have also made noteworthy contributions in food safety risk assessment as well as risk management (McMeekin et al., 1997). Different types of FMCs (Table 4.2) associated with food production are briefly described.

TABLE 4.1 Microorganism in Different Food/Food Products Available in World Market.

Food/food products	Microorganism	References
Tomato-based drink	*Latobacillus acidophilus*	Yoon et al. (2004)
	Lactobacillus plantarum	
	Lactobacillus casei	
	Lactobacillus delbrueckii	

TABLE 4.1 *(Continued)*

Food/food products	Microorganism	References
Acidophilus butter	*Bifidobacterium* spp.	Gomes and Malcata
	L. acidophilus	(1999)
Yoghurts		
Regular full-fat yoghurts	*Streptococcus thermophilus*	Aryana and Mcgrew
	L. casei	(2007)
	L. delbrueckii spp. *bulgaricus*	
Iranian yoghurt drink (Doogh)	*L. acidophilus*	Mortazavian et al.
	Bifidobacterium lactis	(2008)
Stirred fruit yoghurts	*L. acidophilus*	Kailasapathy et al.
	Bifidobacterium animalis spp. *lactis*	(2008)
Argentinian yoghurts	*Bifidobacterium bifidum*	Vinderola et al.
	L. acidophilus	(2000b)
Banana-based yoghurt	*S. thermophilus*	Kumar and Thakur
	L. delbrueckii ssp. *bulgaricus*	(2014)
Corn milk yoghurt	*Lactobacillus bulgaricus*	Supavititpatana et al.
	S. thermophilus	(2008)
Frozen yoghurt	*L. bulgaricus*	Davidson et al. (2000)
	S. thermophilus	
Mango soy fortified probiotic yoghurt	*S. thermophilus*	Kaur et al. (2009)
	L. bulgaricus	
	Bifidobacterium bifidus	
	L. acidophilus	
Rice-based yoghurt	*L. acidophilus*	Helland et al. (2005)
	B. animalis	
	Lactobacillus rhamnosus	
Traditional Greek yoghurt	*L. plantarum*	Maragkoudakisa et al.
	Lactobacillus paracasei ssp. *tolerans*	(2006)
Soy-based stirred yoghurt	*Lactococcus lactis*	Saris et al. (2003)
Desserts		
Frozen synbiotic dessert	*L. bulgaricus*	Davidson et al. (2000)
	S. thermophilus	
	Bifidobacterium longum	
	L. acidophilus	

TABLE 4.1 *(Continued)*

Food/food products	Microorganism	References
Soy-based frozen desserts	*B. lactis* *L. acidophilus* *L. paracasei* subsp. *paracasei* *L. rhamnosus* *Saccharomyces boulardii*	Heenan et al. (2005)
Frozen dairy dessert	*L. bulgaricus* *S. thermophilus*	Hong and Marshall (2001)
Cheese		
Argentine fresco cheese	*Bifidobacterium* spp. *B. longum* *B. bifidum* *L. acidophilus* *L. casei* *L. lactis* *S. thermophilus*	Vinderola et al. (2000a)
Canestrato pugliese hard cheese	*B. bifidum* *B. longum*	Corbo et al. (2001)
Cheddar cheese	*B. animalis* ssp. *lactis B. longum* *L. acidophilus* *L. casei*	Ong and Shah (2009)
Cottage cheese	*Bifidobacterium infantis*	Blanchette et al. (1996)
Cultured cottage cheese	*B. infantis*	Blanchette et al. (1996)
Feta cheese	*B. lactis* *L. acidophilus* *L. lactis* ssp. *lactis* *L. lactis* ssp. *cremoris* *S. thermophilus*	Kailasapathy and Masondole, (2005); Karimi et al. (2012a)
Goat semisolid cheese	*Bifidobacterium* spp. *L. acidophilus*	Gomes and Malcata (1999)
Turkish Beyaz cheese	*B. longum* *L. acidophilus* *L. lactis* ssp. *cremoris* *Lactococcus cremoris* *L. lactis* *L. lactis* subsp. *lactis*	Kiliç et al. (2009); Yangılar and Özdemir (2013)

TABLE 4.1 *(Continued)*

Food/food products	Microorganism	References
Minas fresco cheese	*L. acidophilus* *S. thermophilus*	Souza and Saad, (2009)
Reduced-fat semihard cheese	*L. paracasei* ssp. *paracasei*	Thage et al. (2005)
White-brined cheese	*L. acidophilus*	Özer et al. (2008)
	L. lactis ssp. *cremoris* *B.* bifidum	
	L. lactis ssp. *lactis*	
Juice		
Cabbage juice	*L. delbrueckii*	Yoon et al. (2006)
	L. casei	
	L. plantarum	
Carrot juice	*L. acidophilus*	Nazzaro et al. (2009)
Ginger juice	*B. longum*	Chen et al. (2009)
Grape and passion fruit juices	*B. animalis* subsp. *lactis*	Saarela et al. (2006)
Noni juice	*L. casei*	Wang et al. (2014)
	L. plantarum	
	B. longum	
Cranberry, orange, and pineapple juices	*L. Paracasei* ssp. *paracasei*	Sheehan et al. (2007)
	Lactobacillus salivarius ssp. *salivarius*	
	L. casei	
	L. rhamnosus	
	B. animalis subsp. *lactis*	
Beverage		
Fermented maize beverage	*L. bulgaricus* var *delbrueckii* or *Lactobacillus brevis*	McMaste et al. (2005)
Millet or sorghum flour-based fermented probiotic beverage	Lactic acid bacteria	Muianja et al. (2003)
Non-fermented fruit juice beverages	*Aspergillus oryzae*	Renuka et al. (2009)
Milk beverage	*Leuconostoc mesenteroides*	Shobharani and Agrawal, (2009)

TABLE 4.1   *(Continued)*

Food/food products	Microorganism	References
Whey protein-based drinks	*B. animalis/lactis*	Dalev et al. (2006)
	B. infantis	
	Bifidobacterium breve	
	L. plantarum	
	S. thermophilus	
Low-fat ice cream	*B. animalis*	Davidson et al. (2000); Haynes and Playne (2002); Akalin and Erisir (2008)
	B. lactis	
	L. paracasei ssp. *paracasei*	
	L. delbrueckii ssp. *bulgaricus*	
	L. acidophilus	
	Streptococcus salivarius subsp. *thermophilus*	
Probiotic ice cream	*B. lactis*	Kailasapathy and Sultana, (2003)
	L. acidophilus	
Synbiotic ice cream	*B. lactis*	Homayouni et al. (2008)
	L. casei	
Fermented milk		
High pressure homogenized milk	*L. paracasei*	Patrignani et al. (2009)
	L. acidophilus	
Fermented goat's milk	*S. thermophilus*	Martın-Diana et al. (2003)
	L. acidophilus	
	Bifidobacterium spp.	
Peanut milk	*Bifidobacterium pseudocatenulatum*	Mustafa et al. (2009)
Skim milk	*B. pseudocatenulatum*	Mustafa et al. (2009)
Soy milk drink	*B. longum*	Donkor et al. (2007)
	B. lactis	
	L. acidophilus	
	L. casei	
	L. delbrueckii spp. *bulgaricus*	
	S. thermophilus	
Cereal-based Product		
Oat-based drink	*L. plantarum*	Angelov et al. (2006)

TABLE 4.1   *(Continued)*

Food/food products	Microorganism	References
Oat-based products	*Lactobacillus reuteri*	Mårtensson et al.
	L. acidophilus	(2002)
	B. bifidum	
Oat and barley concentrates	*L. acidophilus*	Lambo et al. (2005)
	L. delbrueckii ssp. *bulgaricus*	
	Pediococcus damnosus	
	S. salivarius ssp. *thermophilus*	
Malt-based drink	*L. reuteri*	Kedia et al. (2007)
Meat products		
Greek dry fermented sausage	*Lactobacillus curvatus* ssp. *curvatus*	Papamanoli et al.
	L. plantarum	(2003)
Salami (traditional Italian dry fermented sausage)	*L. casei*	Rebucci et al. (2007)
	L. rhamnosus	
Dry fermented Sausage	*L. rhamnosus*	Erkkilä and Suihko (2001)
Fermented sausage	*Lactobacillus gasseri*	Arihara et al. (1998)
Herkules (typical Czech fermented sausage)	*L. casei*	Burdychova et al. (2008)
Salami	*L. reuteri*	Muthukumarasamy and Holley (2006; 2007)
	B. longum	
Fifteen different Scandinavian-type fermented meat products	*L. plantarum*	Klingberg et al. (2005)
Dry fermented sausage	*L. rhamnosus*	Erkkilä and Petaja (2000)
		Ammor and Mayo, (2007)
Low-fat fermented Sausage	*L. plantarum*	Kim et al. (2008)
	P. damnosus	
Tempeh	*Rhizopus oligosporus*	Shurtleff and Aoyagi, (2001)

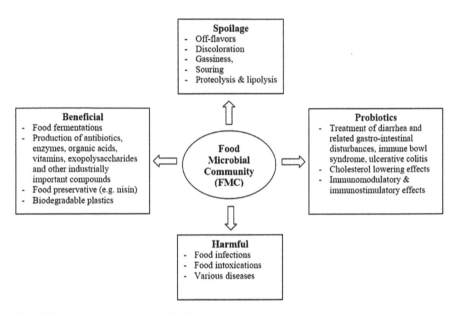

FIGURE 4.2   Role of various FMCs in food and health.

TABLE 4.2   Prevalence Some Important Microbial Communities in Food Production.

Microbial community	Strain
Acetobacter spp.	*Acetobacter polyoxogenes*
Aspergillus spp.	*A. oryzae, A. flavus, A. niger*
Bacillus spp.	*B. subtilis, B. amyloliquefaciens*
Bifidobacterium spp.	*B. animalis, B. bifidum, B. breve, B. infantis, B. lactis, B. longum,. B. pseudocatenulatum*
Brevibacterium spp.	*B. linens, B. sanguinis*
Cladosporium spp.	*Cladosporium fulvum, Cladosporium elegans*
Gluconobacter spp.	*Gluconobacter oxydans, Gluconobacter cerevisiae*
Lactobacillus spp.	*L. bulgaricus, L. brevis, L. acidophilus, L. casei, L. curvatus, L. debrueckii, L. gasseri, L. lactis, L. paracasei, L. plantarum, L. reuteri, L. rhamnosus, L. salivarius*
Lactococcus spp.	*L. cremoris, L. lactis*
Leuconostoc spp.	*L. mesenteroides, L. dextranicum*
Micrococcus spp.	*M. luteus, M. roseus, M. lactis*
Pediococcus spp.	*P. cerevisiae, P. damnosus*
Penicillium spp.	*P. Glabrum, P. Purpurogenum, P. nalgiovense*

TABLE 4.2   *(Continued)*

Microbial community	Strain
Propionibacterium spp.	*P. cyclohexanicum, P. jensenii, P. microaerophilum, P. acidicpropionici, P. thoenii, P. freudenreichii, P. acnes, P. avidum, P. humerusii, P. acidifaciens, P. australiense*
Rhizopus spp.	*R. oligosporus, R. oryzae*
Saccharomyces spp.	*S. boulardii, S. cerevisiae*
Streptococcus spp.	*S. salivarius, S. thermophilus*
Trichothecium spp.	*T. roseum*
Zygosaccharomyces spp.	*Z. rouxii*

4.2.1   LACTIC ACID BACTERIA

4.2.1.1   LACTOBACILLUS SPP.

Lactobacillus genus consists of Gram-positive, nonmotile, nonspore-forming, facultatively anaerobic or microaerophilic, rod-shaped bacteria. The genus contains more than 180 species; being larger genera of lactic acid bacteria (LAB) group, it converts lactose and other sugars to lactic acid, as a major end product with little amounts of acetic acid, ethanol, and acetaldehyde (Hammes and Hertel, 2009).

According to metabolism, *Lactobacillus* spp. has been divided into three groups:

- **The group I:** obligatory homofermentative, for example, *L. delbrueckii, L. acidophilus, L. salivarius, L. helveticus.*
- **The group II:** facultatively heterofermentative, for example, *L. plantarum, Lactobacillus casei,* Lb. sakei, *Lactobacillus curvatus.*
- **The group III:** obligatory heterofermentative, for example, *Lactobacillus brevis, L. fermentum, L. buchneri, L. reuteri.*

Some strains of *Lactobacillus* spp. and other LAB may have possible therapeutic properties or confer health benefits in the host when administered in adequate amounts which are termed as probiotics. A study was conducted by researchers from the Medical Center Beth Israel Deaconess and the University of California in 2009 on the protective effects of some strains of these bacteria to antitumur and anticancer effects in mice (Chen et al., 2009).

Major species of *Lactobacilli*, such as *L. casei, L. acidophilus, L. helveticus*, and so forth, have established probiotic potential through immunostimulatory and immunomodulatory functions. A number of studies have indicated that *L. casei* can amend potentially harmful bacterial activities like those of β-glucuronidase and nitroreductase and exert anticancer activities (Goldin and Gorbach, 1984; Marteau et al., 1990; Uccello et al., 2012). Based on technological performances, in vitro adhesion capacity and intestinal transit tolerance after administration in a rat, Bertazzoni et al. (2004) evaluated *L. casei* strains for their probable use in health-promoting foods. Authors reported that all four *L. casei* strains showed potential for application in probiotic fermented milk.

4.2.1.2   STREPTOCOCCUS SPP.

Streptococcus consist of Gram-positive, cocci-shaped, nonmotile, catalase-negative bacteria. It is classified in the phylum Firmicutes and order LAB. They mainly grow in chains or pairs as in these bacteria cell division takes place along a single axis. *Streptococcus thermophilus* is found in fermented milk products such as yoghurt, beside *Lactobacillus bulgaricus*. The two work synergistically, with *S. thermophilus* providing *L. bulgaricus* with folic and formic acid. Several strains are producing EPS (De Vuyst et al., 2003). Further, β-galactosidase enzyme is produced by *S. thermophilus* isolated from yoghurt after acid whey fermentation (Princely et al., 2013). Some species of *Streptococcus* are pathogenic for human and animals, for example, *S. pneumonia, S. pyogenes* (Lamagni et al., 2008). *Streptococcus agalactis* and *Streptococcus dysgalactis* are most common isolates in clinical mastitis of milch animals (Abd-Elrahman, 2013).

4.2.1.3   LACTOCOCCUS SPP.

Lactococcus genus is one member of the LAB and belongs to Strepto-coccaceae family. Seven species of Lactococcus are *L. lactis, L. chungangensis, L. piscium, L. fujiensis, L. garvieae, L. raffinolactis, and L. plantarum*. *L. lactis* is important species in the genus comprises several types of species such as *L. lactis* ssp. *cremoris, L. lactis* ssp. *lactis, L. lactis* ssp. *tructae*, and *L. lactis* ssp. *hordniae*. In the dairy industry, they are chiefly used in the manufacture of fermented dairy products such as curd (dahi), buttermilk, cheeses, and so forth. Nisin is a bacteriocin produced by *Lactococci* strains that have a wide spectrum inhibition against pathogenic bacteria and legally permitted biopreservative in food products (Şanlibaba et al., 2009).

4.2.1.4   LEUCONOSTOC SPP.

Leuconostoc is Gram-positive, cocci-shaped catalase-negative bacteria; placed within the family Leuconostocaceae. Some strains of *Leuconostocs* are resistant to vancomycin. The genus contains 15 species. All species of *Leuconostoc* are heterofermentative and belong to LAB group. They also found to synthesise EPS biopolymer; mainly dextran from carbohydrate fermentation. *Leuconostoc mesenteroides* sp. *mesenteroides* are also found to produce low-calorie sweeteners (Al-Sahlany, 2008; Patra et al., 2009; Verma et al., 2017a). *Leuconostoc* spp. causes spoilage of refrigerated storage food products (Säde, 2011). Some *Leuconostoc* strains are pathogenic, such as *L. mesenteroides* caused nosocomial urinary tract infection in North India (Taneja et al., 2005).

4.2.1.5   PEDIOCOCCUS SPP.

Pediococcus genus is placed within the family of Lactobacillaceae which are Gram-positive, homofermentative (D and L lactic acid production from glucose), nonspore-forming, and vancomycin-resistant. Genus consist of total nine species and the best known *P. acidolactici* and *P. pentosaceus* are used in the fermentation of cabbage and other vegetables, that is, used to make sauerkraut and kimchi as well as other fermented cereal, meat, and milks products (Olaoya et al., 2008; Patel et al., 2012; Verma et al., 2017a). *P. acidolactici* and *P. pentosaceus* are producing pediocin PA-1/AcH, stable over a large range of pH and wide spectrum of inhibitory action (Patel and Shah, 2016). Pediocin production from the local isolate of *P. acidolactici* AK was studied by Niamah, (2010). The highest inhibition against the test bacteria *Staphylococcus aureus* was reached to 79.55% and its addition into meat and cheese reduced the bacterial numbers and increased shelf life of meat and cheese samples.

4.2.1.6 WEISSELLA SPP.

Weissella genus includes Gram-positive, coccid to rod-shaped, hetero-fermentative bacteria, consists of 19 validly described species. They are associated with natural fermentation flora of sourdough, kimchi, and other Asian and African traditional fermented foods. They are found to improve the body texture and organoleptic qualities of such fermented foods though virtue of their ability to produce EPS (Patel et al., 2012; Patel and Prajapati, 2016; Verma et al., 2017a). These biopolymers are receiving increased attention for extensive industrial applications, predominantly in bakeries (bread, sourdough, and biscuits) and for the development of cereal-based functional fermented beverages and prebiotic candidature (Patel and Prajapati, 2013). Recently, EPS-producing *W. cibaria* and *W. confusa* strains also found to possess antagonistic activities against food-borne pathogens (Patel et al., 2013; Shah et al., 2016). In contrast to this, several *Weissella* strains are associated with bacteraemia infections, sepsis (Lee et al., 2011; Welch and Good, 2013).

4.2.2 OTHER FOOD-GRADE BACTERIA

4.2.2.1 BIFIDOBACTERIUM SPP.

Bifidobacterium was first isolated by Henry Tissier from mother's milk and called the bacterium *B. bifidus* (Guarner et al., 2008). *Bifidobacterium* belongs to the Bifidobacteriaceae family. The genus consists of Gram-positive, nonmotile, anaerobic bacteria, and possesses 60% G+C content of DNA. It has about 33 species and they mainly produce acetic acid and lactic acid in 3:2 ratio from lactose degradation. The bacterium is natural flora of the gastrointestinal tract (GIT), vagina, and mouth of mammals, including humans (Mayo and van Sinderen, 2010). The genus *Bifidobacterium* uses fructose-6-phosphate phosphoketolase pathway to ferment carbohydrates (Vlkova et al., 2002). Some *Bifidobacterium* spp., such as *B. adolescentis, B animalis, B. breve, B. bifidum, B. infantis, B. lactis,* and *B. longum* are used as probiotics.

4.2.2.2   *BREVIBACTERIUM SPP.*

Brevibacterium is Gram-positive, aerobic, rods-shaped soil bacterium. The G+C content of bacterial DNA may range from 60–67 mol.% (Jones and Keddie, 1986). The genus is within the family Brevibacteriaceae that was recorded by Breed 1953 and content 25 species. *B. linen* is present on the human skin ubiquitously and causes foot odor (Gruner et al., 1993). *B. linen* is used for surface ripening of brick cheese. *Brevibacterium* included specifically with each of the biotechnological and the clinical significance. It has been isolated from food, blood, and ear discharge (Gelsomino et al., 2004). From the blood of HIV patient, *B. casei* was isolated after 16S rRNA gene sequencing test; discern the isolates to belong to new bacteria called *B. sanguinis* (Wauters et al., 2004).

4.2.2.3   *ACETOBACTER SPP.*

The *Acetobacter* spp. is one genera of acetic acid bacteria (AAB) group based on their abilities to oxidize glucose or ethanol to acetic acid or lactic acid (De Ley et al., 1984). They are Gram-negative rods classified in the Acetobacteraceae family. Because of their capability to oxidize ethanol stronger than glucose, many strains are used in the manufacturing of vinegar at industrial level (Swings, 1991). Among various *Acetobacter* spp., *A. pasteurianus, A. aceti, A. europaneus*, and *A. polyoxogenes* are the chief strains considered for vinegar factories; moreover, they do not affect acetate in the later stage of fermentation (Entani et al., 1985; Sievers et al., 1992). The strain *A. pasteurianus* of the genus *Acetobacter* reported as dominant AAB in vinegar production (Entani et al., 1985). Multiple species of AAB are capable of incomplete oxidation of carbohydrates and alcohols to organic acids, ketones, and aldehydes (Matsushita et al., 2003; Deppenmeier et al., 2002). AAB or their enzymes are used in the biotransformations of 2-keto-L-gulonic acid (used for the vitamin C production), shikimate (a key intermediate for a number of antibiotics), D-tagatose (used as a bulking agent in foods), and a noncalorific sweetener.

4.2.2.4   GLUCONOBACTER SPP.

The *Gluconobacter* spp. is another genus of AAB group that have an ability to over oxidize acetic acid or lactic acid (De Ley et al. 1984). Most strains of *Gluconobacter* are used for industrial applications, such as fermentation of sorbose, ketogluconic acid, and dihydroxyacetone; they prefer glucose more than ethanol (Swings 1991). The genus *Gluconobacter* was postulated in the year 1935 and include that genus were competent to oxidize glucose into gluconate considerably better than of oxidizing glucose to acetate (Yamada and Yukphan, 2007).

4.2.2.5   PROPIONIBACTERIUM SPP.

Propionibacterium is a Gram-positive, catalase-positive, nonmotile, and nonspore-forming bacilli (Stackebrandt et al., 2006). It produces propionic acid as foremost product of the carbohydrate fermentation and also well-known to biosynthesis vitamin B_{12} and tetrapyrrole compounds (Kiatpapan and Murooka, 2002). Some species of *Propionibacterium* are used as starter cultures in the dairy industry and some possess probiotic status (Table 4.3). *P. Freudenreichii* ssp. *shermanii* is used in manufacturing Swiss cheese, where it leads to eyes formation due to the production of CO_2 in cheese and the characteristic propionic acid provides sweet aroma to the product (Langsrud and Reinbold, 1973). Strains of this genus had shown production of propionic acid from whey and glycerol media (Kośmider et al., 2010). *P. freudenreichii* found to produce antibacterial compound (Irina et al., 2012).

TABLE 4.3   *Propionibacterium* spp. Used in the Dairy Industry.

Dairy industry	Improve health
P. Cyclohexanicum	P. acnes
P. Jensenii	P. avidum
P. Microaerophilum	P. humerusii
P. Acidicpropionici	P. acidifaciens
P. Thoenii	P. australiense
P. Freudenreichii	Pb. granulosum

Source: Adapted from Zárate (2012).

4.2.2.6   BACILLUS SPP.

Bacillus is the genus of the family Bacillaceae (Fischer 1895), the genus includes Gram-positive, rod-shaped, endospore-producing aerobic bacteria; some species grow under anaerobic condition and can grow in poor media naturally. *Bacillus* includes both free-living and parasitic species; pathogenic to animals, plants, and human. Several species of this genus such as *B. cereus, B. weihenstephanensis, B. mycoides, B. Anthracis*, and *B. thuringiensis* cause food poisoning with symptoms of diarrhea and/or vomiting (Didelot et al., 2009). *B. anthracis* is the causative agent of anthrax, a widespread disease of livestock and sporadically of human (Sneath et al., 1986). Some of the *Bacillus* spp. have the ability to produce some useful substances, such as *B. amyloliquefaciens* is synthesizing Iturin A—a natural antibiotic protein (Lin et al., 2007). Moreover, it produces various types of enzymes to use in foods, industrial importance, for example, α-amylase used in hydrolysis of starch, and the protease subtilisins employed in the detergents (Demirkan, 2011). Some *Bacillus* strains are able to polyhydroxyalkonate (PHAs) production (Al-Sahlany, 2015). The PHAs are considered as green biodegradable plastic and can be used in medical, pharmacy, and food-related products. While *B. thuringiensis* is usually known for its use in biological pesticide production.

4.2.2.7   MICROCOCCUS SPP.

Micrococcus genus is belonging to the Micrococcaceae family, found in natural sources as water, dust, soil, human skin, animal, and dairy products. Micrococci are Gram-positive, spherical-shaped cells that are oxidase-positive and catalase-positive and citrate-negative. The genus content seven species including *M. mortus, M. mucilaginosis, M. roseus, M. antarcticus, M. flavus, M. luteus*, and *M. lylae*. Micrococcal DNA is rich in guanine (G) and cytosine (C), usually exhibiting 65–75% G+C content. *M. luteus* was found to produce antimicrobial compound against food-borne pathogens, for example, *Listeria monocytogenes, Salmonella typhimurium*, and *Escherichia coli* (Akbar et al., 2014) and used for biodegradation of plastics in vitro (Sivasankari and Vinotha, 2014).

4.2.3 YEASTS

4.2.3.1 SACCHAROMYCES SPP.

Saccharomyces belongs to genus yeast classified in fungi kingdom, family Saccharomycetaceae. *Saccharomyces* spp. produces ascospores; when stained with Gram's stain, vegetative cells emerge out Gram-positive, while ascospores appear Gram-negative. It includes 23 species of yeast and many members play pivotal role in food production. *Saccharomyces cerevisiae* is an example of yeasts which is used in making bread, beer, wine, and related beverages. Bruno et al. (2009) studied the effect of feeding yeasts on dairy cows during summer; resulted in improved milk yield and milk components in heat-stressed cows.

4.2.3.2 ZYGOSACCHAROMYCES SPP.

Zygosaccharomyces is a genus belong to Saccharomycetaceae family. It was first described under the *Saccharomyces* genus; in 1983, the genus was reclassified to its current name in the work of Barnett and coworker (Barnett et al., 1983). Twelve species of *Zygosaccharomyces* genus were *Z. bailii, Z. fermentati, Z. bisporus, Z. cidri, Z. florentinus, Z. mellis, Z. lentus, Z. kombuchaensis, Z. microellipsoides, Z. pseudorouxii, Z. rouxii,* and *Z. mrakii* (James and Stratford, 2011; Suh et al., 2013). They had a long and well-known role as spoilage organism of the food industry; many of the species in the genus to be markedly resistance to many common foods keeping methods (Merico et al., 2003). *Z. lentus* grew at 25–26°C, pH 2·2–7·0, and it showed resistance to food preservative materials such as dimethyl dicarbonate, sorbic acid, and benzoic acid (Steels et al., 1999).

4.2.3.3 CANDIDA SPP.

Candida is classified in ascomycota division and Saccharomycetaceae family. Numerous *Candida* spp. are harmless endosymbionts or commensals of animals and human while some of them are associated with infections, called candidemia, for example, *Candida albicans*. Several species of this yeast such as *C. kefir* and *C. maris* are found to be associated with the fermentation of acid alcoholic fermented milk products such as kefir, while *C. antarctica* is a source of important lipases at the industrial level.

4.2.4 MOLDS

4.2.4.1 PENICILLIUM SPP.

Penicillium is a genus of ascomycetous fungi which had a great importance in the food and drug industry. It was described in 1809 by Johann Heinrich Friedrich Link. They are employed as starter molds for the manufacture of several cheese types, that is, Camembert and Roquefort cheese and industrially used for the production of antibiotic penicillin. Some strains of *Penicillium* (namely *P. glabrum*) is associated with the production of mycotoxins, which represents a potential health risk to animals and humans (Samson et al., 2004; Nevarez et al., 2008). Apart from that they are considered as spoilage organisms in fruits, vegetables, and dairy products. *Penicillium* spores and other molds were found in beef and pork used as a raw material; in salami production, it added flavor to salami during raw meat fermentation (Mižáková et al., 2002). *P. purpurogenum* was gave the maximum synthesis of red pigment (2.46 g/L) when grown on Czapek-Dox media, pH 5 at 24°C (Méndez et al., 2011). Some *Penicillium* species can produce antimicrobial compound; *P. nalgiovense* showed antimicrobial production and used 0.125 µg/ml for *C. albicans* inhibition (Svahn et al., 2015).

4.2.4.2 ASPERGILLUS SPP.

Aspergillus is a genus consisting of a few hundred mold species found in various climates worldwide. In 1729, Pier Antonio Micheli first recorded this fungus. *Aspergillus* belongs to the Deuteromycetes fungi group, which is a group with no known sexual state. The most common species of *Aspergillus* genus are *A. niger*, *A. flavus*, *A. ochraceus*, *A. parasiticus*, *A. aliaceus*, and *A. carbonarius* (Perrone et al., 2007). Some *Aspergillus* spp. are very important in industrial products, for example, *A. niger* is used for citrate production (Grewal and Kalra, 1995); glucoamylase production from *A. niger* by solid-state fermentation (Slivinski et al., 2011); pectinase production from seven local isolates of *A. niger* by solid-state fermentation

(Akhter et al., 2011). They mainly cause spoilage in dairy products, meat, fruits, and vegetables. Aflatoxins (B_1, B_2, G_1, and G_2) are secondary metabolites that are produced from some species (e.g. *A. flavus*). It affects human and animal health and as it is transferred to foods through feed animals on fodder that contain aflatoxins.

4.2.4.3 CLADOSPORIUM SPP.

Cladosporium is a genus of fungi belonging to family Davidiellaceae. It nearly contains more than 772 species. Some species develop black brown to olive green colonies, and possess dark pigmented conidia, formed in a simple or branched chain. Few species of *Cladosporium* are pathogenic to plants while some may parasitize other fungi. They are rarely human pathogen, but reported to cause skin, toenails, and lungs infections as well as sinuses. The air-borne spores of the fungus are significant allergens; in large amounts can severely affect people with respiratory diseases and asthma. One member of *Cladosporium* spp. is responsible for contamination in grapes; in the grapes, the fungal growth begins if the humidity and temperature are appropriate for its growth. The rot and damage to the grape berries before maturity leads to the early harvesting (Aydogdu and Gucer, 2009).

4.2.4.4 RHIZOPUS SPP.

Rhizopus genus is one member of fungal species widespread in nature and characterized by the pigmented rhizoids and the occurrence of stolons, and usually globose sporangia. It includes mainly eight species, that is, *R. caespitosus, R. oryzae, R. delemar, R. microspores, R. homothallicus, R. stolonifer, R. schipperae,* and *R. reflexus* (Ellis, 1997). The strain *R. microsporus* var. *oligosporus* find its use in preparation of tempeh, a soybean fermented food. Commercially, it is employed for lactic acid production from renewable material (Zhang et al., 2010). Chitosan production has been reported from cheap sources by *R. arrhizus* (Cardoso et al., 2012). *R. stolonifer*, known as black bread mold develops fruit rot on tomato, strawberry, and sweet potato; commercially used for to produce fumaric acid and cortisone. Some *Rhizopus* spp. causes fungal infections (zygomycosis) to human that can be fatal (Chinn and Diamond, 1982).

4.2.4.5   *TRICHOTHECIUM SPP.*

Trichothecium spp. are common, but pose a small percentage of fungal organisms. They developed flat and granular colonies, which are initially white and later turn to light pink in color. They are plant pathogens, but there are now studies of infection in humans and animals. *T. roseum* produces variety of secondary metabolites including mycotoxins, roseotoxins, and trichothecenes, which can spoil a variety of fruit crops. Secondary metabolites of *T. roseum*, specifically Trichothecinol A, are being investigated as potential *antifungal, anticancer, and antimetastatic drugs* (*Batt and Tortorello, 2014; Taware* et al., *2014*). *Trichothecium* spp. is one of the fungal infections found in rice samples in Portugal (Magro et al., 2006). Trichothecin, produced by *T. roseum*, is an inhibiting compound for other fungi, but some fungi had been found to be resistant to ribosomes as *Fusarium oxysporum* and *S. cerevisiae* can be "in vitro" (Iglesias and Ballesta, 1994).

4.3   POTENTIAL USE OF MICROORGANISMS IN DIFFERENT FOOD PRODUCTION

Numerous FMCs are adhering and have been empirically used in the production of different type of foods since ancestral times, but recently, new product development has been directed by FMCs due to consumer demand for healthier foods. Some of the food products are described herein which production, the FMCs have their potential role.

4.3.1   *MILK-BASED PRODUCTS*

4.3.1.1   *BUTTER AND BUTTERMILK*

Buttermilk is the liquid remaining after butter churning from milk or cream. This type of buttermilk is known as traditional buttermilk. Acidophilus butter was produced using *Bifidobacterium* spp. and *L. acidophilus* (Gomes and Malcata, 1999). Gullón et al. (2013) considered buttermilk as one of the most appropriate matrix to sustain the viability of probiotic bacteria and to make healthy beverage/drink with required sensory attributes.

L. lactis and *Lactobacillus* spp. are utilized for the development of flavor/aroma and acid in butter and buttermilk. The organism *L. lactis* ssp. *diacetylactis* converts citrate present in milk to diacetyl, which imparts a special buttery flavor to the product. These bacteria are used to ferment skim milk, cultured buttermilk forms, if the cream is fermented sour cream produces (Prescott et al., 2002). Bio-sour cream was produced by probiotic starter *L. acidophilus, B. Bifidum,* and *S. thermophilus*. In recent experiment, using spray-chilling technology probiotic strains *L. acidophilus* (LAC-04) and *B. animalis* ssp. *lactis* (BI-01) were encapsulated in cocoa butter (Pedroso et al., 2013). Cultured butter with probiotic *L. acidophilus* La-5 was produced by Tsisaryk et al. (2014). Authors reported that probiotic bacterium stimulated the 11-trans isomerization of fatty acid and enhanced the proportion of conjugated linolenic acid in cultured butter.

4.3.1.2 CHEESE

Cheese is believed to be one of the oldest foods developed nearly about 8000 years ago. It is a dairy product made from milk that is necessary for nutrition and development because it provides all micro (vitamins and enzymes) and macronutrients (proteins, lipids, and saccharides) and also support to the normal function of metabolism (Boza and Sanz Sampelayo, 1997; Ataro et al., 2008). About 2000 distinct variety of cheese are fashioned all over the world, representing approximately 20 general types (Prescott et al., 2002).

Brevibacterium linens; Lactobacillus spp. such as *L. bulgaricus, L. plantarum, L. diacetylactis, L. cremoris, L. helveticus, L. lactis*; Streptococcus spp. such as *S. thermophilus*; Propionibacterium spp. such as *P. Shermanii*; and Penicillium spp. such as *P. camemberti, P. candidum, P. freudenreichii,* and *P. roqueforti* are commonly employed in different varieties of cheese (Brüssow et al., 1998; Prescott et al., 2002). Examples of probiotic bacteria containing cheese types are Iranian white-brined cheese (Ghoddusi and Robinson, 1996), goat semisolid cheese (Gomes and Malcata, 1999), Turkish Beyaz cheese (Kiliç et al., 2009), white-brined cheese (Özer et al., 2008), Minas fresco cheese (Souza and Saad, 2009), Scamorza ewe milk cheese (Albenzio et al., 2013), and white cheese (Yerlikaya and Ozer, 2014).

Manufacturers try to produce cheeses enriched by various healthy additives including probiotic cultures. The intake of cheese supplemented with probiotics shown a number of health benefits, such as reinforcement of intestinal immunity, improvements in oral and intestinal health, and upgradation of the immune system in the elder people (Lollo et al., 2012; Albenzio et al., 2013). It is noticed that cheese exhibits buffering action against the acidic environment of the GIT; creates a suitable passage during gastric transit of probiotics and enhance their survival throughout (Ortakci et al., 2012; Karimi et al., 2012a, b).

During manufacturing of fresh white cheese, all probiotics persist at threshold level (above 7 log·cfu/g) throughout the storage period. Cheeses made with *S. thermophilus* and *L. casei*, recorded maximum evalua- tion points in sensory score in compare to other probiotic combinations (Yerlikaya and Ozer, 2014). Minas fresh cheese with *Lactobacillus para- casei* showed a great potential as a functional food (Buriti et al., 2005).

B. longum (strains B1 and B2), *B. bifidum* (strains B3 and B4), *Bifi- dobacterium* spp. strain B5, *L. acidophilus* (strains A1 and A2), *L. casei* (strains C1 and C2), *L. lactis* (strain A6), and *S. thermophilus* (strain A4) were assessed in combinations for manufacturing of Argentinian fresco cheese. All probiotic showed good viability at the end. The probiotics had the tremendous aptitude to remain viable up to 3 h when a cheese homogenate at pH 3 was used to moderately simulate the acidic environ- ment in the stomach; however, cell viability was more affected at pH 2, *B. bifidum* was the most resistant bacterium (Vinderola et al., 2000a). Ong and Shah (2009) reported that manufactured cheddar cheeses were significantly affected ($p < 0.05$) the concentration of lactate and acetate by the type of probiotic strain used, ripening temperatures, ripening time, as well as their interactions. Texture and appearance attributes of Scamorza ewe milk cheese were improved with the incorporation of probiotic strains (Albenzio et al., 2013).

The microencapsulation process keeps the probiotic microorganisms alive in the dairy products. In Feta cheese, free and alginate encapsulated *L. acidophilus* and *B. lactis*, water-holding capability of encapsulated cells were high due to EPS production; however, microencapsulation process did not demonstrate significant effect on probiotic bacteria viability. Therefore, preferred to use the types of probiotic bacteria possessing poly- saccharide production, acidity, and salt-tolerant attributes (Kailasapathy and Masondole, 2005).

4.3.1.3   YOGHURT AND RELATED FERMENTED MILK

Yoghurt is predominantly manufactured from cow's milk employing two homofermentative bacteria *L. delbrueckii* ssp. *bulgaricus* and *S. salivarius* ssp. *thermophilus* (1:1 ratio) due to their protocooperative action (De Brabandere and De Baerdemaeker, 1999; Tamime and Robinson, 1999; Lucey, 2002; Chandan, 2013). With these organisms growing in consorsium, acid is produced by *S. thermophilus* and aroma components are formed by the *L. bulgaricus*. Freshly prepared yoghurt contains 109 cfu/g (Chandan, 2013). A variety of food resources, including a combination of mango pulp–soy milk, fruit pulp, cow milk, soy milk, and buffalo milk have been used to make different yoghurt varieties (Granata and Morr, 1996; Öztürk and Öner, 1999; Kumar and Mishra, 2004).

Probiotic frozen yoghurt was developed using probiotic starter and traditional starter; the addition of probiotic bacteria to frozen yoghurt had no effect on lactose and protein concentration as well as on sensory characteristics (Davidson et al., 2000). Total acidity and pH was 0.15% and 5.6 at the end of the fermentation and stored for 11 weeks at −20°C. Greek set-type yoghurt was produced by probiotic starter cultures (*L. plantarum* and *L. paracasei* subsp. *tolerans*). It had good physicochemical properties and after refrigerated storage for 2 weeks, the viability of starter bacteria was 107 cfu/g (Maragkoudakisa et al., 2006).

Some cereals and plants can be used in yoghurt production for the people who suffer from sensitivity of milk and vegetarians or to improve some of the sensory qualities, for example, corn (Sankhavadhana, 2001), soybean (Saris et al., 2003), rice (Helland et al., 2005), mango, mixed berry, passion fruit and strawberry (Kailasapathy et al., 2008), mango and soy milk (Kaur et al., 2009), and banana (Kumar and Thakur, 2014).

Microencapsulation of *L. acidophilus* LA-5 and *B. lactis* Bb-12 was used during Iranian yoghurt drink (doogh) production and stored at 4°C for 42 days. Final pH of the encapsulated yoghurt at the end of storage was 4.52 while pH of free cells containing yoghurt was 3.78. In the yoghurt containing encapsulated cells, the viable count of *L. acidophilus* and *bifidobacteria* were 5.5 and 4.0 log cycles, respectively comparatively more than yoghurt containing free cells of bacteria (Mortazavian et al. 2008).

Milk treated with high-pressure homogenization (HPH) before manufacturing of probiotic fermented milk product. The agglomeration of HPH milk production is significantly more compact compared with traditional products and did not affect the probiotic bacteria viability. All samples received high scores in sensory analysis. HPH treatment of milk can help to expand the market of probiotic fermented milk, particularly in the context of texture parameters (Patrignani et al., 2009).

Probiotic starter cultures (*L. acidophilus* LA-5, *S. thermophilus* ST-20Y and *Bifidobacterium* BB-12) were used in the production of fermented goat's milk. The outcome of results revealed that whey protein concentrate (WPC) added to milk fermented increased total protein content and viscosity, and improved the sensory qualities of the product. Addition of 1% WPC enhanced the number of *S. thermophilus* and was also a reason for the reduction of the fermentation time (Martın-Diana et al., 2003).

Bifidobacterium pseudocatenulatum G4 was used in peanut milk fermentation. The peanut milk was supplemented with yeast extract and fructooligosaccharides (FOS), stored at 4°C and 25°C for 2 weeks. After storage period, the numbers of viable G4 cells were decreased to < 7 log·cfu/ ml in the samples held at 25°C and while organic acids including lactate, propionate, acetate, and butyric acid were increased. At 4°C storage, the probiotic viability maintained in fermented products was higher exceeding the mean recommended level of 7 log·cfu/ml (Kabeir et al., 2009).

It is important to maintain the viable counts of each probiotic bacterial strain per milliliter or gram of finished functional products greater than a minimum standard level, that is, 106 cfu/ml still consumption to delivered desired health benefit in the host (Mortazavian et al., 2012). Regular consumption of probiotic cultures, accompanied probably by compounds "prebiotic" improves the digestive process of the consumer health (Itsaranuwat et al., 2003).

In order to reduce the fermentation time, soy milk was allowed to ferment at 42°C with yoghurt culture YCX11 together with *B. animalis* ssp. *lactis* Bb12 (Božanić et al., 2011). It was shown to shorten the fermentation time to 4 h, while the viable bifidobacteria count found to increase for the half of the logarithm scale. The viable bacterial count remained invariable and beyond 10^7 cfu/ml during storage at 4°C for 28 days.

4.3.1.4   ICE CREAM AND FROZEN DAIRY DESSERT

Ice cream and related confectionary products are very much liked by all age group people and addition of probiotics and/or prebiotics may add value by providing health beneficiary effects. Storage at freezing temperatures affects theviability of incorporated bacteria. Nevertheless, the addition of prebiotic oligosaccharides and microencapsulation technique appears more attractive in enhancing the viability of probiotics in such frozen food products. Several studies deal with the development of probiotic ice creams and frozen desserts.

The viability of three strains of probiotic bacteria (*L. acidophilus, B. lactis*, and *L. paracasei* ssp. *paracasei*) was estimated in low-fat ice cream during −25°C storage. *B. lactis* were showed the best survival, whereas *L. acidophilus* survived lower than other strains after 52 weeks. The addition of Hi Maize™ to the production was ameliorated the viability of bacteria after 12 months storage period (Haynes and Playne, 2002). The low-fat ice cream mix was allowed to ferment with a starter culture of yoghurt and probiotic bacteria. The product was stored for 11 weeks at −20°C; the presence of probiotic bacteria in the system supplemented had no effects in protein and lactose concentration and sensory characteristics (Davidson et al., 2000).

Akalin and Erisir (2008) studied the effect of inulin and oligofructose addition on the viability of probiotic *B. animalis* Bb-12 and *L. acidophilus* La-5 strains low-fat ice cream during −18°C storage till 90 days. Oligofructose and inulin were considerably improved the survival of probiotic bacteria in ice cream mix, and also improved the rheological characteristics of the end product. The addition of insulin as prebiotic to Yog-ice cream led to improved melting properties, enhance viscosity, and meltdown characteristics. There is also a direct correlation between increased prebiotic concentration and rheological properties of the final product (El-Nagar et al., 2002).

The effect of encapsulation of *L. acidophilus* and *B. lactis* on β-D-galactosidase activity in the ice cream during 24 weeks of storage at −20°C was investigated (Kailasapathy and Sultana, 2003). The encapsulation of the bacteria showed decrease of 2.06 log than with the viability of bacteria without encapsulation was 2.52 log in end storage period. Overall, the results indicated that the encapsulation of probiotics in the ice cream could enhance their viability. The viability of free and encapsulated probiotic bacteria *L. casei (Lc-01)* and *B. lactis (Bb-12)* were enhanced during the entire shelf life study in synbiotic ice cream incorporated with 1% of resistant starch without any significant effect on the organoleptic attributes (Homayouni et al., 2008).

The microfluidization method was used in the milk homogenization of ice cream mixes. Overall, the sensory properties of low-fat ice-cream production by microfluidization method were not significant than the control samples (Olson et al., 2003).

4.3.2   FRUIT AND VEGETABLE-BASED PRODUCTS

Various fermented vegetable and fruit products should have arisen from the time people began to collect and store it. Natural fermentation of individual or mixed vegetables resulted in the development of diverse fermented products namely sauerkraut, kimchi, pickles, and so forth, which can contain carrot, cabbage, radish, turnip, cucumber, beet, pepper, and/or olive. Several researches had incorporated probiotics in various fruits and vegetable substrates (Beganović et al., 2011; Martins et al., 2013).

4.3.2.1   JUICES

Fermented cabbage juice can be a healthy drink for vegetarians and people suffering from lactose sensitivity. Probiotic starters *L. plantarum* C3, *L. delbrueckii* D7, and *L. casei* A4 were used in the cabbage juice production and incubated at 30°C. After storage for 2 weeks at 4°C, survival of *L. delbrueckii* and *L. plantarum* were reduced while *L. casei* did not survive high acidity and low pH (Yoon et al., 2006). The encapsulated *L. acidophilus* was used to ferment carrot juice for 8 weeks of storage at 4°C. The encapsulation significantly strengthened the cell survival after the fermentation and during storage (Nazzaro et al., 2009).

The fermented ginger juice was incorporated with three spices from *Zingiberaceae* plants (highly antioxidant) and *B. longum* as probiotic starter. Period of fermentation was 35–40 h. The pH of juice was 3.5–4.0 and overall antioxidant performances were 68–75% at the end of the fermentation, which suggests positive effect of *Zingiberaceae* spices on fermentation by probiotic bacteria and may help to develop a novel functional food (Chen et al., 2009).

Stability and functionality of freeze-dried cells of *B. animalis* ssp. *lactis* E-2010 (Bb-12) were better in fruit juice than skim milk; propose to predict the functionality of probiotic bacteria in unfavorable environment of GIT (e.g., survival in acidic stomach and bile salt stress) (Saarela et al., 2006). Noni juice that is claimed to cure a number of human diseases was inoculated with LAB and bifidobacteria to prepare probiotic noni juice to enhance its health beneficiary activities. The numbers of bacteria were 109 cfu/ml at the fermentation end; at 4°C storage and low pH had no effect on survival of *B. longum* and *L. plantarum*. Noni juice produced using *B. longum* had a higher antioxidant action (Wang et al., 2014).

Sheehan et al. (2007) examined the suitability of probiotic cultures for fortification in cranberry, pineapple, and orange fruit juice to assess their technological robustness and acid tolerance. The appropriateness of fruit juice samples as delivery vehicles for probiotic cultures was determined by testing the viability of five *Lactobacillus* strains and one *bifidobacterium* strain (*L. salivarius* ssp. *salivarius* UCC118, *L. salivarius* ssp. *salivarius* UCC500, *Lactobacillus rhamnosus GG*, *L. paracasei* ssp. *paracasei* NFBC43338, *L. casei* DN-114 1 and one *B. animalis* ssp. *lactis* Bb-12 strain). The ability of these probiotics to survive for at least 12 weeks in orange and pineapple juice renders them suitable strains for utilization (Sheehan et al., 2007). In another investigation, the aptness of tomato juice as a carrier for four probiotic species of *Lactobacillus* (*L. casei* A4, *L. acidophilus* LA39, *L. delbrueckii* D7, and *L. plantarum* C3) was determined. The viable cell counts of *Lactobacillus* spp. was $1-9 \times 109$ cfu/ml after 72 h fermentation, at pH 4.1 with increased acidity to 0.65% (Yoon et al., 2004).

4.3.2.2 FERMENTED ONION

Sour onions were produced by fermentation with LAB, added with salt and sugar, followed by fermentation at 18°C (Roberts and Kidd, 2005). The pH was 3.25–3.35 and acidity was 1.2–1.5 g of lactic acid per 100 ml at the end of the fermentation and sour onion had a typical tart taste with the characteristic of sauerkraut onion aroma, without raw onions pungency.

4.3.2.3 FERMENTED BANANA

Fermentation of banana puree is suitable for specialized foods with banana flavored. About 16 types of LAB were tested (1% inoculums) for fermentation of banana puree. After 24 h of fermentation, the pH was reduced to 4.5 and *L. plantarum* gave good results compared to other LAB (de Porres et al., 1985).

Apart from the flavor, recently ethanol and vinegar were produced from banana fruit pulp; highest alcohol level production was 7.77% and acidity% was 4.67% at the end of the fermentation with 15% of *Acetobactor aceti* (Saha and Banerjee, 2013). An enzyme, exoglucanase was produced from the banana stalk in solid-state fermentation method using *Bacillus subtilis* (Shafique et al., 2004). Such microbial processes could serve as an economical method to obtain valuable enzyme (s) from banana waste. *L. acidophilus* cells immobilized by κ-carrageenan were used for fermentation of banana medium. The viable count of bacteria was higher than 10^8 cfu/ml in the medium at the fermentation end than free cells which were 10^6 cfu/ml. The effect of FOS in banana was not more significant in immobilized bacteria than with free cell (Tsen et al., 2009). Recently, Gaikwad (2015) developed a simple method to produce ethanol from different deteriorated fruits using Baker's yeast in which banana achieved maximum 6.63% w/v ethanol.

4.3.2.4   TOMATO-BASED DRINK

Tomato juice is one fermented functional foods, which does not give desired taste and only used as a healthy drink. Tomato contains 93.1% water, 4.89% carbohydrate, mineral, vitamins, and low protein and lipid (Suzuki et al., 2005). Tomato juice was utilized to make probiotic juice using four strains of LAB (*L. acidophilus* LA39, *L. casei* A4, *L. delbrueckii* D7, and *L. plantarum* C3). The viable cell counts of the LAB was $1–9 \times 109$ cfu/ml after 3 days of fermentation and after storage at 4°C for 4 weeks, it was $10^6–10^8$ cfu/ml. LAB decreased pH value to 4.1 with corresponding increase in the total acidity to 0.65% (Yoon et al., 2004).

Tomato juice fermented by *B. breve*, *B. longum*, and *B. infantis* was inoculated at 35–37°C for 6 h. The pH value was reduced to 3.51 and 3.80 for juice product by *B. breve* and *B. Longum*, respectively, with corresponding increase in total acidity (13.50 and 12.50, respectively). Heat treatment (100°C) for tomato juice increased the lycopene content from 88 to 113 µg/g (Koh et al., 2010).

4.3.3 CEREAL-BASED PRODUCT

Cereals are analyzed extensively in recent years because of their potential applications in developing healthy foods. Cereals contribute about more than 60% of the food production in the world, offering proteins, dietary fibers, vitamins, minerals, and other growth factors obligatory for human health. Most of cereal-based products involve natural fermentation aid by mixed population of bacteria, yeasts, and molds. During the fermentation process, some of this microflora may act parallel and some in a sequential manner (Blandino et al., 2003; De Vuyst et al., 2009).

Bread is one of the most popular ancient of human food, produced by the help of microorganisms. By using more complex assemblages of microorganisms, bakers can produce special bread such as sourdoughs. The yeast *Saccharomyces exiguus,* together with a *Lactobacillus* spp., produces the characteristic acidic flavor and aroma of such bread (Prescott et al., 2002). *L. bulgaricus*, in combination with *S. cerevisiae* added to sourdough led to enhance the sensory characteristics with increased keeping quality of bread (Rehman et al., 2007). The aflatoxin B was removed from sourdough making use of LAB in fermentation processes of the bread dough (Zinedne et al., 2005). Rollán et al. (2010) suggested that L AFB decrease molds corruption in bread and baked goods and reduce anti-nutritional factors such as phytic acid. Several EPS-producing strains found to influence the sourdough viscoelastic properties and enhance the body texture as well as organoleptic characteristics of sourdough (Ricciardi et al., 2009; Wolter et al., 2014). EPS have the possible aptitude to substitute hydrocolloids, currently that are employed as bread improvers and therefore meet the consumers' demand for a lesser use of additives in foods (Galle and Arendt, 2014).

A new application of cereals is in the development of synbiotic drinks or beverages incorporating probiotics with cereals. The oat seed contains β-glucan (0.31–0.36%), which is prebiotic oligosaccharides that stimulate the viable probiotic cells. Under refrigerated storage, the keeping quality of oat drink was estimated to be 21 days (Angelov et al., 2006).

Mårtensson et al. (2002) mentioned that oat-based products can be used to sustain the growth and viability of intestinal bacteria in humans and during cold storage. The β-glucan rich concentrates from barley and oat showed decrease in insoluble fiber content after lactic fermentation with probiotics (Lambo et al., 2005). Rathore et al. (2012) studied the consequence of single and mixed cereal flour (malt and barley) suspensions on the fermentation of two probiotic strains *L. acidophilus* (NCIMB 8821) and *L. plantarum* (NCIMB 8826) at 30°C for 28 h of incubation. Authors concluded that by changing the substrate or inoculum composition, the functional and organoleptic attributes of cereal-based probiotic drinks could be significantly modified. In another investigation, it was noticed that growth of LAB was accelerated by the presence of yeast in a cereal-based probiotic foods. The pH was reduced with simultaneous increase in the biosynthesis of lactate and ethanol in mixed culture broth to that of pure LAB culture medium (Kedia et al., 2007).

4.3.4   MEAT AND FISH-BASED PRODUCTS

Probiotic bacteria are largely utilized in the fermented dairy products; however, yet their industrial appliance in fermented meat or fish-based foods is not that much popular. A variety of meat and fish products have been developed, especially sausages, Lebanon bologna, fish sauces (processed using halophilic *Bacillus* spp.), katsuobushi, and izushi besides the fermentation of dairy products. Among LAB, *L. plantarum* and *S. cerevisiae* are often involved in fermentation of meat sausages. The product called Izushi is prepared by lactic fermentation of the mixture containing fresh fish, rice, and vegetables by *Lactobacillus* spp.; while katsuobushi is produced from tuna fermentation by *Aspergillus. glaucus*. Both these fermentations were begun in Japan (Prescott et al., 2002).

Meat and meat products can be improved with the naturally occurring bioactive components within the meat matrixes and these compounds were also reloaded with exerting positive effects on human health as reviewed by Arihara (2006). LAB and bifidobacteria are used in the fermentation of meat product and the range of fermented meat products available around the world is almost equal to that of cheese. Rubio et al. (2014) assessed the effect of two probiotic strains *L. plantarum* 299V and *L. rhamnosus* GG during the production of Spanish fermented sausages. *L. plantarum* 299V inoculated at 10^5 cfu/g achieved as well as maintained high counts during ripening and at the end of the storage period (10^8 cfu/g) suppressing (60%) the growth of endogenous microflora and led to produce functional meat sausages with the highly satisfied sensory quality overall.

Inoculation of thermotolerant LAB resulted in more cohesive, but less hard texture of cooked sausage formulations. Further, the microscopic study revealed that *L. plantarum* UAM10a and *P. acidilacti* UAM15c secreted EPSs; thus, in compared to non-inoculated samples, these sausages showed high moisture stability and improved textural properties (Pérez-Chabela et al., 2013).

Papamanoli et al. (2003) revealed presence of 90% lactobacilli, 4% enterococci, 3% pediococci, and 3% other genera such as *Weissella viridescens, Leuconostoc pseudomesenteroides,* and *Leuconostoc* spp. in naturally fermented dry sausages. Among these isolates, *L. curvatus, L. sakei,* and *L. plantarum* showed inhibition of two *S. aureus* strains and some *L. monocytogenes* strains. Lactobacilli isolated from salami contained 38% *L. casei,* 32% *L. paracasei,* 21% *L. rhamnosus,* and 9% *Lb. sakei, L. casei,* and *L. rhamnosus* had antimicrobial activity against *E. coli* and *S. typhimurium* suggesting there probable use as novel probiotic starter in meat products (Rebucci et al., 2007).

Muthukumarasamy and Holley (2006) in salami product used 7 log·cfu/g from *P. pentosaceus* and *Staphylococcus carnosus* as starter cultures and added 1% (w:w) *L. reuteri,* microencapsulated by sodium alginate. After drying, the numbers of microencapsulated *L. reuteri* cells were diminished by only ≤ 0.5 log unit, whereas non-encapsulated cells dropped by 2.6 log units; had no significant difference in sensory evaluation. Further, free *L. reuteri* cells alone or in combination with *B. longum* diminished *E. coli* O157: H7 by 3.0 log·cfu/g while *B. longum* led a 1.9 log·cfu/g decrease, whereas microencapsulation enhanced viability of *L. reuteri* and *B. longum;* the process reduced their antimicrobial activity against the used *pathogen* (Muthukumarasamy and Holley, 2007).

Three strains of probiotic *L. rhamnosus* were used in dry fermented sausage than with *P. pentosaceus* control starter. After fermentation, viable cells of lactobacilli increased from 10^7 to 10^8–10^9 cfu/g, the pH value was 4.9–5.0. Moreover, the sensory parameters suggested that the dry sausages fermented using *L. rhamnosus* strains were superior over the control samples except dry sausage production by *L. rhamnosus* LC-705 (Erkkilä and Suihko, 2001).

About, 22 strains of nonstarter LAB were isolated from 15 different sources of fermented meat products. Strains were identified by random amplified polymorphic DNA, analytical profile index, and 16S rRNA methods. Results indicated that it contained five strains of each *L. sakei, L. plantarum,* and *L. farciminis,* four of *L. alimentarius* strains, two of *L.*

brevis strains, and one strain of *L. versmoldensis*. Some strains were applied as starter bacteria, after fermentation the viable cells of bacteria ranged from 4.7×10^7 to 2.9×10^8 cfu/g, at pH 4.8–4.9 (Klingberg et al., 2005).

The survival of eight LAB strains was evaluated under conditions similar to those in the GIT in meat model as starters Erkkilä and Petaja (2000). Strains of *P. acidilactici* (P2) and *L. sake* (RM10) exhibited the best competence to survive under acidic conditions and high bile salt concentration.

Probiotic starters *L. plantarum* and *Pediococcus damnosus* were used in fermented sausage. After ripening time (1–2 week), 90% fat content was reduced when viable cells of probiotic bacteria were 10^8–109 cfu/g. Low-fat fermented sausage had high number cells of probiotic bacteria and better physicochemical properties (Kim et al., 2008).

4.3.5 SOYA-BASED PRODUCT

The whole grains of soybean possess considerable amounts of flavonoids, terpenoids, α-linolenic acid, isoflavones, and other natural antioxidants such as carotene, ascorbic acid, and tocopherol (Anderson et al., 1995; Das and Deka, 2012). Hence, the therapeutic values of soya bean products are also noteworthy. In "meju," *B. subtilis* utilize soy protein, increase their population and develop unique odor of ammonia during doenjang making (Kim et al., 2009). Soy sauce is the fermented food famous in Japan produced from raw soybeans that are fermented with *Aspergillus oryzae* or *A. soyae, Z. rouxii*, and *L. delbrueckii* (Jay, 2000). Heenan et al. (2005) used probiotic cultures (*L. acidophilus* MJLA1, *L. paracasei* ssp. *paracasei* 1, *L. rhamnosus* 100-C, *B. lactis* BB-12, *B. lactis* BBDB2, *Saccharomyces boulardii*) into a non-fermented frozen soy dessert with initial populations of 106 cfu/g and evaluated survival of probiotics, and sensory acceptance. After 6 months of storage, the survival of all probiotic bacteria was 107 cfu/g, except *S. boulardii* which decreased during the storage time, suggesting that frozen soy dessert was appropriate food product for the probiotic bacteria delivery. Four species of *Lactobacillus* were evaluated for their ability to grow and impact on growth *Rhizopus oligosporus* during barley tempeh fermentation. None of the *Lactobacillus* spp. significantly affected the growth of *R. oligosporus* and the pH of the finished product barley tempeh (Feng et al., 2005). Soybeans and soybean meals fermented with *A. oryzae* GB-107 in a bed-packed solid fermentor led to increase in the protein content, reduced peptide size, and eliminated trypsin inhibitors within 48 h of the period (Hong et al., 2004).

4.3.6 ALCOHOLIC BEVERAGES

Most of the alcoholic beverages are manufactures using yeasts because of their high tolerance to grow and survive in the presence of ethanol. Recently, several studies have been reported dealing with the isolation of LAB and probiotification of alcoholic beverages using probiotic strains of LAB or yeasts; few studies have been briefly discussed in the below section.

4.3.6.1 BEER AND WINE

Beer production uses cereals such as barley, rice, and wheat. Most beers have about 3–6% (v/v) alcohol content (Bamforth 2002). Rare information is available about fermentation of malt-based beverages using LAB (Bernd et al. 2007; Tenge and Geiger 2001). The viability of *B. lactis* BB-12 and *L. acidophilus* LaA-5, probiotic strains were analyzed in low-alcohol beer (having 2.5% alcohol) and non-alcoholic beer (containing <0.5% alcohol) during storage (5°C) for 20 days. The beer was fermented using *S. cerevisiae* 70424 and *Saccharomyces rouxii* 2531and yeast cells were inactivated by heat treatment at the end of the fermentation and inoculated with probiotic cultures. Within 20th days of the storage period, both the probiotics (particularly *L. acidophilus*) dramatically lost their viability, at least three logarithmic cycles. The maximum decrease in the viability was observed in the beer fermented by *S. cerevisiae* containing probiotic *L. acidophilus*, whereas the lowest was observed in the beer fermented by *S. rouxii*, containing bifidobacteria (Sohrabvandi et al., 2010). Outcomes of such studies suggest that beer is not an appropriate medium to retain the probiotic strains viability even at low temperature.

Wine production or the science of enology (Greek *oinos*, wine and *ology*, the science of), begins with the grapes collection, followed by crushing and separation of the liquid portion called must which is subsequently fermented. It concludes various storage and aging steps as depicted in Figure 4.3 (Willey et al., 2008). *Acetobacter* and *Gluconobacter* (Prescott et al., 2002; García-Ruiz et al., 2014) assessed probiotic properties of

11 LAB isolated from wine. Depending on the LAB strain, the adhesion level to Caco-2 cells varied from 0.37 to 12.2%; specifically *P. pentosaceus* CIAL-86 showed higher level (> 12%) of adhesion to intestinal cells in compared to other reference probiotic strains. It showed more anti-adhesion activity against *E. coli* CIAL-153, that is, > 30%. Probiotic *Lactobacillus* spp. isolated from Raffia wine showed cell-mediated immune responses in rats (Flore et al., 2010). According to Passoth et al. (2007), a commercial plant for production of alcohol was dominated by a consortium of yeast *Dekkera bruxellensis* and *Lactobacillus vini* along with high numbers of other LAB. The interaction between LAB and yeasts was investigated by van Beek et al. (2002) on strong alcoholic beverages production in malt whiskey distilleries. LABs are accountable for the lactate and acetate formation in the middle and late stage of the fermentation procedure.

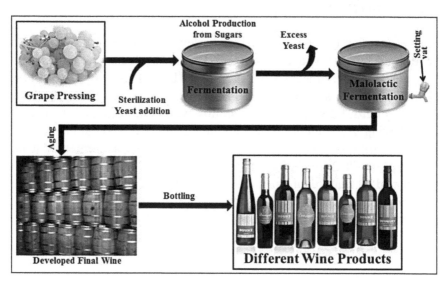

FIGURE 4.3 (**See color insert.**) Process of wine production from grapes (Willey et al. 2008).

4.3.6.2 *MILK AND CEREAL-BASED ALCOHOLIC BEVERAGES*

Beverages are refreshing liquids intended for human consumption, some of which are fermented using microorganisms, mainly LAB. Fermented household bushera was produced by LAB and their viable counts were between 7.1 and 9.4 log cfu/ml (Muianja et al., 2003). Microencapsulation of *B. lactis* was used in tow beverage, amasi, and mahewu. After storage, small differences in viability between immobilized and free cells of *B. lactis* was observed (McMaste et al., 2005). *L. mesenteroides* as starter culture was used in fermented milk beverage with additional supplements; 100 mg/l of tryptone was the best supplement to led increase in viable cells of starter bacteria and maximum bioavailability of iron, magnesium, and zinc. The increased level of beneficial fatty acids equally supports the therapeutic value of the product (Shobharani and Agrawal, 2009).

Nondairy beverages have been developed from variety of fermented cereals and plants, such as wheat and oat bran (Blandino et al., 2003) and green coconut water (Prado et al., 2008). *Lactobacillus* and *Pediococcus* strains were isolated from *Omegisool*, a Korean traditionally fermented alcoholic beverage (Oh and Jung, 2015). A few isolates also revealed good probiotic candidature. Togwa drink is a starch saccharified, alcoholic free, and produced in Tanzania. Usually, maize flour or finger millet malt was used in the production of togwa (Oi and KIitabatake, 2003). Rozada-Sanchez et al. (2008) examined the growth of bifidobacteria in malt hydrolysate enriched with growth promoters, such as yeast extract. *Makgeolli* is known as traditional Korean alcoholic beverage produced using grains and natural starter, *nuruk*. The major bacterial phylum shifted from *γ-Proteobacteria* to *Firmicutes* significantly as the fermentation progresses, and similarly the proportion of *Saccharomyces* fungi increased (Jung et al., 2012). The numbers of bacteria in the finished product was higher than in commercial rice beers, but overall there was no change in fungi content as indicated by the results of quantitative polymerase chain reaction.

4.3.7 MISCELLANEOUS FOOD PRODUCTS

Chocolate is the highest confectionery product with unique sensory and tissue characteristics of its kind. It is also a source of magnesium and many other biologically active substances, such as polyphenols and tocopherols, which are beneficial to humans (Nebesny et al., 2007). Chocolate is naturally rich in antioxidant compounds and its nutritional quality can be further improved by the addition of probiotic bacteria and/or prebiotics

(Gadhiya et al., 2015). The unique taste of chocolate is especially valuable for consumers; importantly, the sensory characteristics should remain unchanged despite the addition of the proportion of LAB. The microencapsulation of probiotic bacteria led to the enhanced survival of *L. helveticus* and *B. longum* in chocolate. Probiotics in the form of coating around the chocolate are the admirable solution to protect the bacteria in environmental stress conditions and did not decrease the numbers of bacteria in the finished food (Possemiers et al., 2010). Homayouni et al. (2014) concluded that the enriched chocolate with *L. paracasei* can be stored at ambient temperature without having any unfavorable changes in its physicochemical properties, such as water activity and pH. The development of foods that confirm health and safety object is one of the main research priorities in the food industry. Betoret et al. (2003) used *S. cerevisiae* and *L. casei* food microorganisms to develop many probiotic-enriched dried fruits. Impregnated apple juice with 107 cfu/g of *L. casei* was led to best stability and ensures the preservation of fruit. Viable cells count of *L. casei* in the end storage product was 106 cfu/g.

4.4 CONCLUSION

FMCs plays crucial role in the development of specific food products. Some of this microbiota also influence human health. Selection of diverse food substrates or their components, such as milk, vegetable, fruits, meat, and beverages and combinations of appropriate microbial strain can help to provide desired physicochemical, technological, and organoleptic effects and development of a wide range of food products. Further, optimization of process parameters can help to enhance the production of organic acids, vitamins, antibiotics, enzymes, or EPSs at the industrial level. With the help of novel biotechnological methodologies now it is possible to select specific strain of microorganism with advance technological or biomedical application in the development of functional foods. Incorporation of probiotic microorganism or prebiotic components novel processing techniques such as ultrahigh temperature treatment, high-pressure homogenization, high-intensity pulsed electric field, antimicrobial packaging, use of recombinant strains, and nanotechnology may help to improve the food safety and quality of various food products and also the shelf life.

KEYWORDS

- acetic acid bacteria
- antifungal
- anti-metastatic drugs
- *Aspergillus*
- *Bifidobacterium*
- bioactive peptides
- biopreservative
- conjugated linolenic acid
- exopolysaccharides
- food microbial community
- functional foods
- microencapsulation
- *Penicillium*
- polyhydroxyalkonate
- probiotification
- protocooperative action
- secondary metabolite
- synbiotic drinks

REFERENCES

Abd-Elrahman, A. H. Mastitis in Housed Dairy Buffaloes: Incidence, Etiology, Clinical Finding, Antimicrobial Sensitivity and Different Medical Treatment Against *E. coli* Mastitis. *Life Sci. J.* **2013,** *10*(1), 531–538.

Akalin, A. S.; Erisir, D. Effects of Inulin and Oligofructose on the Rheological Characteristics and Probiotic Culture Survival in Low-Fat Probiotic Ice Cream. *J. Food Sci.* **2008,** *73*(4), 184–188.

Akbar, A.; Sitara, U.; Ali, I.; Muhammad, N.; Khan, S. A. Isolation and Characterization of Biotechnologically Potent *Micrococcus luteus* Strain from Environment. *Pak. J. Zool.* **2014,** *46*(4), 967–973.

Akhter, N.; Morshed, M. A.; Uddin, A.; Begum, F.; Sultan, T.; Azad, A. K. Production of Pectinase by *Aspergillus niger* Cultured in Solid State Media. *Int. J. Biosci.* **2011,** *1*(1), 33–42.

Albenzio, M.; Santillo, A.; Caroprese, M.; Ruggieri, D.; Napolitano, F.; Sevi, A. Physicochemical Properties of Scamorza Ewe Milk Cheese Manufactured with Different Probiotic Cultures. *J. Dairy Sci.* **2013,** *96*(5), 2781–2791.

Al-Sahlany, S. T. G. Production of Mannitol from a Local Isolate of *Leuconostoc mesenteroides.* M.Sc. Thesis, College of Agriculture, Basrah University, Basrah, Iraq, 2008, p 108.

Al-Sahlany, S. T. G. Production of Biodegradable Plastic (Polyhydroxybutyrate) by the Local Isolate of *Bacillus cereus* B5 and using in Packaging of Some Foods. Ph.D. Thesis, College of Agriculture, Basrah University, Basrah, Iraq, 2015, p 186.

Ammor, M. S.; Mayo, B. Selection Criteria for Lactic Acid Bacteria to be used as Functional Starter Cultures in Dry Sausage Production: An Update. *Meat Sci.* **2007,** *76,* 138–146.

Anderson, J. W.; Johnstone, B. M.; Cook-Newell, M. E. Meta-Analysis of the Effects of Soy Protein Intake on Serum Lipids. N. Engl. J. Med. **1995,** *333,* 276–282.

Angelov, A.; Gotcheva, V.; Kuncheva, R.; Hrstozova, T. Development of a New Oat Based Probiotic Drink. *Int. J. Food Microbiol.* **2006,** *112*(1), 75–80.

Arihara, K. Strategies for Designing Novel Functional Meat Products. *Meat Sci.* **2006,** *74,* 219–229.

Arihara, K.; Ota, H.; Itoh, M.; Kondo, Y.; Sameshima, T.; Yamanaka, H.; Akimoto, M.; Kanai, S.; Miki, T. *Lactobacillus acidophilus* Group Lactic Acid Bacteria Applied to Meat Fermentation. *J. Food Sci.* **1998,** *63,* 544–547.

Aryana, K. J.; Mcgrew, P. Quality Attributes of Yogurt with *Lactobacillus casei* and Various Prebiotics. *LWT-Food Sci. Technol.* **2007,** *40,* 1808–1814.

Ataro, A.; Mc Crindle, R. I.; Botha, B. M.; Mc Crindle, C. M. E.; Ndibewu, P. P. Quantification of Trace Elements in Raw Cow's Milk by Inductively Coupled Plasma Mass Spectrometry (ICP-MS). *Food Chem.* **2008,** *111*(1), 243–248.

Aydogdu, H.; Gucer, Y. Microfungi and Mycotoxins of Grapes and Grape Products. *Trakia J. Sci.* **2009,** *7,* 211–214.

Bamforth, C. W. Nutritional Aspects of Beer—A Review. *Nutr. Res.* **2002,** *22,* 227–237.

Barnett, J. A.; Payne, R. W.; Yarrow, D. *Yeasts: Characteristics and Identification*; Cambridge University Press: Cambridge, 1983.

Batt, C. A.; Tortorello, M. *Encyclopedia of Food Microbiology,* 2nd Ed.; Elsevier: London UK, 2014; p 1014.

Beganović, J.; Pavunc, A. L.; Gjuračić, K.; Špoljarec, M.; Šušković, J.; Kos, B. Improved Sauerkraut Production with Probiotic Strain *Lactobacillus plantarum* L4 and *Leuconostoc mesenteroides* LMG 7954. *J. Food Sci.* **2011,** *76*(2), M124–M129.

Bernd, S.; Lutz-Guenther, F.; Frank, I.; Diana, M.; Bianka, S. Procedure for the Production of Probiotic Wort Extracts to Produce Malt-Based Beverage, Comprises Isolating Special Mash-Acidifying Bacterial Strains from Acidified Malt Mash and Culturing Autoclaved Malt Wort Over Several Culture Lines. Office EP, 2007; DE102005047899.

Bertazzoni, M. E.; Benini, A.; Marzotto, M.; Sbarbati, A.; Ruzzenente, O.; Ferrario, R.; Hendriks, H.; Dellaglio, F. Assessment of Novel Probiotic *Lactobacillus casei* Strains for the Production of Functional Dairy Foods. *Int. Dairy J.* **2004,** *14*(8), 723–736.

Betoret, N.; Puente, L.; Diaz, M. J.; Pagan, M. J.; Garcia, M. J.; Gras, M. L.; Martinez-Monzo, J.; Fito, P. Development of Probiotic-Enriched Dried Fruits by Vacuum Impregnation. *J. Food Eng.* **2003,** *56,* 273–277.

Blanchette, L.; Roy, D.; Belanger, G.; Gauthier, S. F. Production of Cottage Cheese using Dressing Fermented by Bifidobacteria. *J. Dairy Sci.* **1996**, *79,* 8–15.

Blandino, A.; Al-Aseeri, M. E.; Pandiell, S. S.; Cantero, D.; Webb. C. Cereal-Based Fermented Foods and Beverages. *Food Res. Int.* **2003**, *36,* 527–543.

Boza, J.; Sanz Sampelayo, M. R. Aspectos nutricionales de la leche de cabra. Anales de la Real Academia de Ciencias Veterinarias de Andalucia Oriental. **1997**, *10,* 109–139.

Božanić, R.; Lovković, S.; Jeličić, I. Optimising Fermentation of Soymilk with Probiotic Bacteria. *Czech J. Food Sci.* **2011**, *29*(1), 51–56.

Bruno, R. G. S.; Rutigliano, H. M.; Cerri, R. L.; Robinson, P. H.; Santos, J. E. P. Effect of Feeding *Saccharomyces Cerevisiae* on Performance of Dairy Cows During Summer Heat Stress. *Anim. Feed Sci. Technol.* **2009**, *150*(3), 175–186.

Brüssow, H.; Bruttin, A.; Desiere, F.; Lucchini, S.; Foley, S. Molecular Ecology and Evolution of *Streptococcus thermophilus* Bacteriophages—A Review. *Virus Genes* **1998**, *16*(1), 95–109.

Burdychova, R.; Dohnal, V.; Hoferkova, P. Biogenic Amines Reduction by Probiotic *L. casei* During Ripening of Fermented Sausages. *Chemickélisty* **2008**, *102*(Suppl.), S265–S1311.

Buriti, F. C.; da Rocha, J. S.; Assis, E. G.; Saad, S. M. Probiotic Potential of Minas Fresh Cheese Prepared with the Addition of *Lactobacillus paracasei*. *LWT-Food Sci. Technol.* **2005**, *38*(2), 173–180.

Caplice, E.; Fitzgerald, G. F. Food Fermentations, Role of Microorganisms in Food Production and Preservation. *Int. J. Food Microbiol.* **1999**, *50,* 131–149.

Cardoso, A.; Lins, C. I. M.; dos Santos, E. R.; Silva, M. C. F.; Campos-Takaki, G. M. Microbial Enhance of Chitosan Production by *Rhizopus arrhizus* using Agro Industrial Substrates. *Molecules* **2012**, *17,* 4904–4914.

Chandan, R. C. History and Consumption Trends. In *Manufacturing Yogurt and Fermented Milks,* 2nd Ed.; Chandan, R. C., Kilara, A., Eds.; John Wiley & Sons, Inc. 2013.

Chen, I. N. C. C.; Wang, C. Y.; Chang, T. L. Lactic Fermentation and Antioxidant Activity of Zingiberaceae Plants in Taiwan. *Int. J. Food Sci. Nutr.* **2009**, *60*(2S), 57–66.

Chinn, R.Y.; Diamond, R. D. Generation of Chemotactic Factors by *Rhizopus oryzae* in the Presence and Absence of Serum: Relationship to Hyphal Damage Mediated by Human Neutrophils and Effects of Hyperglycemia and Ketoacidosis. *Infect. Immun.* **1982**, *38*(3), 1123–1129.

Corbo, M. R.; Albenzio, M.; De Angelis, M.; Sevi, A.; Gobbetti, M. Microbiological and Biochemical Properties of Canestrato Pugliese Hard Cheese Supplemented with Bifidobacteria. *J. Dairy Sci.* **2001**, *84,* 551–561.

Dalev, D.; Bielecka, M.; Troszynska, A.; Ziajka, S.; Lamparski, G. Sensory Quality of New Probiotic Beverages Based on Cheese Whey and Soy Preparation. *Pol. J. Food Nutr. Sci.* **2006**, *15,* 65–70.

Das, A. J.; Deka, S. C. Fermented Foods and Beverages of the North-East India. *Int. Food Res. J.* **2012**, *19*(2), 377–392.

Davidson, R. H.; Duncan, S. E.; Hackney, C. R.; Eigel, W. N.; Boling, J. W. Probiotic Culture Survival and Implications in Fermented Frozen Yogurt Characteristics. *J. Dairy Sci.* **2000**, *83,* 666–673.

De Brabandere, A. G.; De Baerdemaeker, J. G. Effects of Process Conditions on the pH Development During Yogurt Fermentation. *J. Food Eng.* **1999**, *41,* 221–227.

De Ley, J.; Gillis, M.; Swings, J. Family VI. Acetobacteraceae. In *Bergey's Manual of Systematic Bacteriology;* Krieg, N. R., Holt, J. G., Eds.; Williams & Wilkins: Baltimore, 1984; Vol. 1, pp 267–278.

de Porres, E.; de Arriola, M. C.; Garcia, R.; Rolz, C. Lactic Acid Fermentation of Banana Puree. *Lebensm. Wiss. u. Techno.* **1985,** *18,* 379–382.

De Vuyst, L.; Vrancken, G.; Ravyts, F.; Rimaux, T.; Weckx, S. Biodiversity, Ecological Determinants, and Metabolic Exploitation of Sourdough Microbiota. *Food Microbiol.* **2009,** *26*(7), 666–675.

De Vuyst, L.; Zamfira, M.; Mozzia, F.; Adrianya, T.; Marshalld, V.; Degeesta, B.; Vaningelgem, F. Exopolysaccharide-Producing *Streptococcus thermophilus* Strains as Functional Starter Cultures in the Production of Fermented Milks. *Int. Dairy J.* **2003,** *13,* 07–717.

Demirkan, E. Production, Purification, and Characterization of α-Amylase by *Bacillus subtilis* and its Mutant Derivate. *Turk. J. Biol.* **2011,** *35,* 705–712.

Deppenmeier, U.; Hoffmeister, M.; Prust, C. Biochemistry and Biotechnological Applications of *Gluconobacter* Strains. *Appl. Microbiol. Biotechnol.* **2002,** *60,* 233–242.

Didelota , X.; Barkerb, M.; Falushc, D.; Priest, F. G. Evolution of Pathogenicity in the *Bacillus cereus* Group. *Syst. Appl. Microbiol.* **2009,** *32,* 81–90.

Donkor, O. N.; Henriksson, A.; Vasiljevic, T.; Shah, N. P. α-Galactosidase and Proteolytic Activities of Selected Probiotic and Dairy Cultures in Fermented Soymilk. *Food Chem.* **2007,** *104*(1), 10–20.

Ellis, D. H. Zygomycetes, Chapter 16. In *Topley and Wilson's Microbiology and Microbial Infections*, 9th ed.; Edward Arnold: London, 1997; pp 247–277.

El-Nagar, G.; Clowes, G.; Tudorica, C. M.; Kuri, V. Rheological Quality and Stability of Yogh-Ice Cream with Added Inulin. *Int. J. Dairy Technol.* **2002,** *55*(2), 89–93.

Entani, E.; Ohmori, S.; Masai, H.; Suzuki, K.-I. Acetobacter *Polyoxogenes* sp. nov., a New Species of an Acetic Acid Bacterium useful for Producing Vinegar with High Acidity. J. Gen. Appl. Microbiol. **1985,** *31,* 475–490.

Erkkilä, S.; Petaja, E. Screening of Commercial Meat Starter Cultures at Low pH and in the Presence of Bile Salts for Potential Probiotic use. *Meat Sci.* **2000,** *55,* 297–300.

Erkkilä, S.; Suihko, M.-L.; Eerola, S.; Petaja, E.; Mattila-Sandholm, T. Dry Sausage Fermented by *Lactobacillus rhamnosus* Strains. *Int. J. Food Microbiol.* **2001,** *64,* 205–210.

Feng, X. M.; Eriksson, A. R. B.; Schnurer, J. Growth of Lactic Acid Bacteria and *Rhizopus oligosporus* During Barley Tempeh. *Int. J. Food Microbiol.* **2005,** *104,* 249–256.

Flore, T. N.; François, Z. N.; Félicité, T. M. Immune System Stimulation in Rats by *Lactobacillus* sp. Isolates from Raffia Wine (*Raphia vinifera*). *Cell. Immunol.* **2010,** *260*(2), 63–65.

Fratamico, P. M.; Bayles, D. O. *Foodborne Pathogens: Microbiology and Molecular Biology;* Fratamico, P. M., Bayles, D. O., Eds.; Caister Academic Press: UK, 2005.

Gadhiya, D.; Patel, A.; Prajapati, J. B. Current Trend and Future Prospective of Functional Probiotic Milk Chocolates and Related Products—A Review. *Czech J. Food Sci.* **2015,** *33*(4), 295–301.

Gaikwad, B. G.. Production of Ethanol from Fruits and Waste Food. *LS–An Int. J. Life Sci.* **2015,** *18,* 41.

Galle, S.; Arendt, E. K. .Exopolysaccharides from Sourdough Lactic Acid Bacteria. *Crit. Rev. Food Sci. Nutr.* **2014**, *54*(7), 891–901.

García-Ruiz, A.; de Llano, D. G.; Esteban-Fernández, A.; Requena, T.; Bartolomé, B.; Moreno-Arribas, M. V. Assessment of Probiotic Properties in Lactic Acid Bacteria Isolated from Wine. *Food Microbiol.* **2014** *44*, 220–225.

Gelsomino, R.; Vancanneyt, M.; Vandekerckhove, T. M.; Swings, J. Development of a 16s RRNA Primer for the Detection of *Brevibacterium* spp. *Lett. Appl. Microbiol.* **2004**, *38*, 532–535.

Ghoddusi, H. B.; Robinson, R. K. The Test of Time. *Dairy Ind. Int.* **1996** *61*, 25–28.

Goldin, B. R.; Gorbach, S. L. The Effect of Milk and Lactobacillus Feeding in Human Intestinal Bacterial Enzyme Activity. *Am. J. Clin. Nutr.* **1984**, *39*, 756–761.

Gomes, A. M. P.; Malcata, F. X. *Bifidobacterium* spp. and *Lactobacillus acidophilus*: Biological, Biochemical, Technological and Therapeutic Properties Relevant for use as Probiotics. *Trends Food Sci. Technol.* **1999**, *10*, 139–157.

Granata, L. A.; Morr, C. V. Improved Acid, Flavor and Volatile Compound Production in a High Protein and Fiber Soymilk Yogurt-Like Product. *J. Food Sci.* **1996**, *61*(2), 331–336.

Grewal, H. S.; Kalra, K. L. Fungal Production of Citric Acid. *Biotechnol. Adv.* **1995**, *13*, 209–234.

Gruner, E.; Pfyffer, G. E.; von Graevenitz, A. Characterization of *Brevibacterium* spp. from Clinical Specimens. *J. Clin. Microbiol.* **1993**, *31*(6), 1408–1412.

Guarner, F.; Khan, A. G.; Garisch, J.; Eliakim, R.; Gangl, A.; Thomson, A.; Krabshuis, J.; Mair, T. L.; Kaufmann, P.; de Paula, J. A.; Fedorak, R.; Shanahan, F.; Sanders, M. E.; Szajewska, H. Probiotics and Prebiotics. World Gastroenterology Organisation Practice Guideline. 2008. http://doctor-ru.org/main/1100/1105.pdf (accessed June 21, 2016).

Gullón, P.; Souza, P. Z. D.; Gullón, B.; Pintado, M. E.; Gomes, A. Suitability of Buttermilk as Culture Medium to Support the Survival of Probiotic Strains. *Abstract Book, Portuguese Congress of Microbiology and Biotechnology*, December 6–8, 2013, Portugal.

Hammes, W. P.; Brandt, M. J.; Francis, K. L.; Rosenheim, J.; Seitter, M. F. H.; Vogelmann, S. A. Microbial Ecology of Cereal Fermentations. *Trends Food Sci. Technol.* **2005**, *16*, 4–11.

Hammes, W. P.; Hertel, C. Genus I. Lactobacillus Beijerink 1901. In: De Vos P, Garrity GM, Jones D, Krieg NR, Ludwig W, Rainey FA, Schleifer KH, Whitman WB, editors. Bergey's manual of systematic bacteriology, vol. 3, 2nd ed., Springer, Berlin, 2009. pp 465–510.Haynes, I. N.; Playne, M. J. Survival of Probiotic Cultures in Low-Fat Ice Cream. *Aust. J. Dairy Technol.* **2002**, *57*, 10–14.

Heenan, C. N.; Adams, M. C.; Hosken, R. W.; Fleet, G. H. Survival and Sensory Acceptability of Probiotic Microorganisms in a Nonfermented Frozen Vegetarian Dessert. *LWT-Food Sci. Technol.* **2005**, *37*(4), 461–466.

Helland, M. H.; Wciklund, T.; Narvhus, J. A. Growth and Metabolism of Selected Strains of Probiotic Bacteria in Milk- and Water-Based Cereal Puddings. *Int. Dairy J.* **2005**, *14*, 957–965.

Homayouni, A.; Azizi, A.; Ehsani, M. R.; Yarmand, M. S.; Razavi, S. H. Effect of Microencapsulation and Resistant Starch on the Probiotic Survival and Sensory Properties of Synbiotic Ice Cream. *Food Chem.* **2008**, *111*(1), 50–55.

Homayouni, A.; Roudbaneh, M.; Aref Hosseyni, S. R. Filled Chocolate Supplemented with *Lactobacillus paracasei*. *Int. Res. J. Appl. Basic Sci.* **2014**, *8*(11), 2026–2031.

Hong, K.-J.; Lee, C.-H.; Kim, S. W. *Aspergillus oryzae* GB-107 Fermentation Improves Nutritional Quality of Food Soybeans and Feed Soybean Meals. *J. Med. Food* **2004,** *7*(4), 430–435.

Hong, S. H.; Marshall, R. T. Natural Exopolysaccharides Enhance Survival of Lactic Acid Bacteria in Frozen Dairy Desserts. *J. Dairy Sci.* **2001,** *84,* 1367–1374.

Iglesias, M.; Ballesta, J. P. G. Mechanism of Resistance to the Antibiotic Trichothecin in the Producing Fungi. *Eur. J. Biochem.* **1994,** *223*(2), 447–453.

Irina, V. D.; Lee, H.; Tourova, T. P.; Ryzhkova, E. P.; Netrusov, A. I. *Propionibacterium freudenreichii* Strains as Antibacterial Agents at Neutral pH and their Production on Food-Grade Media Fermented by Some Lactobacilli. *J. Food Saf.* **2012,** *32,* 48–58.

Itsaranuwat, P.; Al-Haddad, K. S. H.; Robinson, R. K. The Potential Therapeutic Benefits of Consuming 'Health-Promoting' Fermented Dairy Products: A Brief Update. *Int. J. Dairy Technol.* **2003,** *56,* 203–210.

James, S.; Stratford, M. *Zygosaccharomyces* Barker 1901. In *The Yeasts: A Taxonomic Study;* Kurtzman, C. P., Fell, J. W., Boekhout, T., Eds.; Elsevier: London, United Kingdom, 2011; Vol. 1, pp 937–947.

Jay, J. M. *Modern Food Microbiology,* 6th Ed.; An Aspen Pub., Aspen Publishers, Inc.: Gaithersburg, Maryland, 2000; p 679.

Jones, D.; Keddie, R. M. Genus Brevibacterium. In *Bergey's Manual of Systematic Bacteriology;* Holt, J. G., Ed.; Williams & Wilkins: Baltimore, 1986; p 1301.

Jung, M. J.; Nam, Y. D.; Roh, S. W.; Bae, J. W. Unexpected Convergence of Fungal and Bacterial Communities During Fermentation of Traditional Korean Alcoholic Beverages Inoculated with Various Natural Starters. *Food Microbiol.* **2012,** *30*(1), 112–123.

Kabeir, B. M.; Yazid, A. M.; Hakim, M. N.; Khahatan, A.; Shaborin, A.; Mustafa, S. Survival of Bifidobacterium pseudocatenulatum G4 During the Storage of Fermented Peanut Milk (PM) and Skim Milk (SM) Products. *Afr. J. Food Sci.* **2009,** *3*(6), 151–155.

Kailasapathy, K.; Masondole, L. Survival of Free and Microencapsulated *Lactobacillus acidophilus* and *Bifidobacterium lactis* and their Effect on Texture of Feta Cheese. *Aust. J. Dairy Technol.* **2005,** *60*(3), 252–258.

Kailasapathy, K.; Sultana, K. Survival of β-D-Galactosidase Activity of Encapsulated and Free *Lactobacillus acidophilus* and *Bifidobacterium lactis* in Ice-Cream. *Aust. J. Dairy Technol.* **2003,** *58*(3), 223–227.

Kailasapathy, K.; Harmstorf, I.; Phillips, M. Survival of *Lactobacillus acidophilus* and *Bifidobacterium animalis* spp. *Lactis* in Stirred Fruit Yogurts. *LWT-Food Sci. Technol.* **2008,** *41,* 1317–1322.

Karimi, R.; Mortazavian, A. M.; Karimi, M. Incorporation of *Lactobacillus casei* in Iranian Ultrafiltered Feta Cheese Made by Partial Replacement of NaCl with KCl. *J. Dairy Sci.* **2012a,** *95*(8), 4209–4222.

Karimi, R.; Mortazavian, A.; Cruz, A. Viability of Probiotic Microorganisms in Cheese During Production and Storage: A Review. *Dairy Sci. Technol.* **2011,** *91,* 283–308.

Karimi, R.; Sohrabvandi, S.; Mortazavian, A. M. Review Article: Sensory Characteristics of Probiotic Cheese. *Compr. Rev. Food Sci. Food Saf.* **2012b,** *11*(5), 437–452.

Kaur, H.; Mishra, H. N.; Umar, P. Textural Properties of Mango Soy Fortified Probiotic Yogurt: Optimization of Inoculum Level of Yogurt and Probiotic Culture. *Int. J. Food Sci. Technol.* **2009,** *44,* 415–424.

Kedia, G.; Wang, R.; Patel, H.; Pandiella, S. S. Use of Mixed Cultures for the Fermentation of Cereal-Based Substrates with Potential Probiotic Properties. *Process Biochem.* **2007**, *42*(1), 65–70.

Kiatpapan, P.; Murooka, Y. Genetic Manipulation System in Propionibacteria. *J. Biosci. Bioeng.* **2002**, *93*, 1–8.

Kiliç, G. B.; Kuleansan, H.; Eralp, I.; Karahan, A. G. Manufacture of Turkish Beyaz Cheese Added with Probiotic Strains. *LWT-Food Sci. Technol.* **2009**, *42*(5), 1003–1008.

Kim, T. W.; Lee, J. H.; Kim, S. E.; Park, M. H.; Chang, H. C.; Kim, H. Y. Analysis of Microbial Communities in Doenjang, a Korean Fermented Soybean Paste, using Nested PCR-Denaturing Gradient Gel Electrophoresis. *Int. J. Microbiol.* **2009**, *131*, 265–271.

Kim, Y. J.; Lee, H. C.; Park, S. Y.; Park, S. Y.; Oh, S.; Chin, K. B. Utilization of Probiotic Starter Cultures for the Manufacture of Low-Fat Functional Fermented Sausages. *Korean J. Food Sci. Anim. Resour.* **2008**, *28*(1), 51–58.

Klingberg, T. D.; Axelsson, L.; Naterstadt, K.; Elsser, D.; Budde, B. B. Identification of Potential Probiotic Starter Cultures for Scandinavian Type Fermented Sausages. *Int. J. Food Microbiol.* **2005**, *105*, 419–431.

Koh, J.-H.; Kim, Y.; Oh, J.-H. Chemical Characterization of Tomato Juice Fermented with Bifidobacteria. *J. Food Sci.* **2010**, *75*(5), 428–432.

Kośmider, A.; Drożdżyńska, A.; Blaszka, K.; Leja, K.; Czaczyk, K. Propionic Acid Production by *Propionibacterium freudenreichii* subsp. *Shermanii* using Crude Glycerol and Whey Lactose Industrial Wastes. *Pol. J. Environ. Stud.* **2010**, *19*(6), 1249–1253.

Kumar, D.; Thakur, S. N. Sensory and Physico-Chemical Analysis of Banana Based Probiotic Soy Yoghurt Toward Consumer's Acceptance. *Int. J. Sci. Res.* **2014**, *3*(6), 96–98.

Kumar, P.; Mishra, H. N. Mango Soy Fortified Set Yoghurt: Effect of Stabilizer Addition on Physicochemical, Sensory and Textural Properties. *Food Chem.* **2004**, *87*, 501–507.

Lamagni, T. L.; Neal, S.; Keshishian, C.; Alhaddad, N.; George, R.; Duckworth, G.; Vuopio-Varkila, J.; Efstratiou, A. Severe *Streptococcus pyogenes* Infections, United Kingdom, 2003–2004. *Emerging Infect. Dis.* **2008**, *14*(2), 202–209.

Lambo, A. M.; Oste, R.; Nyman, M. G. E. L. Dietary Fiber in Fermented Oat and Barley β-Glucan Rich Concentrates. *Food Chem.* **2005**, *85*(2), 283–293.

Langsrud, T.; Reinbold, G. W. Flavour Development and Microbiology of Swiss Cheese. A Review. II Starters, Manufacturing Process and Procedure. *J. Milk Food Technol.* **1973**, *36*, 531–542.

Lee, M. R.; Huang, Y. T.; Liao, C. H.; Lai, C. C.; Lee, P. I.; Hsueh, P. R. Bacteraemia Caused by *Weissella confusa* at a University Hospital in Taiwan 1997–2007. *Clin. Microbiol. Infect.* **2011**, *17*, 1226–1231.

Lin, H-Y.; Rao, Y. K.; Wu, W.-S.; Tzeng, Y.-M. Ferrous Ion Enhanced Lipopeptide Antibiotic Iturin a Production from *Bacillus amyloliquefaciens* B128. *Int. J. Appl. Sci. Eng.* **2007**, *5*(2), 123–132.

Lollo, P. C. B.; Cruz, A. G.; Morato, P. N.; Moura, C. S.; Carvalho-Silva, L.; Oliveira, C. A. F.; Faria, J. A. F.; Amaya-Fartan, J. Probiotic Cheese Attenuates Exercise-Induced Immune Suppression in Wistar Rats. *J. Dairy Sci.* **2012**, *5*(7), 3549–3558.

Lucey, J. A. Foundation Scholar Award Formation and Physical Properties of Milk Protein Gels. *J. Dairy Sci.* **2002**, *85*, 281–294.

Magro, A.; Carvalho, M. O.; Bastos, M. S. M.; Carolino, M.; Adler, C. S.; Timlick, B.; Mexia, A. Mycoflora of Stored Rice in Portugal. In *Proceedings of the 9th International*

Working Conference on Stored Product Protection, Campinas, São Paulo, Brazil, Oct 15–18, 2006; Lorini, I., Bacaltchuk, B., Beckel, H., Deckers, D., Sundfeld, E., dos Santos, J. P., Biagi, J. D., Celaro, J. C., Faroni, L. R. D'A., Bortolini, L. de O. F., Sartori, M. R., Elias, M. C., Guedes, R. N. C., da Fonseca, R. G., Scussel V. M., Eds.; Brazilian Post-harvest Association—ABRAPOS, Passo Fundo: RS, Brazil, 2006. (ISBN 8560234004). pp 128–134.

Maragkoudakisa, P. A.; Miarisa, C.; Rojeza, P.; Manalisb, N.; Magkanarib, F.; Kalantzopoulosa, G.; Tsakalidou, E. Production of Traditional Greek Yogurt using Lactobacillus Strains with Probiotic Potential as Starter Adjuncts. *Int. Dairy J.* **2006,** *16*(1), 52–60.

Marteau, P.; Pochart, P.; Flourie, B.; Pellier, P.; Santos, L.; Desjeux, J. F. Effect of Chronic Ingestion of Fermented Dairy Product Containing *Lactobacillus acidophilus* and *Bifidobacterium bifidum* on Metabolic Activities of the Colonic Flora in Humans. *Am. J. Clin. Nutr.* **1990,** *52,* 685–688.

Mårtensson, O.; Öste, R.; Holst, O. The Effect of Yoghurt Culture on the Survival of Probiotic Bacteria in Oat-Based, Non-Dairy Products. *Food Res. Int.* **2002,** *35*(8), 775–784.

Martin-Diana, A. B.; Janer, C.; Pelaez, C.; Requena, T. Development of a Fermented Goat's Milk Containing Probiotic Bacteria. *Int. Dairy J.* **2003,** *13*(10), 827–833.

Martins, E. M. F.; Ramos, A. M.; Vanzela, E. S. L.; Stringheta, P. C.; de Oliveira Pinto, C. L.; Martins, J. M. Products of Vegetable Origin: A New Alternative for the Consumption of Probiotic Bacteria. *Food Res. Int.* **2013,** *51*(2), 764–770.

Matsushita, K.; Fujii, Y.; Ano, Y.; Toyama, H.; Shinjoh, M.; Tomiyama, N.; Miyazaki, T.; Sugisawa, T.; Hoshino, T.; Adachi, O. 5-Keto-D-Gluconate Production is Catalyzed by a Quinoprotein Glycerol Dehydrogenase, Major Polyol Dehydrogenase in *Gluconobacter* Species. *Appl. Environ. Microbiol.* **2003,** *69,* 1959–1966.

Mayo, B.; van Sinderen, D. *Bifidobacteria: Genomics and Molecular Aspects;* Caister Academic Press, 2010; p 260.

McMaste, L. D.; Kokott, S. J.; Abratt, V. R. Use of Traditional African Fermented Beverages as Delivery Vehicles for *Bifidobacterium lactis* DSM 10140. *Int. J. Food Microbiol.* **2005,** *102*(2), 231–237.

McMeekin, T. A.; Brown, J.; Krist, K.; Miles, D.; Neumeyer, K.; Nichols, D.S.; Olley, J.; Presser, K.; Ratkowsky, D. A.; Ross, T.; Salter, M.; Soontranon, S. Quantitative Microbiology: A Basis for Food Safety. *Emerging Infect. Dis. J.* **1997,** *3*(4), 541–549.

Méndez, A.; Pérez, C.; Montañéz, J. C.; Martínez, G.; Aguilar, C. N. Red Pigment Production by *Penicillium purpurogenum* GH2 is Influenced by pH and Temperature. *J. Zhejiang Univ. Sci. B* **2011,** *12*(12), 961–968.

Meroth, C. B.; Walter, J.; Hertel, C.; Brandt, M. J.; Hammes, W. P. Monitoring the Bacterial Population Dynamics in Sourdough Fermentation Processes by using PCR-Denaturing Gradient Gel Electrophoresis. *Appl. Environ. Microbiol.* **2003,** *69,* 475–482.

Mižáková, A.; Pipová, M.; Turek, P. The Occurrence of Moulds in Fermented Raw Meat Products. *Czech J. Food Sci.* **2002,** *20*(3), 89–94.

Mortazavian, A. M.; Mohammadi, R.; Sohrabvandi, S. Delivery of Probiotic Microorganisms into Gastrointestinal Tract by Food Products. In *New Advances in the Basic and Clinical Gastroenterology;* Brzozowski, T., Ed.; InTech Pub.: London, 2012; pp 121–146.

Mortazavian. A. M.; Ehsani, M. R.; Azizi, A.; Razavi, S. H.; Mousavi, S. M.; Sohrabvandi, S.; Reinheimer, J. A. Viability of Calcium-Alginate-Microencapsulated Probiotic Bacteria in Iranian Yogurt Drink (DOOGH) During Refrigerated Storage and under Simulated Gastrointestinal Conditions. *Aust. J. Dairy Technol.* **2008,** *63,* 24–29.

Muianja, C. M. B.; Narvhus, J. A.; Treimo, J.; Langsrud, T. Isolation, Characterisation and Identification of Lactic Acid Bacteria from Bushera: A Ugandan Traditional Fermented Beverage. *Int. J. Food Microbiol.* **2003,** *80*(3), 201–210.

Mustafa, S.; Shaborin, A.; Kabeir, B. M.; Yazid, A. M.; Hakim, M. N.; Khahtanan, A. Survival of *Bifidobacterium pseudocatenulatum* G4 During the Storage of Fermented Peanut Milk (PM) and Skim Milk (SM) Products. *Afr. J. Food Sci.* **2009,** *3*(6), 150–155.

Muthukumarasamy, P.; Holley, R. A. Microbiological and Sensory Quality of Dry Fermented Sausages Containing Alginate-Microencapsulated *Lactobacillus reuteri. Int. J. Food Microbiol.* **2006,** *111,* 164–169.

Muthukumarasamy, P.; Holley, R. A. Survival of *Escherichia coli* O157:H7 in Dry Fermented Sausages Containing Micro-Encapsulated Probiotic Lactic Acid Bacteria. *Food Microbiol.* **2007,** *24,* 82–88.

Nazzaro, F.; Fratinni, F.; Coppola, R.; Sada, A.; Orlando, P. Fermentative Ability of Alginate Prebiotic Encapsulated *Lactobacillus acidophilus* and Survival under Simulated Gastrointestinal Conditions. *J. Funct. Foods* **2009,** *1*(3), 319–323.

Nebesny, E.; Żyżelewicz, D.; Moty, I. Dark Chocolates Supplemented with *Lactobacillus* Strains. *J. Eur. Food Res. Technol.* **2007,** *225,* 33–42.

Nevarez, L.; Vasseur, V.; Le Drean, G.; Tanguy, A.; Guisle-Marsollier, I.; Houlgatte, R.; Barbier, G. Isolation and Analysis of Differentially Expressed Genes in *Penicillium glabrum* Subjected to Thermal Stress. *Microbiology* **2008,** *154,* 3752–3765.

Niamah, A. K. Production of Pediocin from a Local Isolate of *Pediococcus acidilactici* and Studying its Properties and using it in Some Foods. Ph.D. Thesis, College of Agriculture, Basrah University, Basrah, Iraq. 2010, p 179.

Oh, Y. J.; Jung, D. S. Evaluation of Probiotic Properties of *Lactobacillus* and *Pediococcus* Strains Isolated from Omegisool, a Traditionally Fermented Millet Alcoholic Beverage in Korea. *LWT-Food Sci. Technol.* **2015,** *63*(1), 437–444.

Oi, Y.; KIitabatake, N. Chemical Composition of an East African Traditional Beverage, Togwa. *J. Agric. Food Chem.* **2003,** *51,* 7024–7028.

Olson, D. W.; White, C. H.; Watson, C. E. Properties of Frozen Dairy Desserts Processed by Microfluidization of their Mixes. *J. Dairy Sci.* **2003,** *86,* 1157–1162.

Ong, L.; Shah, N. P. Probiotic Cheddar Cheese: Influence of Ripening Temperatures on Survival of Probiotic Microorganisms, Cheese Composition and Organic Acid Profiles. *LWT-Food Sci. Technol.* **2009,** *42,* 1260–1268.

Ortakci, F.; Broadbent, J. R.; Mcmanus, W. R.; and Mchahon, D. J. Survival of Microencapsulated Probiotic *Lactobacillus paracasei* LBC-1e During Manufacture of Mozzarella Cheese and Simulated Gastric Digestion. *J. Dairy Sci.* **2012,** *95*(11), 6274–6281.

Özer, B.; Kirmaci, H. A.; Senel, E.; Atamer, M.; Hayaloğlu, A. Improving the Viability of *Bifidobacterium bifidum* BB-12 and *Lactobacillus acidophilus* LA-5 in White-Brined Cheese by Microencapsulation. *Int. Dairy J.* **2008,** *19,* 22–29.

Öztürk B. A.; Öner, M. D. Production and Evaluation of Yogurt with Concentrated Grape Juice. *J. Food Sci.* **1999,** *64*(3), 530–532.

Papamanoli, E.; Tzanetakis, N.; Litopoulou-Tzanetaki, E.; Kotzekidou, P. Characterization of Lactic Acid Bacteria Isolated from a Greek Dry Fermented Sausage in Respect of their Technological and Probiotic Properties. *Meat Sci.* **2003**, *65*, 859–867.

Passoth, V.; Blomqvist, J.; Schnürer, J. *Dekkera bruxellensis* and *Lactobacillus vini* Form A Stable Ethanol-Producing Consortium in a Commercial Alcohol Production Process. *Appl. Environ. Microbiol.* **2007**, *73*(13), 4354–4356.

Patel, A.; Lindström, C.; Patel, A.; Prajapati, J. B.; Holst, O. Screening and Isolation of Exopolysaccharide Producing Lactic Acid Bacteria from Vegetables and Indigenous Fermented Foods of Gujarat, India. *Int. J. Fermented Foods* **2012**, *1*(1), 77–86.

Patel, A.; Prajapati, J. B. Food and Health Applications of Exopolysaccharides Produced by Lactic Acid Bacteria. *Adv. Dairy Res.* **2013**, *1*, 107.

Patel, A.; Prajapati, J. B. Partial Characterization of Exopolysaccharides Obtained from Novel Isolates of *Weissella* Spp. *Indian J. Dairy Sci.* **2016**, *69*(3), 310–315.

Patel, A.; Shah, N. Recent Advances in Antimicrobial Compounds Produced by Food Grade Bacteria in Relation to Enhance Food Safety and Quality. *J. Innovative Biol.* **2014**, *1*(4), 189–194.

Patel, A.; Shah, N.; Ambalam, P.; Prajapati, J. B.; Holst, O.; Ljungh, A. Antimicrobial Profile of Lactic Acid Bacteria Isolated from Vegetables and Indigenous Fermented Foods of India Against Clinical Pathogens using Microdilution Method. Biomed. Environ. Sci. 2013, 26(9), 759–764.

Patel, A.; Shah, N. Bioactive Components of Fermented Foods. In *Fermented Foods, Part I: Biochemistry and Biotechnology;* Montet, D., Ray, R. C., Eds.; CRC Press, 2016; pp 282–303.

Patra, F.; Tomar, S. K.; Arora, S. Technological and Functional Applications of Low-Calorie Sweeteners from Lactic Acid Bacteria. *J. Food Sci.* **2009**, *74*, 16–21.

Patrignani, F.; Burns, P.; Serrazanetti, D.; Vinderola, G.; Reinheimer, J.; Lanciotti, R.; Guerzoni, M. E. Suitability of High Pressure-Homogenized Milk for the Production of Probiotic Fermented Milk Containing *Lactobacillus paracasei* and *Lactobacillus acidophilus*. *J. Dairy Res.* **2009**, *76*(1), 74–82.

Pedroso, D. L.; Dogenski, M.; Thomazini, M.; Heinemann, R. J. B.; Favaro-Trindade, C. S. Microencapsulation of *Bifidobacterium animalis* Subsp. Lactis and *Lactobacillus acidophilus* in Cocoa Butter using Spray Chilling Technology. *Braz. J. Microbiol.* **2013**, *44*(3), 777–783.

Pérez-Chabela, M. L.; Díaz-Vela, J.; Reyes-Menéndez, C. V.; Totosaus, A. Improvement of Moisture Stability and Textural Properties of Fat and Salt Reduced Cooked Sausages by Inoculation of Thermotolerant Lactic Acid Bacteria. *Int. J. Food Prop.* **2013**, *16*(8), 1789–1808.

Perrone, G.; Susca, A.; Cozzi, G.; Ehrlich, K.; Varga, J.; Frisvad, J. C.; Meijer, M.; Noonim, P.; Mahakarnchanakul, W.; Samson, R. A. Biodiversity of *Aspergillus* Species in Some Important Agricultural Products. *Stud. Mycol.* **2007**, *59*, 53–66.

Possemiers, S.; Marzorati, M.; Verstraete, W.; de Wiele, T. V. Bacteria and Chocolate: A Successful Combination for Probiotic Delivery. *Int. J. Food Microbiol.* **2010**, *141*, 97–103.

Prado, F. C. J. L; Parada, J. C.; Carvalho, C.; Soccol, R. Isolation and Characterization of Lactic Acid Bacteria from Green Coconut Microbiota for us in Non-Dairy Probiotic

Beverage. 18th International Congress of Chemical and Process Engineering. Proceedings: 18th ICCPE. Praga: CHISA 2008. CD. 2008, pp 1–2.

Prescott, L. M.; Harley, J. P.; Klein, D. A. Microbiology of Food. In *Microbiology*, 5th ed.; Prescott, L. M., Harley, J. P., Klein, D. A., Eds.; McGraw–Hill: New York City, 2002; pp 963–990.

Princely, S.; Basha, N. S.; Kirubakaran, J. J.; Dhanaraju, M. D. Biochemical Characterization, Partial Purification, and Production of an Intracellular Beta-Galactosidase from *Streptococcus thermophilus* Grown in Whey. *Eur. J. Exp. Biol.* **2013,** *3*(2), 242–251.

Rathore, S.; Salmerón, I.; Pandiella, S. S. Production of Potentially Probiotic Beverages using Single and Mixed Cereal Substrates Fermented with Lactic Acid Bacteria Cultures. *Food Microbiol.* **2012,** *30*(1), 239–244.

Rebucci, R.; Sangalli, L.; Fava, M.; Bersani, C.; Cantoni, C.; Baldi, A. Evaluation of Functional Aspects in *Lactobacillus* Strains Isolated from Dry Fermented Sausages. *J. Food Qual.* **2007,** *30*(2), 187–201.

Rehman, S.; Nawaz, H.; Hussain, S.; Ahmad, M. M.; Murtaza, M. A.; Ahmad, M. S. Effect of Sourdough Bacteria on the Quality and Shelf Life of Bread. *Pak. J. Nutr.* **2007,** *6*(6), 562–565.

Renuka. B.; Kulkarni, S. G.; Vijayanand, P.; Prapulla, S. G. Fructooligosaccharide Fortification of Selected Fruit Juice Beverages: Effect on the Quality Characteristics. *LWT—Food Sci. Technol.* **2009,** *43*(5), 1031–1033.

Ricciardi, A.; Parente, E.; Zotta, T. Modelling the Growth of *Weissella cibaria* as a Function of Fermentation Conditions. *J. Appl. Microbiol.* **2009,** *107,* 1528–1535.

Rivera-Espinoza, Y.; Gallardo-Navarro, Y. Non-Dairy Probiotic Products. *Food Microbiol.* **2010,** *27,* 1–11.

Roberts, J. S.; Kidd, D. R. Lactic Acid Fermentation of Onions. *LWT—Food Sci. Technol.* **2005,** *38*(2), 185–190.

Rollán, G.; Gerez, C. L.; Dallagnol, A. M.; Torino, M. I.; Font, G. Update in Bread Fermentation by Lactic Acid Bacteria. In *Current Research, Technology and Education Topics in Applied Microbiology and Biotechnology;* Méndez-Vilas, A., Ed.; Formatex Research Center: Spain, 2010; Vol. 2, pp 1168–1174.

Rozada-Sanchez, R.; Sattur, A. P.; Thomas, K.; Pandiella, S. S. Evaluation of *Bifidobacterium* spp. for the Production of a Potentially Probiotic Malt-Based Beverage. *Process Biochem.* **2008,** *43,* 848–854.

Rubio, R.; Martín, B.; Aymerich, T.; Garriga, M. The Potential Probiotic *Lactobacillus rhamnosus* CTC1679 Survives the Passage Through the Gastrointestinal Tract and its use as Starter Culture Results in Safe Nutritionally Enhanced Fermented Sausages. *Int. J. Food Microbiol.* **2014,** *186,* 55–60.

Saarela, M.; Virkajarvi, I.; Alakomi, H. L. Stability and Functionality of Freeze-Dried Probiotic Bifidobacterium Cells During Storage in Juice and Milk. *Int. Dairy J.* **2006,** *16*(12), 1477–1482.

Säde, E. *Leuconostoc* Spoilage of Refrigerated, Packaged Foods. Ph.D. Thesis, Faculty of Veterinary Medicine, University of Helsinki, Helsinki, Finland. 2011, pp 57.

Saha, P.; Banerjee, S. Optimization of Process Parameters for Vinegar Production using Banana Fermentation. *Int. J. Res. Eng. Technol.* **2013,** *2*(9), 501–514.

Samson, R. A.; Hoekstra, E. S.; Frisvad, J. C.; Filtenborg, O. *Introduction to Food and Airborne Fungi,* 7th Ed.; Centraalbureau voor Schimmelcultures (CBS): Utrecht, Netherlands, 2004.

Sankhavadhana, P. Development of Yoghurt-Like Fermented Milk Product using Probiotic Cultures. MS Thesis. Chiang Mai University, Thailand, 2001.

Şanlibaba, P.; Akkco, N.; Akcelik, M. Identification and Characterization of Antimicrobial Activity of Nisin a Produced by *Lactococcus lactis* Subsp. *Lactis* LL27. *Czech J. Food Sci.* **2009,** *27*(1), 55–64.

Saris, P. E. J.; Beasley, S.; Tourila, H. Fermented Soymilk with a Monoculture of *Lactococcus lactis. Int. J. Food Microbiol.* **2003,** *81,* 159–162.

Shafique, S.; Asgher, M.; Sheikh, M. A.; Asad, M. J. Solid State Fermentation of Banana Stalk for Exoglucanase Production. *Int. J. Agric. Biol.* **2004,** *3,* 488–491.

Shah, N.; Patel, A.; Ambalam, P.; Holst, O.; Ljungh, A.; Prajapati, J. Determination of an Antimicrobial Activity of *Weissella confusa, Lactobacillus fermentum,* and *Lactobacillus plantarum* against Clinical Pathogenic Strains of *Escherichia coli* and *Staphylococcus aureus* in Co-Culture. *Ann. Microbiol.* **2016,** *66,* 1137–1143.

Sheehan, V. M.; Ross, P.; Fitzgerald, G. F. Assessing the Acid Tolerance and the Technological Robustness of Probiotic Cultures for Fortification in Fruit Juices. *Innov. Food Sci. Emerg.Technol.* **2007,** *8*(2), 279–284.

Shobharani, P.; Agrawal, R. Supplementation of Adjuvants for Increasing the Nutritive Value and Cell Viability of Probiotic Fermented Milk Beverage. *Int. J. Food Sci. Nutr.* **2009,** *60*(S6), 70–83.

Shurtleff, W.; Aoyagi, A. *The Book of Tempeh;* Ten Speed Press: Berkeley, California, 2001.

Silveira, A. C.; Conesa, A.; Aguayo, E.; Artes, F. Alternative Sanitizers to Chlorine for use on Freshcut "Galia" (*Cucumis melo* var. *catalupensis*) Melon. *J. Food Sci.* **2008,** *73,* 405–411.

Singh, B. K.; Campbell, C. D.; Sorenson, S. J.; Zhou, J. Soil Genomics. *Nat. Rev. Microbiol.* **2009,** *7*(10), 756.

Sivasankari, S.; Vinotha, T. In-Vitro Degradation of Plastics (Plastic Cup) using *Micrococcus luteus* and *Masoniella* Sp. *Scholars Acad. J. Biosci.* **2014,** *2*(2), 85–89.

Slivinski, C. T.; Machado, A. V. L.; Iulek, J.; Ayub, R. A.; de Almeida, M. M. Biochemical Characterisation of a Glucoamylase from *Aspergillus niger* Produced by Solid-State Fermentation. *Braz. Arch. Biol. Technol.* **2011,** *54*(3), 559–568.

Sneath, P. H. A.; Mair, N. S.; Sharpe, M. E.; Holt, J. G. *Bergey's Manual of Systematic Bacteriology;* Williams and Wilkins: Baltimore, Maryland, USA, 1986; Vol. 2, pp 1105.

Sohrabvandi, S.; Razavi, S. H.; Mousavi, S. M.; Motazavian, A. M. Viability of Probiotic Bacteria in Low-Alcohol-and Non-Alcoholic Beer During Refrigerated Storage. *Philipp. Agric. Sci.* **2010,** *93,* 24–28.

Souza, C. H. B.; Saad, S. M. I. Viability of *Lactobacillus acidophilus* La-5 Added Solely or in Co-Culture with a Yogurt Starter Culture and Implications on Physicochemical and Related Properties of Minas Fresh Cheese During Storage. *LWT-Food Sci. Technol.* **2009,** *42,* 633–640.

Stackebrandt, E.; Cummins, C. S.; Johnson, J. L. Family Propionibacteriacea: The Genus *Propionibacterium.* In *The Prokaryotes: A Handbook on the Biology of Bacteria;* Dworkin, M., Falkow, S., Rosemberg, E., Schleifer, K. H., Stackebrandt, E., Eds.; Springer: Singapur, 2006; pp 400–418.

Steels, H.; James, S. A.; Roberts, I. N.; Stratford, M. Zygosaccharomyces Lentus: A Significant New Osmophilic, Preservative-Resistant Spoilage Yeast, Capable of Growth at Low Temperature. *J. Appl. Microbiol.* **1999**, *87*(4), 520–527.

Suh, S. O.; Gujjari, P.; Beres, C.; Beck, B.; Zhou, J. Proposal of *Zygosaccharomyces parabailii* Sp. Nov. and *Zygosaccharomyces pseudobailii* Sp. Nov., Two New Species Closely Related to *Zygosaccharomyces bailii*. *Int. J. Syst. Evol. Microbiol.* **2013**, *63*, 1922–1929.

Supavititpatana, P.; Wirjantoro, T. I.; Arunee Apichartsrangkoon, A.; Raviyan, P. Addition of Gelatin Enhanced Gelation of Corn–Milk Yogurt. *Food Chem.* **2008**, *106*, 211–216.

Suzuki, M.; Yamamoto, T.; Kawai, Y.; Inoue, N.; Yamazaki, K. Mode of Action of Piscicocin CS526 Produced by *Carnobacterium piscicola* CS526. *J. Appl. Microbiol.* **2005**, *98*(5), 1146–1151.

Svahn, S. K.; Chryssanthou, E.; Olsen, B.; Bohlin, L.; Göransson, U. *Penicillium nalgiovense* Laxa Isolated from Antarctica is a New Source of the Antifungal Metabolite Amphotericin B. *Fungal Biol. Biotechnol.* **2015**, *2*, 1.

Swings, J. The Genera *Acetobacter* and *Gluconobacter*. In *The Prokaryotes;* Balows, A., Truper, H. G., Dworkin, M., Harder, W., Schleifer, K.-H., Ed.; Springer Verlag: New York, Vol. III, 1991; pp 2268–2286.

Tamime, A. Y.; Robinson, R. K. *Yoghurt Science and Technology,* 2nd Ed.; TJ. International: Cornwall, UK, 1999.

Taneja, N.; Rani, P.; Emmanue, R.; Khudaier, B. Y.; Sharma, S. K.; Tewari, R.; Sharma, M. Nosocomial Urinary Tract Infection Due to *Leuconostoc mesenteroides* at a Tertiary Care Center in North India. *Indian J. Med. Res.* **2005**, *122*, 178–179.

Taware, R, Abnave, P.; Patil, D.; Rajamohananan, P.; Raja, R.; Soundararajan, G.; Kundu, G.; Ahmad, A. Isolation, Purification and Characterization of Trichothecinol-A Produced by Endophytic Fungus *Trichothecium* sp. and its Antifungal, Anticancer and Antimetastatic Activities. Sustainable Chem. Processes **2014**, *2*(1), 8.

Tenge, C.; Geiger, E. Alternative Fermented Beverages-Functional Drinks. *MBAA Tech. Q.* **2001**, *38,* 33–35.

Thage, B. V.; Broe, M. L.; Petersen, M. H.; Petersen, M. A.; Bennedsen, M. A.; Ardö, Y. Aroma Development in Semi-Hard Reduced-Fat Cheese Inoculated with *Lactobacillus paracasei* Strains with Different Aminotransferase Profiles. *Int. Dairy J.* **2005**, *15*, 795–805.

Tsen, J. H.; Lin, Y. P.; King, V. Response Surface Methodology Optimisation of Immobilised Lactobacillus Acidophilus Banana Puree Fermentation. *Int. J. Food Sci. Technol.* **2009**, *44*(1), 120–127.

Tsisaryk, O.; Musiy, L.; Golubets, O.; Shkaruba, S. Fatty Acid Composition of Cultured Butter with Probiotic Lbc. Acidophilus La-5 Produced in Winter Time. Conference paper: 2014 ADSA-ASAS-CSAS Joint Annual Meeting, 2014.

Uccello, M.; Malaguarnera, G.; Basile, F.; D'agata, V.; Malaguarnera, M.; Bertino, G.; Vacante, M.; Drago, F.; Biondi, A. Potential Role of Probiotics on Colorectal Cancer Prevention. *BMC Surg.* **2012**, *12*(Suppl 1), S35.

van Beek, S.; Priest, F. G. Evolution of the Lactic Acid Bacterial Community During Malt Whisky Fermentation: A Polyphasic Study. *Appl. Environ. Microbiol.* **2002**, *68*, 297–305.

Verma, D. K.; Srivastav, P. P. *Microorganisms in Sustainable Agriculture, Food and the Environment.* as part of book series on *Innovations in AGRICULTURAL MICROBIOLOGY*; Verma, D. K., Srivastav, P. P., Eds.; Apple Academic Press: USA, 2017.

Verma, D. K.; Mahato, D. K.; Billoria, S.; Kapri, M.; Prabhakar, P. K.; Ajesh Kumar, V.; Srivastav, P. P. Microbial Approaches in Fermentations for Production and Preservation of Different Food. In *Microorganisms in Sustainable Agriculture, Food and the Environment, as part of book series on Innovations in Agricultural Microbiology;* Verma, D. K., Srivastav, P. P., Eds.; Apple Academic Press: USA, 2017a.

Verma, D. K.; Mahato, D. K.; Billoria, S.; Kapri, M.; Prabhakar, P. K.; Ajesh Kumar, V.; Srivastav, P. P. Microbial Spoilage in Milk Products, Potential Solution, Food Safety and Health Issues. In *Microorganisms in Sustainable Agriculture, Food and the Environment. as part of book series on Innovations in AGRICULTURAL MICROBIOLOGY;* Verma, D. K., Srivastav, P. P., Eds.; Apple Academic Press: USA, 2017b.

Vinderola, C. G.; Reinheimer , J. R. lactic Acid Bacteria: A Comparative "In-Vitro" Study of Probiotic Characteristics and Biological Barrier Resistance. *Food Res. Int.* **2003,** *36,* 895–904.

Vinderola, C. G.; Prosello, W.; Ghiberto, T. D.; Reinheimer, J. A. Viability of Probiotic (*Bifidobacterium, Lactobacillus acidophilus and Lactobacillus casei*) and Non-Probiotic Microflora in Argentinean Fresco Cheese. *J. Dairy Sci.* **2000a,** *83,* 1905–1911.

Vinderola, C. G.; Bailo, N.; Reinheimer, J. A. Survival of Probiotic Microflora in Argentinian Yoghurts During Refrigerated Storage. *Food Res. Int.* **2000b,** *33,* 97–102.

Vlkova, E.; Medkova, J.; Rdad, V. Comparison Four Method Identification of Bifidobacteria to the Genus Level. *Czech J. Food Sci.* **2002,** *20*(5), 171–174.

Wang, C. Y.; Wu, S. J.; Shyu, Y. T. Antioxidant Properties of Certain Cereals as Affected by Food-Grade Bacteria Fermentation. *J. Biosci. Bioeng.* **2014,** *117*(4), 449–456.

Wauters, G.; Haase, G.; Avesani, V.; Charlier, J.; Janssens, M.; Broeck, G. V.; Delmee, M. Identification of a Novel *Brevibacterium* Species Isolated from Humans and Description of *Brevibacterium sanguinis* sp. nov. *J. Clin. Microbiol.* **2004,** *42*(6), 2829–2832.

Welch, T. J.; Good, C. M. Mortality Associated with Weissellosis (*Weissella* sp.) in USA Farmed Rainbow Trout: Potential for Control by Vaccination. *Aquaculture* **2013,** 388–391, 122–127.

Whitman, W. B.; Coleman, D. C.; Wiebe, W. J. Prokaryotes: The Unseen Majority. *Proc. Natl. Acad. Sci. USA.* **1998,** *95,* 6578–6583.

Willey, J. M.; Sherwood, L. M.; Woolverton, C. J. Microbiology of Food. In *Prescott, Harley, and Klein's Microbiology,* 7th Ed.; McGraw–Hill, New York, 2008; p 1023.

Wolter, A.; Hager, A. S.; Zannini, E.; Galle, S.; Gänzle, M.; Waters, D. M.; Arendta, E. K. Evaluation of Exopolysaccharide Producing *Weissella cibaria* Mg1strain for the Production of Sourdough from Various Flours. *Food Microbiol.* **2014,** *37,* 44–50.

Yamada, Y.; Yukphan, P. Genera and Species in Acetic Acid Bacteria. *Int. J. Food Microbiol.* **2007,** *125,* 15–24.

Yangılar, F.; Özdemir, S. Microbiological Properties of Turkish Beyaz Cheese Samples Produced with Different Probiotic Cultures. *Afr. J. Microbiol. Res.* **2013,** *7*(22), 2808–2813.

Yerlikaya, O.; Ozer, E. Production of Probiotic Fresh White Cheese using Co-Culture with *Streptococcus thermophilus. Food Sci. Technol. (Campinas),* **2014,** *34*(3), 471–477.

Yoon, K. Y.; Woodamns, E. E.; Hang, Y. D. Probiotication of Tomato Juice by Lactic Acid Bacteria. *J. Microbiol.* **2004,** *42,* 315–318.

Yoon, K. Y.; Woodams, E. E.; Hang, Y. D. Production of Probiotic Cabbage Juice by Lactic Acid Bacteria. *Bioresour. Technol.* **2006,** *97*(12), 1427–1430.

Zárate, G. Dairy Propionibacteria: Less Conventional Probiotics to Improve the Human and Animal Health. In *Probiotic in Animals;* Rigobelo, E. C., Ed.; InTech Publishers: London, 2012; pp 153–202.

Zhang, J.; Liu, G.; Li, P.; Qu, Y. Pentocin 31–1, a Novel Meat-Borne Bacteriocin and its Application as Biopreservative in Chill-Stored Tray-Packaged Pork Meat. *Food Control* **2010,** *21,* 198–202.

Zinedine, A.; Faid, M.; Benlemlih, M. In-Vitro Reduction of Aflatoxin B1 by Strains of Lactic Acid Bacteria Isolated from Moroccan Sourdough Bread. *Int. J. Agric. Biol.* **2005,** *7*(1), 67–71.

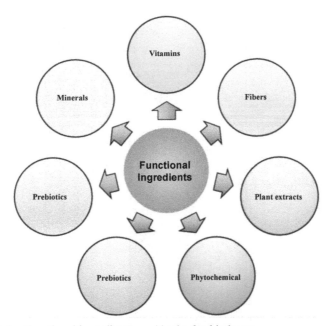

FIGURE 1.1   Functional ingredients used in the food industry.

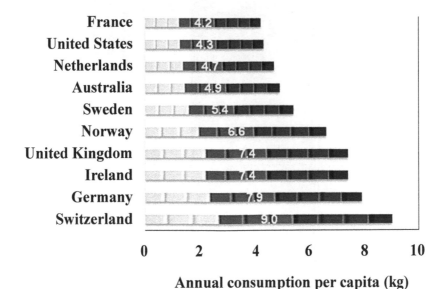

FIGURE 1.3   Annual chocolate consumption of some countries per capita.
Source: Adapted from Ascarelli (2015).

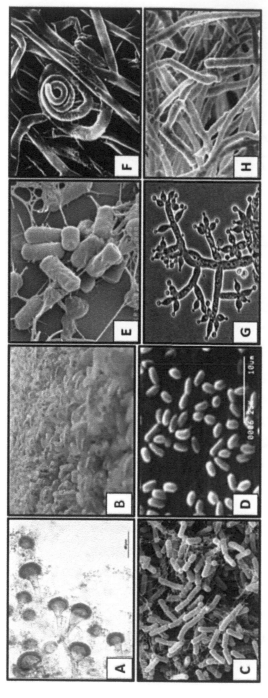

FIGURE 3.1 Microscopic image of microorganisms. (a) *Aspergillus niger*, (b) *Aspergillus oryzae*, (c) *Bacillus amyloliquefaciens*, (d) *Bacillus licheniformis*, (e) *Bacillus subtilis*, (f) *Streptomyces lividans*, (g) *Trichoderma harzianum*, and, (h) *Trichoderma reesei*.

Source: Adapted from a: http://medicinembbs.blogspot.in/2013/02/microscopic-morphology-of-aspergillus.html; b: https://en.wikipedia.org/wiki/Aspergillus_oryzae; c: http://www.gettyimages.in/detail/photo/bacillus-amyloliquefaciens-is-a-gram-high-res-stock-photography/128628136; d: http://www.sciencephoto.com/media/13170/view; e: Reprinted from Zweers et al. (2008). ©Zweers et al; licensee BioMed Central Ltd. 2008. https://creativecommons.org/licenses/by/2.0/; f: (to come); g: http://www.mycolog.com/CHAP4a.htm; h: Public domain.

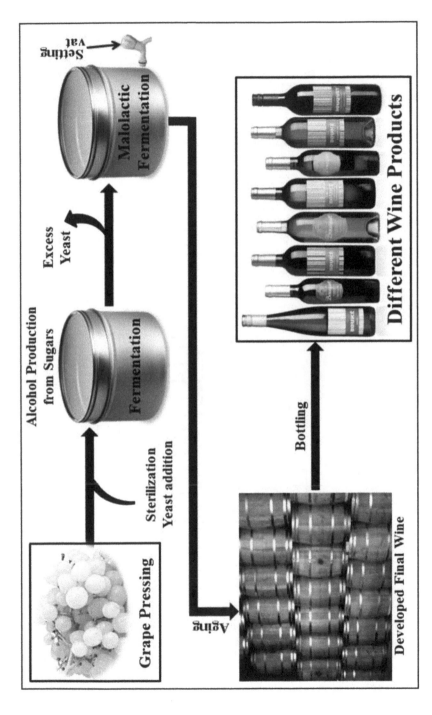

FIGURE 4.3 Process of wine production from grapes (Willey et al. 2008).

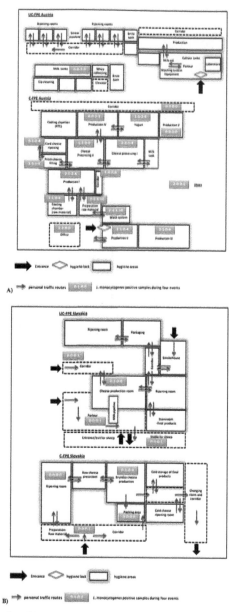

FIGURE 6.1 Example of *Listeria monocytogenes* spreading in Austrian industrial dairy (A) and Slovakian farmhouse cheese making (B) food processing environments.

Reprinted with permission from Muhterem-Uyar, M.; Dalmasso, M.; Bolocan, A. S.; Hernandez, M.; Kapetanakou, A. E.; Kuchta, T.; Manios, S. G.; Melero, B. and Minarovičová, J. Environmental sampling for Listeria monocytogenes control in food processing facilities reveals three contamination scenarios. Food Control, 2015, 51, 94–107. © 2015 Elsevier.

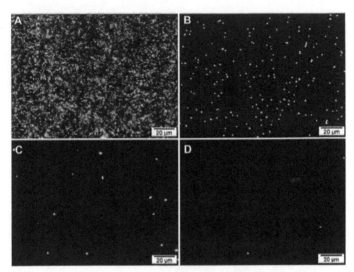

FIGURE 6.3 Vital status of *L. monocytogenes* in biofilms on stainless steel surfaces treated with (A) 5 log·pfu/ml phage P100, (B) 6 log·pfu/ml phage P100, (C) 7 log· pfu/ml phage P100, and (D) 8 log· pfu/ml of phage P100 at 48 h. Epifluorescence digital images of the stain with LIVE/DEAD.

Sources: Reprinted from Montañez-Izquierdo et al. (2012).

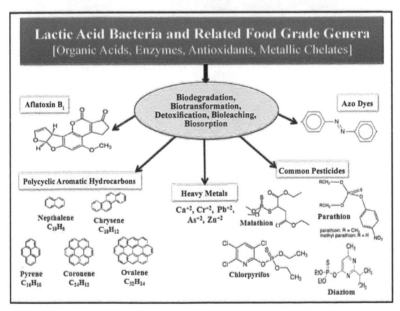

FIGURE 7.1 Schematic representation of various mechanisms of food-grade bacteria for biological control of toxic compounds.

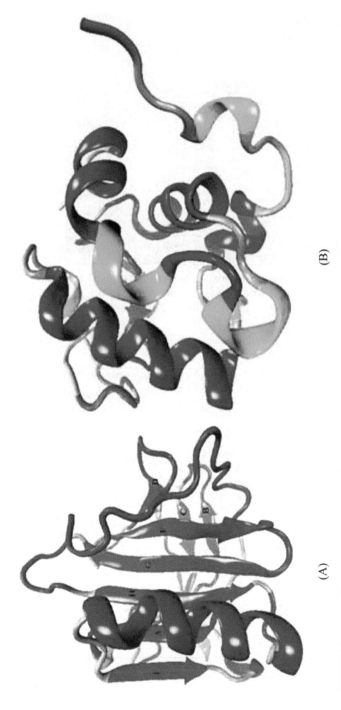

(A)

(B)

FIGURE 8.1 Details on the structure of (a) bovine β-LG (model 3NPO.pdb from RCSB protein data bank) and (b) α-LA (1F6S.pdb). (The proteins are represented in new cartoon style using VMD software [Humphrey et al., 1996]. The secondary structure elements are represented as follows: α-helices in purple; 3–10 helices in green; extended β-sheets in violet; turns in red; and coils in pink.)

Source: (a) Adapted from Loch et al. (2011) and (b) adapted from Chrysina et al. (2000).

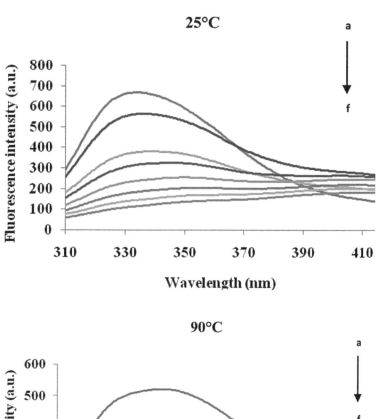

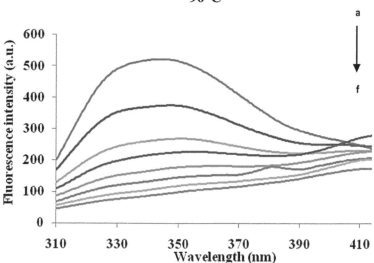

FIGURE 8.2   The emission spectra of the β-LG and sour cherries extract (the anthocyanins concentration in the extract [from a–f] varied from 0 to 0.093 μm cyanidin glycoside equivalents).

Source: Adapted from Oancea et al. (2017) (with permission from Taylor & Francis).

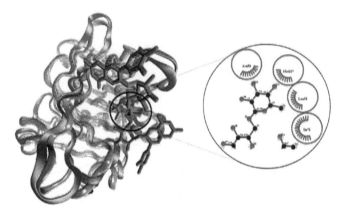

FIGURE 8.3 Superposition of the β-LG-CYR docking models at different temperatures. The image was prepared using the VMD software. The protein at 25 and 90°C is represented in new ribbons style in gray and blue, respectively, whereas, the ligand is represented in orange and red, respectively, in licorice style. In detail are marked the amino acids located in the hydrophobic binding site which are in direct contact with atoms of the CYR at both temperatures. The image presented in the inset was prepared using LigPlot software.

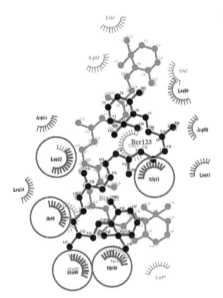

FIGURE 8.4 Superposition of the α-LA–β-carotene (BC) models equilibrated at 25°C (gray) and at 90°C (colored). The hydrophobic contacts directly involving α-LA amino acids are represented by arcs radiating toward the ligand atoms they are in contact with. The amino acids interacting with the BC molecule at both temperatures are circled. The representation was prepared using LigPlot+.

Source: Adapted from Laskowski and Swindells (2011).

CHAPTER 5

PRODUCTION OF EXTRACELLULAR POLYMERIC SUBSTANCES FROM LACTIC ACID BACTERIA: THEIR ANTIMICROBIAL EFFECT AND POTENTIAL APPLICATION IN FOOD INDUSTRY

GOKSEN GULGOR[1,*] and MIHRIBAN KORUKLUOGLU[1,2]

[1]Department of Food Engineering, Faculty of Agriculture, Uludag University, Gorukle, 16059 Bursa, Turkey, Tel.: +90 224 294 15 6

[2]E-mail: mihriban@uludag.edu.tr

*Corresponding author. E-mail: goksengulgor@uludag.edu.tr

5.1 INTRODUCTION

Lactic acid bacteria (LAB) and their metabolites are known as natural bio-preservatives used in food industry. In fermented foods, LAB displays numerous antimicrobial activities (Patel et al., 2017). They are commonly used industrially and their applications could be by a majority due to their metabolites, especially organic acids, which cause rapid acidification of the raw material and also other components which are known as bacteriocins and antibacterial–antifungal peptides. Furthermore, the production of aroma compounds, ethanol, various types of enzymes and exopolysaccharides by them have a remarkable importance for food production (De Vuyst and Leroy, 2007; Verma and Srivastav, 2017).

This chapter is an overview of the main aspects of the general properties of LAB, extracellular polymeric substance (EPS) biosynthesis mechanism, and opportunities to use of EPS and EPS-like secretions biosynthesized by LAB strains. Furthermore, this chapter will also offer an overview of recent researches and studies about metabolites produced by LAB and new usage fields of these metabolites to provide natural improvement of industrial foods.

5.2 LACTIC ACID BACTERIA (LAB): GENERAL INFORMATION

LAB are a group of genera known as nonpathogenic, fermentative, nonspore-forming, Gram-positive, usually catalase-negative bacteria, including *Pediococcus, Lactococcus, Lactobacillus, Leuconostoc, Streptococcus*, and *Weissella* (Mahony et al., 2014).

LAB are generally associated with an environment which is rich in nutrients such as dairy products, meat, beverages, and vegetables. In addition, it has been reported that LAB could be also isolated from several environments such as intestinal tract of animals and humans, soil, lakes, and so forth. LAB have been used for the fermentation of food and feed products since ancient times. Nowadays, the primary applications of LAB strains are still eligible and profitable in the food and feed industry as starter cultures (Wassie and Wassie, 2016; Verma et al., 2017a).

LAB are defined as widespread microorganisms which are members of a heterogeneous family that have the ability to ferment carbohydrates into mainly lactic acid. Lactic acid fermentation is an energy source for LAB which can be classified into two groups as homo- and heterofermentative, according to how they catabolize carbohydrate source. Hexoses are metabolized by glycolytic Embden–Meyerhof pathway for homofermentative LAB genera such as *Lactococcus* spp. and *Pediococcus* sp. More than 90% of the carbohydrate source is converted to lactic acid with the metabolism. Heterofermentative LAB such as *Leuconostoc, Lactobacillus, Weissella*, and so forth, utilize hexoses by Warburg-Dickens pathway. In this pathway, 50% of the carbohydrate source is converted to lactic acid; however, on the other hand, CO_2, some organic acids such as formic acid, acetic acid, and also ethanol are produced during the fermentation period (Kocková et al., 2011; Stoyanova et al., 2012; Saranraj et al., 2013; Sun et al., 2014).

LAB are known as "Generally Regarded as Safe (GRAS)" microorganisms and both the viable cells and their metabolites can be applied to foods as natural additives (Baruah et al., 2016). LAB and their metabolites, especially lactic acid and bacteriocins have been used as natural food preservatives and also EPSs are applied to the food products as thickening/gelling agents because they improve the texture, rheology, mouthfeel, aroma, and odor of many fermented food products (Hassan, 2008; Mostefaoui et al., 2014; Baruah et al., 2016). It is thought that the cell wall composition and structure of LAB have a key role for technological and health applications of LAB in connection with their GRAS status as well as their existence in the human gut microbiota. The cell wall components of LAB can be ordered as peptidoglycan (PG), teichoic acids (TAs), polysaccharides (PSs), and proteins. According to a study about the cell walls of LAB strains (*Lactobacillus rhamnosus, Lactococcus lactis, Lactobacillus casei,* and *Lactobacillus plantarum*), PG variations have roles in taking shape of most important structural and functional properties, such as lysozyme resistance and nisin, inhibition of autolysis, increase in the activity of carboxypeptidase; furthermore, PG structure has an importance in the growth of the bacteria culture. On the other hand, TAs have various essential roles in keeping the cell stable and occurrence of structural–functional properties such as cell morphology, response to UV stress, secretion of protein, nisin resistance, adhesion to epithelial cell surfaces, colonization of intestinal tract, and also recognizing bacteriophages. TAs can bind cations such as Mg^{2+} and protons which create pH balance on the cell wall; therefore, the enzymatic activities could be enhanced with TAs activation. PSs have immunosuppressive function and they play a role in biofilm formation, decreasing adhesion, protection against antimicrobial peptides, cell morphology, and recognizing bacteriophages. Proteins vary in structure as well as function of cells (Chapot-Chartier and Kulakauskas, 2014).

In addition, LAB are commonly used as probiotic in the food industry. Probiotic LAB strains can attach to an intestinal system and proliferate easily on the surface of intestinal epithelium cells. The definition of the term "probiotic" is approved as "live micro-organisms which when administered in adequate amounts, confer a health benefit on the host" (Arroyo-Lopez et al., 2012; Reis et al., 2016). Probiotic LAB colonize and predominate in the newborn's intestine due to their physiological characteristics. Besides, growth promoters are found in human milk such as oligosaccharides which are called "prebiotics" (a nondigestible food ingredient that promotes the growth of beneficial microorganisms in the intestines). Therefore, the probiotic effect of LAB could be observed in the intestinal system of breastfed infants (Reis et al., 2016).

LABs are associated with fermentation process under anaerobic conditions. However, it is indicated that some of them are capable of growing better under aerobic conditions. This demonstrates that some LABs are genetically coded to grow and proliferate under aerobic conditions as they have an aerobic metabolism. Therefore, breathing process could enhance the performance of LAB to be more effective starter cultures. Furthermore, aerobic conditions may improve the flavor and production of vitamin K2 in some cases. It has been reported that the intracellular pyruvate molecule has a key role in metabolic pathways of LAB; therefore, it is significant in metabolic studies. Furthermore, it facilitates the optimization of LAB. These properties are common for almost all products secreted by LAB. Besides, in many cases, the greater or lesser secretion of different metabolites may depend on the regulation between metabolic fluxes and the pyruvate (Valenzuela et al., 2015).

LABs are mostly isolated from dairy and meat products, alcoholic beverages, some plants, fermented food products, and living organisms. LAB is known as starter culture to produce or/and improve the last product of food; however, on the other hand, it is recognized as spoilage bacteria for some food products such as some meat products, mayonnaise, salad dressing, miso (fermented soybean paste), sake (rice wine), soy sauce, and wine (Kubota et al., 2008; Azcarate-Peril and Klaenhammer, 2010). In the food industry, the transport chain of last product is the most important step to hold the shelf-life period between expected time interval. Some factors including transporting period and storage time/conditions after sale affect the shelf life of the food product. Modified atmosphere and the storage temperatures are two main applications to provide hurdling conditions that create negative selective pressure to aerobic spoilage microorganisms but cold-tolerant, facultative anaerobic LAB prevail in nutrient-rich foods. Organoleptic properties of the food product change more slowly by spoilage LAB, whereas aerobic Gram-negative [G-(−)] bacteria particularly reach the logarithmic growth phase fast. The spoilage of the food product by the LAB is more preferred because the results of spoilage by G-(−) bacteria could be observed earlier (Johansson et al., 2011).

It has been demonstrated that some LAB species form biofilm layer under various conditions. A biofilm is known as a microbial community formed by surface-associated cells embedded in the extracellular polymeric matrix. Microorganisms change their behaviors under inadequate conditions by cooperating with each other to survive in the changing environment. The attachment to the surfaces is followed by a period of growth and thickness of biofilm layer depends on the conditions. Biofilm layer is built by EPS-producing microorganisms. LAB–EPS correlation is the key to biofilm formation and adhesion to the surfaces. Moreover, these two factors let colonization of LAB into the different environments.

LAB can produce slime, especially on the nutrient-rich foods; therefore, the spoilage could be observed with slimy area on the food. In addition, it is generally reported that LAB cause changes in natural color and odor of food (Johansson et al., 2011). Some strains of *Lactobacillus acetotolerans* and *Lactobacillus fructivorans* are able to survive in high concentrations of acetic acid and therefore they are known as spoilage strains in vinegar process (Kubota et al., 2008). *L. plantarum* is a typical spoilage bacterium that forms biofilms well and previous studies showed that it has acid and ethanol resistance in biofilms isolated from acidic foods, more particularly in pickled cabbage. Furthermore, it has been reported that *L. plantarum* strains isolated from red wines showed resistance to low pH values and they could proliferate in the pH range from 3.2 to 3.6 under growing condition (Kubota et al., 2009).

5.3 LAB EMPLOYED TO PRODUCE EXTRACELLULAR POLYMERIC SUBSTANCES (EPSs)

EPSs (Fig. 5.1) are metabolic compounds produced by the LAB that are well-known mesophilic group of EPS-producing microorganisms. The most studied genera are *Lactobacillus bulgaricus, Lactobacillus helveticus, Lactobacillus brevis, L. lactis, Leuconostoc mesenteroides,* and *Streptococcus* spp. among the LAB and it is indicated that they are the good EPS producers (Singha, 2012).

Microbial EPSs are divided into two groups, homopolysaccharides (HoPSs; cellulose, dextran, mutan, alternan, pullulan, levan) and heteropolysaccharides (HePSs; gellan and xanthan). HoPSs are composed of a single type of sugar residue, whereas HePSs are composed of various sugar residues (Baruah et al., 2016). There is an increasing demand in food industries for microorganisms producing EPS and many LAB can produce EPSs in all conditions. However, the amount and composition of EPSs depend on microbial strain and also nutritional and environmental conditions that the bacteria are in (Mostefaoui et al., 2014; Abdellah et al., 2015). EPSs and oligosaccharides produced by LAB have found wide usage in the food industry. According to the studies, some bacterial strains found in dairy products such as *Streptococcus thermophilus, Lactobacillus delbrueckii* ssp. *bulgaricus* and *L. lactis* ssp. *cremoris* are capable of producing EPS (Abdellah et al., 2015). Both biosynthesis and secretion of heteropolysaccharides from LAB take place at different phases of growth. Moreover, the type of the HePS and its amount vary from the growth conditions of the LAB. The structure of some HePS could be rope-like or mucoid, whereas some of them showed non-ropy structure (Patel et al., 2012).

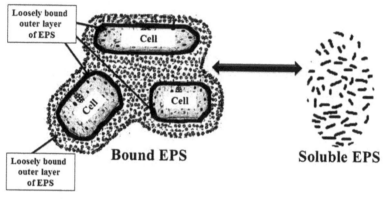

FIGURE 5.1   Sketch of structure of extracellular polymeric substances.

Adapted with permission from Nielsen, P. H.; Jahn, A. Extraction of EPS (Chapter 3). In *Microbial Extracellular Polymeric Substances: Characterization, Structure and Function;* Wingender, J., Neu, T. R., Flemming, H. C. Eds.; Springer Berlin Heidelberg: Berlin, Germany, 1999; pp 49–72. © 1999 Springer.

Rope-like structure in EPS could be observed in the culture medium with long-strand forms (Abdellah et al., 2015). Some LAB are able to produce capsular EPSs, whereas some produce only ropy/slime form. Under some conditions, microorganisms can produce both capsular and slimy forms of EPSs (Ciszek-Lenda, 2011; Patel and Prajapati, 2013).

There are some functional properties of EPS. They play a part in cellular structure because glycocalyx is mainly composed of EPS (Donot et al., 2012). EPSs are used as prebiotics which could be digested only by probiotic bacteria, and therefore, they promote the human gastrointestinal system. Other prebiotic substances are various forms of oligosaccharides which are degraded products of EPSs (Hassan et al., 2008; Baruah et al., 2016). A prebiotic effect of EPS has been observed as it can be metabolized or utilized by probiotic strains because of their specific enzyme systems (Caggianiello et al., 2016).

Another study shows that LAB strains produce various EPSs. According to the study, *Pediococcus parvulus*, isolated from wine, secreted a kind of EPS, β-glucans, which induce the adhesion of bacteria to the surfaces (García-Ruiz et al., 2014). On the other hand, another kind of EPS, α-glucan polymers, was exclusively synthesized by microbial glucansucrase enzymes. The presence of these polymers has been observed in the medium including LAB, especially *Streptococcus, Leuconostoc, Weissella, and Lactobacillus* (van Hijum, 2006).

Microbial metabolites are the main work area for bioprocess engineering and thus microbial PSs could be processed for different targets. In this case, the most important hurdle is the properties of microorganisms. Because, industrially, use of EPSs produced by LAB is preferable to improve the textural and sensorial properties of food product but pathogenic microorganisms and their metabolites should be removed from the food-related materials and surfaces. It should be considered that EPS production and biofilm formation are observed commonly in pathogenic microorganisms and they must be removed from the food-contact surfaces (Baruah et al., 2016).

5.3.1 BIOSYNTHESIS OF EPS FROM LAB

The production of EPSs by the LAB is directly related to the expression of specific gene clusters, which are tagged as *eps* or *cps* that act in surface PS production. Researches showed that the cps/eps clusters have a big role in encoding regulatory enzymes responsible for EPS biosynthesis, polymerization, and secretion. It is known that the eps/cps clusters are located in chromosomes of *S. thermophilus, L. plantarum,* or in plasmids of *L. lactis* and *Pediococcus damnosus* (Caggianiello et al., 2016).

HoPSs secreted by LAB composed of repeating units of a short monosaccharide, such as D-glucose or D-fructose. Glucans and fructans are exopolysaccharides that could be produced by the LAB and their molecular weights range from 10^5 to 10^6 Da (Caggianiello et al., 2016). Almost all glucans produced by LAB are composed of α-glucopyranosyl moieties and bacterial fructans are known as polymers of fructose residues and they are synthesized by the fructosyltransferases (van Hijum, 2004; van Hijum, 2006).

Heteropolysaccharides secreted by LAB are formed from glucose, rhamnose, *N*-acetyl-D-glucosamine, and *N*-acetyl-D-galactosamine and their molecular weights range from 10^4 to 10^6 Da (van Hijum, 2006; Caggianiello et al., 2016).

The secretion period, yield, and variety of EPSs are connected with the temperature, pH, and incubation conditions of the microbial culture. The most known EPS-producing LAB can be ordered as *L. rhamnosus, Lactobacillus kefiranofaciens, L. lactis* ssp. *cremoris, S. thermophilus, L. delbrueckii* ssp. *bulgaricus, L. casei,* and *L. plantarum. L. rhamnosus* is one of the largest EPS producers followed by *L. kefiranofaciens.* Its EPS production rate ranges between 2450 and 2800 mg/l (Caggianiello et al., 2016).

5.3.2   METHODS FOR EXTRACTION OF EPS

EPS production and accumulation are observed around microbial cells. The extraction and separation of EPSs from microbial cells or/and sludge are the most important steps to obtain purified EPS because, in the food industry, the metabolites should be used when they are purified completely. There are two methods for purification (Table 5.1), physical and chemical methods. These methods can be used individually or together. The quantity and quality of EPSs are regarded as significant criteria in extraction methods (Donot et al., 2012).

TABLE 5.1   Methods of Extracellular Polymeric Substance Extraction.

Method	Mechanism
1. Physical methods	Sonication
	Sonication/centrifugation
	High-speed centrifugation
	Heating
2. Chemical methods	Acidic treatment
	Alkaline treatment
	Cation-exchange resin (CER)
	Crown ether
	Ethylenediaminetetraacetic acid (EDTA)
	Enzymatic extraction
	Ethanol extraction
	Glutaraldehyde
	Formaldehyde (HCHO)/sodium hydroxide (NaOH)
	Sodium chloride (NaCl)
	Ammonium hydroxide (NH_4OH)/ethylenediaminetetraacetic acid (EDTA)
	Sulfide

Reprinted with permission from Sheng, G.-P.; Yu, H.-Q.; Li, X.-Y. Extracellular Polymeric Substances (EPS) of Microbial Aggregates in Biological Wastewater Treatment Systems: A Review. *Biotechnol. Adv.* **2010,** *28*(6), 882–894. © 2010 Elsevier.

The physical extraction methods can be ordered as ultrasonic, microwave treatment, centrifugation, and heating. There are three chemical extraction methods which include EDTA, formaldehyde + NaOH, and glutaraldehyde. It has been indicated that the yield of EPSs is more when extracted by chemical methods compared with physical methods. However, with a slight optimization—such as temperature, time, pH—may increase the yield of EPS output. Moreover, combined and repeated extractions should be used to purify and obtain all EPS fractions (Donot et al., 2012).

5.3.3 APPLICATION AND OPPORTUNITIES OF EPS AND EPS-LIKE SUBSTANCES

PSs have some functions in food as water-binding agents, stabilizers, emulsifiers, viscosifying agents, and flavor enhancers. Food-grade microorganisms are used for many years in the food industry and the most important microbial group is LAB, which are able to produce two types of PSs—extracellular and intracellular PSs—according to their location in the cell (Singha, 2012; Patel and Prajapati, 2013). Microorganisms usually adhere to the surfaces easily. After the attachment is completed, they compose microcolony and finally form a biofilm layer which is known as "cell community" glued to each other embedded in a self-produced matrix composed from EPSs (Abe et al., 2013). Microbial EPSs are formed of repeated mono- or oligosaccharides which are linked to each other by glycosidic bonds. The EPS production was first characterized in pathogenic microorganisms because EPS and biofilm formations were related directly with virulence of pathogens. However, nowadays, the EPS production could be used as different and various targets to provide natural preservation of foods because it is demonstrated that food-grade microorganisms produce EPSs as well (Ciszek-Lenda, 2011; Bottacini et al., 2014).

In the food industry, some EPS-producing LAB have already been used as a food additive for various purposes depending on their properties. On the other hand, researchers have been studying the EPS production and stimulation. For this purpose, different microbial strains are kept and grown together. According to the previous studies, synergistic effects have been observed between various microbial species in mix cultures. In this case, various LAB and other microbial species such as some yeast strains—which have an important role in food fermentation or/ and improvement of food properties—could be used together to promote their EPS production (Abe et al., 2013; Derkx et al., 2014). According to the research of a study, *L. plantarum* isolated from vinegar and *Saccharomyces cerevisiae* formed thicker biofilm layer when they were grown together in the medium compared to the thickness of their biofilm layer they formed individually (Abe et al., 2013). These synergistic effects have been observed in another study. According to the study, a lot of LAB and yeast strains were isolated and identified from different traditional/nontraditional fermented products and their biofilm formation was measured individually and together to determine their synergistic/antagonistic effects on each other. The analyses were carried out to determine their biofilm-producing capacity of 380 combinations of yeast and LAB strains. According to the 96-well-plate assay, 36 combinations showed biofilm-positive. In this regards, it is indicated that the combination of *L. plantarum* ML11-11 and *S. cerevisiae* Y11-4 showed the most distinctive biofilm layer at the bottom of the wells, whereas there was no biofilm formation observed inside the wells including their single culture. Another study showed that the remarkable biofilm formation has been observed by the combination of both *L. plantarum–S. cerevisiae* and *L. mesenteroides–S. cerevisiae* (Kawarai et al., 2007; Furukawa et al., 2014; Furukawa et al., 2015). These combinations of beneficial microorganisms should be used for various goals in the food industry.

The use of LAB in industrial fermented foods represents an expanding industry with the dairy products. The most produced commodities are cheese, yoghurt, drinking yoghurt, fermented cream, and milk-based desserts (Douillard and de Vos, 2014). In dairy products, one of the major sensorial expectations significant for the consumer is firmness and creaminess. EPS can bind a hydration water, provide decreasing syneresis, and interact with other compounds found in the dairy product such as micelles and proteins; therefore, EPSs contribute to the rigidity of the casein network. Furthermore, lactic acid accumulation is observed during the fermentation process which lowers the pH in the milk, thereby the solubility of calcium and phosphate contained in micelles. An increase in acidity and EPS in the product promotes the formation of structure desired by consumers (Duboc and Mollet, 2001).

According to the previous studies, purified and crude EPSs, produced by *L. helveticus* MB2-1, which is isolated from traditional Sayram ropy fermented milk in Southern Xinjiang of China, had strong scavenging activities on three free radicals, and, it has also been observed that the bacterial EPS had ion-chelating activity. These results indicate that EPS purified from some LAB strains may find a usage area as being natural antioxidant because of its scavenging and chelating activities. The studies showed that EPSs have a potential value for therapeutics and food industry. The studies provide new food product design without free radicals. Moreover, EPSs obtained from LAB may promote the quality of food product (Li et al., 2014).

S. thermophilus and *L. delbrueckii* sp. *bulgaricus* are the most commonly known starter cultures required for yoghurt process and both bacteria improve the viscosity and mouthfeel of the end product. Besides, according to the previous and recent studies in situ production of EPS by *Lactobacillus mucosae* promotes the rheological, textural, and organoleptic properties by regulating viscosity and also flavor without affecting the yoghurt starter cultures (London et al., 2015).

Kyiad pyrsi is an ethnically fermented beverage which is popular in India. *Leuconostoc lactis* and *L. mesenteroides* have been isolated from the beverage. Its traditional flavor and texture is provided by EPS-producing *Leuconostoc* sp. (Joshi and Koijam, 2014).

"Nordic ropy milk" is a name of fermented traditional milk in Scandinavia, Iceland, Russia, and Mongolia. The fermentation is carried out by mesophilic cocci which produce slime. The fermented milk has been produced traditionally at home and also manufactured industrially. The major LAB are *L. lactis* ssp. *cremoris, L. lactis,* and *L. lactis* biovar *diacetylactis, L. mesenteroides* ssp. *cremoris,* and *Leuconostoc dextranicum, S. thermophilus, L. delbrueckii* ssp. *bulgaricus,* and *L. helveticus.* Moreover, lactose-fermenting yeast strains and mold strain belong to *Geotrichum candidum,* which enhances the sensorial properties of the end product. Ropy bacteria found in the traditional Nordic dairy product are known as *L. lactis* ssp. *cremoris* and *Leuconostoc* spp., which are responsible for the quality of the product although non-ropy strains of these species also exist in the product (Duboc and Mollet, 2001).

Kefir is a traditional, refreshing, and probiotic beverage and it is very popular in Turkey. Kefir grain possesses EPS-producing LAB and various yeast strains combined with casein and complex carbohydrates in a matrix which consist of several PSs. Non-lactose-fermenting yeast and acetic acid bacteria are also found in kefir grains. It is indicated that the microbial flora found in kefir grains can protect the human gastrointestinal system from toxins formed by pathogens as beneficial microorganisms in the grain adhere to the epithelial tissue and also compete with pathogens for nutrition and place that kept them alive. Furthermore, it is reported that regular consumption of kefir beverage provides many benefits to human health; also, due to including coculture, the inhibition of *Helicobacter pylori*—causes ulcer in the gastrointestinal system—has been observed in previous studies (Kandylis et al., 2016).

Table olives is another common fermented vegetable of the food industry and according to a study, the yeast and LAB strains responsible for olive fermentation form biofilm layer. There is a correlation between biofilm formation and EPS production because in biofilm layer not only LAB and yeast strains were observed but also all the species of microorganisms submerged in EPS matrix. It is thought that the direct cell-to-cell communication could exist in the matrix. Microorganisms need a surface to adhere easily to it and after the adherence, they come together as a microcolony by communicating with each other. According to the study, the olive surface is one of the most convenient areas for the adherence of the microorganisms. The adherence and attachment capacities of some LAB such as *Lactobacillus pentosus* strains and also some yeast strains which are isolated from a biofilm matrix formed on the surface of olive skin have been observed and the results showed that the olive skin is very efficient to be vehicle of entry of probiotic microorganisms in the human body (Arroyo-Lopez et al., 2012). During the table olive processing, drupes are immersed in brine which are significantly enriched by nutrients from the olive mesocarp serving as a substrate for the initiation of microbial fermentation and a result of fermentation is developing the sensory characteristics of table olive. Biofilm developments between LAB and yeast strains on Spanish-style green olives have been reported. A research indicated *that L. pentosus* has an ability to survive in high numbers on the olive surface forming biofilm together. This property of the strain pointed to its in vitro probiotic potential. In this case, *L. pentosus* could be a good candidate for natural black olive production with its functional characteristics. The yeast strains cannot recover the surface of drupes; thus, their biofilm formation remains weak in contrast to the LAB. However, the presence of yeast starter resulted in a proper fermentation process. Furthermore, yeast starters affect the quality of end product to provide the desired sensory attributes (Grounta et al., 2016).

In winemaking process, *S. cerevisiae* is an indispensable yeast strain for alcoholic fermentation, often followed by malolactic fermentation (MLF) carried out by *Oenococcus oeni*. MLF promotes the aroma, color, textural properties, and organoleptic profile of wine; therefore, for some types of wine, MLF enhances the desired characteristics and oenological properties (Bastard et al., 2016). According to the study of some LAB isolated from Australian wines of several vintages, some strains have esterase activity, some have the ability to degrade lipids, and it was observed that none of the bacterial strains examined in the study had shown proteolytic activity but during the MLF, the amino acid quantity increased. Wines contain various esters that originated from grape and yeast metabolism and it is understood that LAB could degrade such esters found in the wine. This biochemical reaction could either improve or depreciate wine quality depending on the ester metabolized (Davis et al., 1988).

O. oeni secretes some proteins which make the bacteria more resistant to adverse conditions. These proteins are heat-shock proteins, stress proteins for adaptation of ethanol stress. Biofilm formation is another way to be resistant to environmental stress conditions. The resistance is stabilized in biofilm layer because of metabolic cross-feeding in micro-colonies, cell-to-cell interactions, and chemical–physical balance. The existence of a correlation between bacteriocin resistance and survival in vinous stress conditions has been reported in the previous researches. Within the study, biofilm formation, EPS production, and resistance of *O. oeni* could be observed by scanning electron microscopy (SEM). In this case, the EPS-producing vinous *O. oeni* strains should be preferred to keep the viability of the culture in the stress conditions and therefore prolong MLF period (Bastard et al., 2016). *Weissella confusa, Weissella cibaria,* and *Lactobacillus reuteri* are commonly used as texture-improver in processing bakery products such as volume and softness of loaf because LAB contributes actively to rheological properties of sourdough by providing viscosity and colloidal structure as desired (Caggianiello et al., 2016; Dertli and Yılmaz, 2016).

Various EPS types could be found and extracted from different LAB cultures, and they can have strong preservative and enhancing proper-ties that very valuable for the food industry. In a study, the optimum culture conditions for a novel EPS production by probiotic *L. helveticus* MB2-1 were identified. It is demonstrated that the EPS produced by *L. helveticus* MB2–1 includes galactose, glucose, and mannose. According to in vitro analyses, EPS showed perfect and distinctive emulsifying activities in different oils and hydrocarbon substrates such as olive oil (64%), n-decane (58%), benzene (58%), cedarwood oil (57%), xylene (56%), oleic acid (55%), and castor oil (54%). High antioxidant activity has been observed and also that EPS could inhibit the biofilm-forming pathogens. *L. helveticus* MB211 strain can be used as a natural emulsifying, antioxidant, and biofilm-degradating agent in the food industry and medical science.

Microbial exopolysaccharides provide a lot of physicochemical and biological properties that are applied to the food product. There are various roles of EPS produced by microorganisms, such as hiding the surface of the microorganism, stabilizing the biofilm structure, providing the interactions with other microorganisms to keep them together strictly like an adhesive, and providing the protection from environmental conditions (Badel et al., 2011). EPSs do not appear to be the source of energy because the microorganisms do not have the capability to catabolize the EPSs (Donot et al., 2012). EPSs constitute the structural properties of biofilm layer, and thus, they enhance the firmness of biofilm matrix by acting as a scaffold (Bai et al., 2016).

Food and agriculture industry always look for a new EPS-producing microbial strain which could compete with other established EPS producers such as plant and algae. The most studied microorganisms are *Lactobacillus* spp. LAB are well-known EPS producers, besides dextran can be obtained commercially from *L. mesenteroides*. The most important obstacle to exploiting EPS produced by LAB is inadequate production of the metabolite. *Lactobacillus* spp. are known as the major producers of EPS (Badel et al., 2011). According to another study results, LAB strains have an ability to inhibit the biofilm layer formed by *Staphylococcus aureus* and *Escherichia coli* (Kaur et al., 2015). Besides, when they are used as probiotics, they promote the intestinal system. However, their EPS production was inefficient compared with Gram-negative EPS-producing bacteria and the immobilized cell technology (ICT) could be an innovative and efficient way to solve this handicap of manufacture of EPS obtainment from the LAB. Effective EPS production by LAB means supplying a new industrial sector for the application of EPS as a food additive. Cells were immobilized by adsorption on the food-grade porous rubber and EPSs produced by the cells were collected until they became a mass (Bergmaier et al., 2005).

5.4 ANTIMICROBIAL EFFECTS OF LAB AND THEIR METABOLITES

LAB and their metabolites have been known and used as a natural preservative in the food industry for many years. The antimicrobial effect of LAB is associated with the synthesis of some organic acids which are mainly lactic acid and limitedly formic acid and acetic acid. Moreover, other metabolites of the LAB have been used as antimicrobial substances such as acetaldehyde, reuterin, bacteriocins, hydrogen peroxide, and so forth (Guerrieri et al., 2009; Gómez et al., 2016).

Antimicrobial substances produced by LAB are divided into two main groups which are known as low-molecular mass compounds (molecular weight is <1000 Da) and high-molecular-mass compounds (molecular weight is >1000 Da), such as bacteriocins. All nonbacteriocin antimicrobial compounds produced by LAB strains are of low molecular mass (Suskovic et al., 2010). Antimicrobial compounds could be classified according to Stoyanova et al., (2012) as follows: low-molecular-weight antimicrobial LAB metabolites (organic acids, hydrogen peroxide, pyrrolidone-5-carboxylic acid, diacetyl, reuterin [β-OH–propionic aldehyde]), antifungal compounds of LAB (diketopiperazines, hydroxy derivatives of fatty acids, 3-phenyllactate), and bacteriocins.

5.4.1 MECHANISM OF ANTIMICROBIAL EFFECT OF LAB SUBSTANCES

"Organic acids" provide the increasing of acidity and at pH below 5.0, lactic acid—the main metabolite from fermentation—inhibits the growth of spore-forming bacteria significantly. The antimicrobial activity of "hydrogen peroxide" is connected with its strong oxidizing effect. LABs are able to produce H_2O_2 and the accumulation of peroxides by *Lactococcus* and *Lactobacillus* inhibits food spoilage microorganisms. "Pyrrolidone-5-carboxylic acid" can be produced by some LAB strains and it has bactericidal effect against *Bacillus* and *Enterobacter* spp. "Diacetyl" is mainly active against Gram-negative bacteria. "Reuterin" is produced under anaerobic conditions from glycerol by some *Lactobacillus* sp. and it is effective to inhibit pathogenic microorganisms including enterobacteria. "Diketopiperazines" is formed from protein degradation and also some LAB strains have the ability to produce this antifungal compound. "Hydroxy derivatives of fatty acids" are known as possessing strongly antifungal activity and they are formed by some *L. plantarum* strains. "3-phenyllactate" exhibits antimicrobial activity against a broad range of Gram-positive [G-(+)] and Gram-negative [G-(−)] bacteria and also has antifungal effect. "Bacteriocins" have a specific spectrum of activity, and each bacteriocin has its own specialized immune protein (Lubelski et al., 2008; Stoyanova et al., 2012).

The mechanism of antimicrobial effects of metabolites of LAB changes according to the structure of the antimicrobial agents. Some of them damage permeability of the cell membrane, whereas some inhibit the synthesis of cell wall (Lubelski et al., 2008; Stoyanova et al., 2012).

5.4.2   RESISTANCE MECHANISMS OF LAB AGAINST STRESS CONDITIONS

It is the general acceptance that microbial polymeric substances play a role to protect of the cell from toxic compounds, antibiotics, and most adverse conditions (Patel and Prajapati, 2013). The physiological role of EPSs in the ecology of LAB has not been understood yet but it is thought that EPSs protect the cell from extreme biotic/abiotic conditions such as light, temperature, pH, and osmotic stress (Caggianiello et al., 2016). Microbial EPSs are thought to cover the cell surface and this coating can change the diffusion properties both into and out of the cell (Patel and Prajapati, 2013). EPSs are also thought to have a role in adhesion and recognition mechanism of microbial cells (Caggianiello et al., 2016).

Besides, EPSs produced by LAB play an important role in stimulating immune system. There are a lot of researches that prove the existence of an interaction between biofilm formation and EPS production because they promote the adherence to the surfaces and this attachment is the first step of biofilm formation (Ciszek-Lenda, 2011; Zhang and Zhang, 2014). Biofilm layer has been formed under various conditions but stress factors especially accelerate the formation of biofilm. Microorganisms become more resistant when they are in biofilm layer. The cell-to-cell signaling changes the behavior of microbial cells to increase the ability of survival under adverse conditions. The biofilm formation is the most effective resistance mechanism of the LAB strains.

Ropy phenotype of LAB makes them more tolerant against stress conditions. Some EPS-producing strains of LAB display ropy phenotype, whereas some have phenotype without ropiness. The ropiness of EPSs structure provides desired textural properties to some traditional cheeses.

For example, the slimy structure of EPSs can enhance the colonization of probiotic bacteria easily in gastrointestinal tract; therefore, pathogenic microorganisms could not compete and adhere to the epithelial surface of intestine (Ciszek-Lenda, 2011; Zhang and Zhang, 2014).

5.5 VARIOUS APPLICATIONS OF LAB AND THEIR METABOLITES IN FOOD INDUSTRY

LAB have been used as starter culture in the food industry for many years. The starter culture usage is necessary especially in fermentation processes to provide an end product having desired physicochemical and flavor properties. There are many fermented food and beverages which gain their final form by LAB (Azhari Ali, 2010).

Lactic acid fermentation is the safest and economical method to preserve and extend the shelf life of food products which is based on the use of starter culture and LAB is responsible for the fermentation process (Verma et al., 2017b). LAB initiates the acidification and improves the nutritional value of raw food material by their metabolic activity. During the fermentation period, a lot of beneficial effects have been observed such as stability, color, taste, and aroma properties of raw material are shaped completely. These beneficial effects could be provided by fermentation, but on the other hand, undesired and toxic compounds/metabolites such as biogenic amines (BAs) and D-lactate can exist during the fermentation period. The main metabolite is lactate of lactic acid fermentation and there are two types of lactate according to their isomeric forms which are known as D-lactate and L-lactate. Increasing of D-lactate by fermentation of food causes the toxic effects such as intestinal ischemia, short bowel syndrome, and appendicitis with the intake of the fermented product including D-lactate. The metabolic mechanism demonstrates that choosing of starter culture is the most important step to produce new fermented product. The target of LAB application to the food must increase/promote the organo-leptic, technological, and nutritional value of the end product. Furthermore, the end product must be beneficial to human health (Bergmaier et al., 2005; Li et al., 2014; Bartkiene et al., 2016; Betteridge et al., 2015).

According to another research, *Pediococcus acidilactici, Pediococcus pentosaceus,* and *Lactobacillus sakei* cultures were applied to the surface of meat product before smoking of the meat. Smoking process is an old preservation method of meat. Smoke gives special aromatic taste and color to the meat. Moreover, the shelf life of the end product could be extended because of its dehydrating property. However, a huge amount of BA formation in food was observed after the smoking process and all these BAs threaten human health because they are carcinogenic compounds. In this research, meat samples were coated with LAB cultures, *P. acidilactici, P. pentosaceus,* and *L. sakei* strains before they were smoked and it is observed that BA formation decreased slightly at the meat sample covered with *L. sakei* and *P. acidilactici* were applied together and BA formation was prevented completely in another meat sample covered with *P. pentosaceus* (Bartkiene et al., 2017).

The fermentation processes of traditional food have been based on spontaneous fermentation driven by natural microbiota that exists in the fermentable raw material. This method is known and used traditionally at home but the outcome of such processes is unpredictable because of unknown microbial load and composition of the raw material (Smid et al., 2014). Industrially fermentation processes of raw food materials are significant in food safety management so as to avoid spoilage and/or pathogenic microorganisms which could be found in the raw material rich in nutrients.

Yoghurt, various cheeses, butter, kefir, pickle, wine, beer, and so forth, are some fermented food products and *L. mesenteroides, L. sakei, Lactobacillus kefiri, L. lactis, L. plantarum, O. oeni, S. thermophilus* are some starter cultures which play a role in fermentation period (Azhari Ali, 2010).

In addition, LAB—as natural preservatives—produce lactic acid and other organic acids during the fermentation period that pathogenic microorganisms are inhibited by acidification of the medium with organic acids (Mahony et al., 2014).

In recent years, genomic interest has been developed into the starter LAB cultures and nonstarter LAB (NSLAB) as well. One of the reasons of this interest is that NSLAB and their genomic structures are present naturally in the fermented dairy products and they also contribute to the development of flavor and quality of end product; therefore, NSLAB and similar cultures are called adjunct starters. According to the recent genomic studies about NSLAB, it is determined that *L. helveticus* strain CNRZ 32, which is already used as adjunct starter to reduce bitterness of milk products, possesses four different cell–envelope proteinases, in contrast to other *Lactobacilli* found in dairy products that have one or none (Douillard and de Vos, 2014). Moreover, several strains of *L. helveticus* are used in the production of some types of cheeses such as mozzarella because the EPS produced by *L. helveticus* contributes to water retention in the product (Duboc and Mollet, 2001). The properties of the strain have been shaped according to their genomic structures. Moreover, EPS production is also regulated by genetic expressions, and in this case, EPSs produced by NSLAB may improve the textural and organoleptic properties of fermented products, and it is similar to fermentation by starter LAB culture. Furthermore, *L. helveticus* strains have been used as starter to produce semihard cheeses and some fermented dairy products released to the market around the world (Li et al., 2014). According to the genetic studies about the LAB, it has been indicated that development in mathematic models could be feasible to estimate metabolic fluxes at a genomic scale in the LAB due to having all information derived from genomic sequence (Valenzuela et al., 2015).

It has been reported that several bacteriocins are used in biocontrol treatments to inhibit/prevent biofilm layer formed by foodborne pathogens. In this regard, among the bacteriocins, pediocin PD-1, plantaricin 423, and nisin successfully killed all viable cells in a biofilm layer formed by *O. oeni* found in the grape medium. The research showed that the most effective bacteriocin among the three examined bacteriocins is pediocin PD-1 to prevent potential reformation of malolactic biofilms on stainless steel surfaces during wine processing in food industry (Nel et al., 2002; Rojo-Bezares et al., 2007; Gün and Ekinci, 2009; Sudağıdan and Aydın, 2013). By binding to the PG precursor lipid II, nisin damages permeability of the membrane and inhibits the cell wall synthesis (Lubelski et al., 2008).

Food spoilage microorganisms and also pathogens have to be inhibited by food preservatives. An example, *Listeria monocytogenes, E. coli* O157:H7, and *Salmonella* spp., which are foodborne pathogens can be isolated from meat, dairy, and various food products. They also can be found on filling/packaging food processing equipment, walls, floor drains, conveyors, and racks for transporting products. Besides, *Listeria* can survive and grow at refrigeration temperature. It is indicated in some previous studies that pathogens are capable to adhere to the surfaces and they can form biofilm layer when they are exposed to adverse conditions such as low nutrient medium (Guerrieri et al., 2009; Gómez et al., 2016). It has been reported that bacteriocin-producing LAB have especially showed remarkable anti-*listerial* effect. *Listeria* can survive at increasing pH degrees; therefore, it is indicated in the study that the behavior of *L. plantarum* singly had been more effective to inhibit *L. monocytogenes* compared to coculture (*Pseudomonas putida* and *L. plantarum*) behaviors remained inefficient because *P. putida* had blocked the decreasing of pH. In addition, it is indicated that by the researchers that comparing the anti-*listerial* activity of LAB biofilm against both planktonic and adherent cells, the *L. monocytogenes* adherent cells have exerted higher resistance (Guerrieri et al., 2009). The most effective metabolites produced by LAB are various bacteriocins against foodborne pathogens but there is a disadvantage in extracting and keeping the bacteriocin since the extracted substance can lose the activity over time. It can be offered to employ the bacteriocin-producing microorganisms in biofilm formed by the LAB; therefore, a continuance of antipathogenic activity can be provided (Guerrieri et al., 2009; Gutiérrez et al., 2015). Another important property of bacteriocins produced by the LAB is precluding bacterial resistance against antibiotics because, recently, antibiotic resistance in pathogen microorganisms has been considered a problem, which is linked to the extensive use of classical antibiotics in the treatment of human and animal diseases. In this regard, alternative antimicrobial agents could be researched and such antimicrobial agents play a part in medical and industrial area that is required for food safety. Bacteriocins and other metabolites produced by LAB have become potential alternative compounds which can be applied to the food products safety (Parada et al., 2007).

Disinfection of the food-contact surfaces clean away the pathogens from the food-related environments as it is the most remarkable point during food processing. In recent years, there is a great interest in the development of innovative strategies to keep public healthy because pathogens have become resistant to use of antimicrobial agents and it led to a search of alternatives for disinfection in the food industry. LAB, their metabolites and also among them biofilm-forming strains may offer the alternative preservation methods to use them industrially as natural food additives. It is indicated that LAB successfully reduced *L. monocytogenes* on abiotic surfaces. Furthermore, a lot of studies have demonstrated that bacteriocin-producing LAB could degrade the biofilm layer formed by pathogens in/on the food-related area (Gómez et al., 2016). Current biofilm preventive strategies by LAB species against pathogenic bacteria are carried out by the production of antimicrobial metabolites such as bacteriocin, organic acid, hydrogen peroxide, or by covering the surface of pathogenic bacteria with inhibitory EPS to inhibit the cells and to cut off communication of the cell with external environment; therefore, the cell-to-cell signaling mechanism among pathogenic microorganisms is blocked. Furthermore, in recent years, the competition for nutrients and adhesive sites is another prominent subject (Jalilsood et al., 2015). In this case, the metabolites of LAB or bacteriocin, organic acids, EPS-producing strains, and biofilm-forming LAB shed light on alternative disinfection methods of food-related surfaces (Gómez et al., 2016).

Kawarai et al., (2007) demonstrated that *L. casei* ssp. *rhamnosus* had produced a yeast-biofilm-forming factor(s) with low molecular weight. On the other hand, it is indicated that bacteriocins produced by LAB are divided into the groups according to their chemical structures, molecular weights, and mode of actions. Lantibiotics, one of the bacteriocins, are known as small peptides (<5 kDa) that possess unusual posttranslationally modified residues (Saranraj et al., 2013; Perez et al., 2014). It has been reported that lantibiotics produced by *L. casei* ssp. *rhamnosus* were found in similar molecular weight to what the researchers found for the biofilm-forming factor(s). In the study, it is indicated that, most bacteriocins are heat resistant but the biofilm-forming factor(s) is affected by increased temperatures. Therefore, it seems likely the biofilm-forming factor(s) found in the research would be different from lantibiotics (Kawarai et al., 2007; Parada et al., 2007).

In addition, another metabolic compound from the LAB is surfactant substances. Biosurfactant formed by LAB can be used in food industry as an emulsifier. Surfactants are amphipathic (amphiphilic) compounds having an ability to decrease the surface tension of the water; thus, they are titled "surface-active" compounds because they bring different surfaces of compounds together. They can be produced chemically or by microorganisms. Biosurfactants have been produced by some microorganisms that can be used in the processing of hydrocarbons, carbohydrates, oils, or their mix as a substrate. These surfactant compounds contribute to the quality of food products because they can be used as an emulsifier in the food industry. The main physiological role of biosurfactants is reducing the surface tension at the phase boundary. This reduction in surface tension let microorganisms grow on water-immiscible substrates because of the slight increase in the ability of intake and metabolization. Thus, the substrate is available and ready for uptake (Yılmaz et al., 2010; Fakruddin, 2012). Biosurfactant-producing capability of some LAB such as *Lactobacillus fermentum, Lactobacillus acidophilus, L. plantarum, Lactococcus* spp. strains were determined in a study. A total of 19 LAB isolates were screened to determine their biosurfactant production by drop-collapse and oil-spreading methods. The results showed that 73% isolates decreased the surface activity and they are recognized as biosurfactant producers (Kaur et al., 2015).

LAB and their metabolites provide a lot of benefits to human health and also they are well known to be natural food additives; thus, food quality can be reinforced with these natural enhancers in the food industry. The use of vitamin-producing LAB is both natural and economical, which can be alternative to chemically synthesized vitamins (Salvucci et al., 2016). It has been reported that various LAB strains are able to produce folate (vitamin B_{11}) and this production may be allowed in food products with elevated concentrations of folate. For example, the use of LAB-producing B vitamins in cereal-based products makes the baked goods more preferable and this application could be an interesting alternative to improve the quality of products worldwide because, particularly, during processing and storage period of baked products, much of vitamin B originally contained in grains is lost. The use of mutant overproducing strains of vitamin B in food industry could brush up against legal restriction in most countries because it poses a risk to human health. However, wild-type strains could be easily used because they do not have the same legal restriction compared with a modified organism. This production can be a significant alternative method for intake of folate. *L. lactis, L. reuteri, Enterococcus mundtii, L. plantarum, L. pentosus,* and *P. acidilactici* strains have been reported as the producers of folate (vitamin B_{11}) and, in the case of *L. lactis*, riboflavin (B_2) and cobalamin (B_{12}) vitamins. Biosynthetic pathways of vitamins in LAB are usually complex and require multienzymes (Valenzuela et al., 2015; Salvucci et al., 2016). The biosyntheses of vitamins show differences from each other because some strains produce vitamin B intracellularly, whereas some produce the vitamin extracellularly. These researches showed that vitamin B production was different among the strains, not only in quantitative but also in its cellular location. This sheds light on the elaboration of different vitamin-B-enriched foods. In such foods, vitamin

B could either be released into the food product or be protected inside the cells (Salvucci et al., 2016). Therefore, in the future, filling the deficiency of several vitamins can happen as expected for public health. Furthermore, it seems that the production of vitamin B by viable LAB in food material is unlike to cause undesirable side effects compared to chemically processed vitamin B_{12} such as deficiencies or deactivation of liver enzymes (Burguess et al., 2009; Valenzuela et al., 2015; Salvucci et al., 2016).

5.6 CONCLUSION

EPS-producing LAB play an important role in the manufacture of fermented products such as milk products (yoghurt, fermented creams, milk-based desserts), fermented olives, pickled vegetables, and fermented alcoholic beverages (beer, wine). The fermentation is carried out by LAB which are standardized as starter cultures. Fermented food products meet the consumer expectation because LAB strains—responsible for fermentation—promote texture, mouthfeel, and stability of the end product. Furthermore, the beneficial bacteria strains can contribute to human health with their various properties. It has been known that EPSs obtained from LAB have antiulcerogenic, anticarcinogenic, antioxidant properties, cholesterol-lowering activities, and also immune-stimulating effects. Moreover, fermentation process makes the end product more resistant, and LAB cultures provide the variety within food products. Considering all these contributions of LAB, food shelf life could be prolonged because of preservation properties of the LAB.

There is another issue that LAB are commonly known as probiotic bacteria. Functional, dietary, and intestine stimulator foods are released to the market by using probiotic LAB during the fermentation. Most beneficial effects of LAB result from their EPS production because, in the fermentation process, almost all desired properties of end product could be provided completely by EPS produced by the LAB. Most LAB strains produce intra- or extracellular EPSs. In the food industry, EPSs are known as enhancers of food products because of their thickening, viscosifying, emulsifying, water-binding (gelling), and stabilizing properties. EPSs which are especially produced by LAB have gained increasing attention in recent years due to their health benefits to the consumer. In addition to the beneficial effects to human health, they are used as natural food additives or film-coating agents to cover the food surface. They can be used for film coating of the food products. Although some EPSs form a biofilm that causes diseases and hygiene problems, other EPSs derived from the LAB play critical role in improving structural properties of last product. Moreover, EPSs contribute to the formulation of fermented food. Biosurfactants, bioabsorbents, bioflocculants, heavy metal scavenging agents, drug delivery agents, and others are the new applications of EPSs.

5.7 SUMMARY

LAB are nonpathogenic, fermentative, nonspore-forming, Gram-positive, usually catalase-negative, acid-tolerant bacteria which are found in nutrition-rich environments and almost all of them are known as "GRAS." Most representative genera of LAB are *Lactococcus, Lactobacillus, Leuconostoc, Streptococcus, Pediococcus, Weissella, Oenococcus, Aerococcus, Carnobacterium, Enterococcus,* and *Tetragenococcus.* LAB strains have been used in the food industry for a long time as starter cultures, probiotics, and also natural preservatives. In the food industry, LAB cultures change and improve organoleptic, textural, rheological, and nutritional properties of the food product by enzymatic activities. The metabolites produced by LAB give a variety in fermented foods. Moreover, LAB strains, which are found especially in traditional fermented foods, provide specific flavor, odor, and structure to the end product. It is known that antimicrobial activity is observed when LAB strains grow efficiently. The antimicrobial effect is associated with the synthesis of some organic acids which are mainly lactic acid and occasionally formic acid and acetic acid. Moreover, other metabolites of the LAB have been used as antimicrobial substances which are acetaldehyde, reuterin, bacteriocins, hydrogen peroxide, and so forth. Antimicrobial effect mechanisms of organic acids and metabolites produced by the LAB and the fermentation pathways of carbohydrate sources are mentioned in this chapter. Some LAB strains have a resistance mechanism against the stress conditions such as lack of nutrient, presence of toxic compounds, bacteriophages, and antagonist microorganisms. This resistance mechanism is known to be directly correlated with the biosynthesis of EPSs, long-chain PSs, produced extracellularly and formed from branched repeating units of sugars and their derivatives. EPS has important role in protection of microbial cells because it covers the cell surface weakly. This attachment of the surface is not permanent; thus, EPS can be collected easily and used in various fields of food industry. The importance of microbial metabolites, by-products, and extracellular secretions has been appreciated consistently because of ever-expanding industrial fields.

KEYWORDS

- antagonistic effect
- antimicrobial peptides
- bacteriocins
- biofilm
- biopreservatives
- biosurfactant
- exopolysaccaharide
- organic acids
- prebiotic
- probiotic
- rheology
- ropiness
- stabilizer
- synergistic effect
- texture
- viscosity

REFERENCES

Abdellah, M.; Ahcene, H.; Benalia, Y.; Saad, B.; Abdelmalek, B. Evaluation of Biofilm Formation by Exopolysaccharide-Producer Strains of Thermophilic Lactic Acid Bacteria Isolated from Algerian Camel Milk. *Emirates J. Food Agric.* **2015,** *27*(6), 513–521.

Abe, A.; Furukawa, S.; Watanabe, S.; Morinaga, Y. Yeasts and Lactic Acid Bacteria Mixed-Specie Biofilm Formation is a Promising Cell Immobilization Technology for Ethanol Fermentation. *Appl. Biochem. Biotechnol.* **2013,** *171*, 72–79.

Arroyo-López, F. N.; Bautista-Gallego, J.; Domínguez-Manzano, J.; Romero-Gil, V.; Rodriguez-Gómez, F.; García-García, P.; Garrido-Fernández, A.; Jiménez-Díaz, R. Formation of Lactic Acid Bacteriae-yeasts Communities on the Olive Surface During Spanish-Style Manzanilla Fermentations. *Food Microbiol.* **2012,** *32*, 295–301.

Azcarate-Peril, M. A.; Klaenhammer, T. R. Genomics of Lactic Acid Bacteria: The Post-genomics Challenge—From Sequence to Function (Chapter 2). In *Biotechnology of Lactic Acid Bacteria Novel Applications*; Mozzi, F., Raya, R. R., Vignolo, G. M. Eds.; Wiley-Blackwell: USA, 2010; pp 35–56. ISBN: 978-0-813-81583-1.

Azhari Ali, A. Beneficial Role of Lactic Acid Bacteria in Food Preservation and Human Health. *Res. J. Microbiol.* **2010**, *5*(12), 1213–1221.

Badel, S.; Bernardi, T.; Michaud, P. New Perspectives for *Lactobacilli* Exopolysaccharides. *Biotechnol. Adv.* **2011**, *29*, 54–66.

Bai, Y.; Dobruchowska, J. M.; van der Kaaij, R. M.; Gerwig, G. J.; Dijkhuizen, L. Structural Basis for the Roles of Starch and Sucrose In Homo-exopolysaccharide Formation by *Lactobacillus reuteri* 35-5. *Carbohydr. Polym.* **2016**, *151*, 29–39.

Bartkiene, E.; Bartkevics, V.; Mozuriene, E.; Krungleviciute, V.; Novoslavskij, A.; Santini, A.; Rozentale, I.; Juodeikiene, G.; Cizeikiene, D. The Impact of Lactic Acid Bacteria with Antimicrobial Properties on Biodegradation of Polycyclic Aromatic Hydrocarbons and Biogenic Amines in Cold Smoked Pork Sausages. *Food Control* **2017**, *71*, 285–292.

Bartkiene, E.; Bartkevics, V.; Rusko, J.; Starkute, V.; Bendoraitiene, E.; Zadeike, D.; Juodeikiene, G. The Effect of *Pediococcus acidilactici* and *Lactobacillus sakei* on Biogenic Amines Formation and Free Amino Acid Profile in Different Lupin During Fermentation. *LWT–Food Sci. Technol.* **2016**, *74*, 40–47.

Baruah, R.; Das, D.; Goyal, A. Heteropolysaccharides from Lactic Acid Bacteria: Current Trends and Applications. *J. Probiotics Health* **2016**, *4*(2), 1–6.

Bastard, A.; Coelho, C.; Briandet, R.; Canette, A.; Gougeon, R.; Alexandre, H.; Guzzo, J., Weidmann, S. Effect of Biofilm Formation by *Oenococcus oeni* on Malolactic Fermentation and the Release of Aromatic Compounds in Wine. *Front. Microbiol.* **2016**, *7*(613), 1–14.

Bergmaier, D.; Champagne, C. P.; Lacroix, C. Growth and Exopolysaccharide Production During Free and Immobilized Cell Chemostat Culture of *Lactobacillus Rhamnosus* RW-9595M. *J. Appl. Microbiol.* **2005**, *98*, 272–284.

Betteridge, A.; Grbin, P.; Jiranek, V. Improving *Oenococcus oeni* to Overcome Challenges of Wine Malolactic Fermentation. *Trends Biotechnol.* **2015**, *33*(9), 547–553.

Bottacini, F.; Ventura, M.; van Sinderen, D.; O'Connell Motherway, M. Diversity, Ecology and Intestinal Function of Bifidobacteria. *Microb. Cell Fact.* **2014**, *13*(Suppl. 1), 1–15. DOI: 10.1186/1475-2859-13-S1-S4.

Burguess, C.; Smid, E.; van Sinderen, D. Bacterial Vitamin B2, B11 and B12 Overproduction: An Overview. *Int. J. Food Microbiol.* **2009**, *133*, 1–7.

Caggianiello, G.; Kleerebezem, M.; Spano, G. Exopolysaccharides Produced by Lactic Acid Bacteria: from Health-Promoting Benefits to Stress Tolerance Mechanisms. *Appl. Microbiol. Biotechnol.* **2016**, *100*, 3877–3886.

Chapot-Chartier, M. P.; Kulakauskas, S. Cell Wall Structure and Function in Lactic Acid Bacteria. *Microb. Cell Fact.* **2014**, *13*(Suppl 1), 1–23. DOI: 10.1186/1475-2859-13-S1-S9.

Cıszek-Lenda, M. Biological Functions of Exopolysaccharides from Probiotic Bacteria. *Cent. Eur. J. Immunol.* **2011**, *36*(1), 51–55.

Davis, C. R.; Wibowo, D.; Fleet, G. H.; Lee, T. H. Properties of Wine Lactic Acid Bacteria: Their Potential Enological Significance. *Am. J. Enol. Vitic.* **1988**, *39*(2), 137–142.

De Vuyst L.; Leroy, F. Bacteriocins from Lactic Acid Bacteria: Production, Purification, and Food Applications. *J. Mol. Microbiol. Biotechnol.* **2007**, *13*, 194–199.

Derkx, P. M. F.; Janzen, T.; Sørensen, K. I.; Christensen, J. E.; Stuer-Lauridsen, B.; Johansen, E. The Art of Strain Improvement of Industrial Lactic Acid Bacteria Without the Use of Recombinant DNA Technology. *Microb. Cell Fact.* **2014,** *13*(Suppl. 1), 1–13. DOI: 10.1186/1475-2859-13-S1-S5.

Dertli, E.; Yılmaz, M. T. Functional Properties of Lactic Acid Bacteria (LAB) Playing Crucial Roles on Sourdough Biotechnology. *J. Biotechnol.* **2016,** S4–S109.

Donot, F.; Fontana, A.; Baccou, J. C.; Schorr-Galindo, S. Microbial Exopolysaccharides: Main Examples of Synthesis, Excretion, Genetics and Extraction. *Carbohydr. Polym.* **2012,** *87*, 951–962.

Douillard, F. P.; de Vos, W. M. Functional Genomics of Lactic Acid Bacteria: from Food to Health. *Microb. Cell Fact.* **2014,** *13*(Suppl 1), 1–21. DOI: 10.1186/1475-2859-13-S1-S8

Duboc, P.; Mollet, B. Applications of Exopolysaccharides in the Dairy Industry. *Int. Dairy J.* **2001,** *11*, 759–768.

Fakruddin, M. Biosurfactant: Production and Application. *J. Pet. Environ. Biotechnol.* **2012,** *3*(4), 1–5.

Furukawa, S.; Isomae, R.; Tsuchiya, N.; Hirayama, S.; Yamagishi, A.; Kobayashi, M.; Suzuki, C.; Ogihara, H.; Morinaga, Y. Screening of Lactic Acid Bacteria That Can form Mixed-Species Biofilm with *Saccharomyces Cerevisiae. Biosci. Biotechnol. Biochem.* **2015,** *79*(4), 681–686.

Furukawa, S.; Yoshida, K.; Ogihara, H.; Yamasaki, M.; Morinaga, Y. Mixed-Species Biofilm Formation by Direct Cell-Cell Contact between Brewing Yeasts and Lactic Acid Bacteria. *Biosci. Biotechnol. Biochem.* **2014,** *74*(11), 2316–2319.

García-Ruiz, A.; de Llano, D. G.; Esteban-Fernández, A.; Requena, T. B. B.; Moreno-Arribas, M. V. Assessment of Probiotic Properties in Lactic Acid Bacteria Isolated from Wine. *Food Microbiol.* **2014,** *44*, 220–225.

Gómez, N. C.; Ramiro, J. M. P.; Quecan, B. X. V.; de Melo Franco, B. D. G. Use of Potential Probiotic Lactic Acid Bacteria (LAB) Biofilms for the Control of *Listeria monocytogenes, Salmonella typhimurium,* and *Escherichia Coli* O157: H7 Biofilms Formation. *Front. Microbiol.* **2016,** *7*(863), 1–15.

Grounta, A.; Doulgeraki, A. I.; Nychas, G. J. E.; Panagou, E. Z. Biofilm formation on Conservolea Natural Black Olives During Single and Combined Inoculation with a Functional Lactobacillus Pentosus Starter Culture. *Food Microbiol.* **2016,** *56*, 35–44.

Guerrieri, E.; de Niederhäusern, S.; Messi, P.; Sabia, C.; Iseppi, R.; Anacarso, I.; Bondi, M. Use of Lactic Acid Bacteria (LAB) Biofilms for the Control of *Listeria Monocytogenes* in a Small-Scale Model. *Food Control,* **2009,** *20*, 861–865.

Gün, İ.; Ekinci, F. Y. Biyofilmler: YüzeylerdekiMikrobiyalYaşam. *Gıda* **2009,** *34*(3), 165–173.

Gutiérrez, S.; Martínez-Blanco, H.; Rodríguez-Aparicio, L. B.; Ferrero, M. A. Effect of Fermented Broth from Lactic Acid Bacteria on Pathogenic Bacteria Proliferation. *J. Dairy Sci.* **2015,** *99*, 2654–2665.

Hassan, A. N. ADSA Foundation Scholar Award: Possibilities and Challenges of Exopolysaccharide Producing Lactic Cultures in Dairy Foods. *J. Dairy Sci.* **2008,** *91*(4), 1282–1298.

Jalilsood, T.; Baradaran, A.; Song, A. A. L.; Foo, H. L.; Mustafa, S.; Saad, W. Z.; Yusoff, K.; Rahim, R. A. Inhibition of Pathogenic and Spoilage Bacteria by a Novel Biofilm-Forming *Lactobacillus* Isolate: A Potential Host for the Expression of Heterologous Proteins. *Microb. Cell Fact.* **2015,** *14*(96), 1–14.

Johansson, P.; Paulin, L.; Sade, E.; Salovuori, N.; Alatalo, E. R.; Björkroth, K. J.; Auvinen, P. Genome Sequence of a Food Spoilage Lactic Acid Bacterium, *Leuconostoc gasicomitatum* LMG 18811T, in Association with Specific Spoilage Reactions. *Appl. Environ. Microbiol.* **2011,** *77*(13), 4344–4351.

Joshi, S. R.; Koijam, K. Exopolysaccharide Production by a Lactic Acid Bacteria, *Leuconostoc lactis* Isolated from Ethnically Fermented Beverage. *Natl. Acad. Sci. Lett.* **2014,** *37*(1), 59–64.

Kandylis, P.; Pissaridi, K.; Bekatorou, A.; Kanellaki, M.; Koutinas, A. A. Dairy and Non-Dairy Probiotic Beverages. *Curr. Opin. Food Sci.* **2016,** *7*, 58–63.

Kaur, S.; Amrita; Kaur, P.; Nagpal, R. In Vitro Biosurfactant Production and Biofilm Inhibition by Lactic Acid Bacteria Isolated from Fermented Food Products. *Int. J. Probiotics Prebiotics*, **2015,** *10*(1), 17–22.

Kawarai, T.; Furukawa, S.; Ogihara, H.; Yamasaki, M. Mixed-Species Biofilm Formation by Lactic Acid Bacteria and Rice Wine Yeasts. *Appl. Environ. Microbiol.* **2007,** *73*(14), 4673–4676.

Kocková, M; Gereková, P.; Petruláková, Z.; Hybenová, E.; Šturdík, E.; Valík, L. Evaluation of Fermentation Properties of Lactic Acid Bacteria Isolated from Sourdough. *Acta Chim. Slov.* **2011,** *4*(2), 78–87.

Kubota, H.; Senda, S.; Nomura, N.; Tokuda, H.; Uchiyama, H. Biofilm Formation by Lactic Acid Bacteria and Resistance to Environmental Stress. *J. Biosci. Bioeng.* **2008,** *4*, 381–386.

Kubota, H.; Senda, S.; Tokuda, H.; Uchiyama, H.; Nomura, N. Stress Resistance of Biofilm and Planktonic *Lactobacillus plantarum* subsp. *plantarum* JCM 1149. *Food Microbiol.* **2009,** *26*, 592–597.

Li, W.; Ji, J.; Chen, X.; Jiang, M.; Rui, X.; Dong, M. Structural Elucidation and Antioxidant Activities of Exopolysaccharides from *Lactobacillus Helveticus* MB2–1. *Carbohydr. Polym.* **2014,** *102*, 351–359.

Li, W.; Ji, J.; Rui, X.; Yu, J.; Tang, W.; Chen, X.; Jiang, M.; Dong, M. Production of Exopolysaccharides by *Lactobacillus Helveticus* MB2–1 and its Functional Characteristics in Vitro. *LWT–Food Sci. Technol.* **2014,** *59*, 732–739.

London, L. E.; Chaurin, V.; Auty, M. A.; Fenelon, M. A.; Fitzgerald, G. F.; Ross, R. P.; Stanton, C. Use of *Lactobacillus Mucosae* DPC 6426, an Exopolysaccharide Producing Strain, Positively Influences the Techno-Functional Properties of Yoghurt. *Int. Dairy J.* **2015,** *40*, 33–38.

Lubelski, J.; Rink, R.; Khusainov, R.; Moll, G. N.; Kuipers, O. P. Biosynthesis, Immunity, Regulation, Mode of Action and Engineering of the Model Lantibiotic Nisin. *Cell. Mol. Life Sci.* **2008,** *65*, 455–476.

Mahony, J.; Bottacini, F.; van Sinderen, D.; Fitzgerald, G. F. Progress in Lactic Acid Bacterial Phage Research. *Microb. Cell Fact.* **2014,** *13*(Suppl. 1), 1–12. DOI: 10.1186/1475-2859-13-S1-S1.

Mostefaoui, A.; Hakem, A.; Yabrir, B.; Boutaiba, S.; Badis, A. Screening for Exopolysaccharide-Producing Strains of Thermophilic Lactic Acid Bacteria Isolated from Algerian Raw Camel Milk. *Afr. J. Microbiol. Res.* **2014,** *8*(22), 2208–2214.

Nel, H. A.; Bauer, R.; Wolfaardt, G. M.; Dicks, L. M. T. Effect of Bacteriocins Pediocin PD-1, Plantaricin 423, and Nisin on Biofilms of *Oenococcus oeni* on a Stainless Steel Surface. *Am. J. Enol. Vitic.* **2002**, *53*(3), 191–196.

Nielsen, P. H.; Jahn, A. Extraction of EPS (Chapter 3). In *Microbial Extracellular Polymeric Substances: Characterization, Structure and Function;* Wingender, J., Neu, T. R., Flemming, H. C. Eds.; Springer Berlin Heidelberg: Berlin, Germany, 1999; pp 49–72.

Parada, J. L.; Caron, C. R.; Medeiros, A. B. P.; Ricardo, C. Bacteriocins from Lactic Acid Bacteria: Purification, Properties and use as Biopreservatives. *Braz. Arch. Biol. Technol.* **2007**, *50*(3), 521–542.

Patel, A.; Prajapati, J. B. Food and Health Applications of Exopolysaccharides produced by Lactic acid Bacteria. *Adv. Dairy Res.* **2013**, *1*(2), 1–7.

Patel, A.; Shah, N. and Verma, D. K. Lactic Acid Bacteria (LAB) Bacteriocins: An Ecological and Sustainable Biopreservative Approach to Improve the Safety and Shelf-Life of Foods. In *Microorganisms in Sustainable Agriculture, Food and the Environment* as part of book series on *Innovations in Agricultural Microbiology*; Verma, D. K., Srivastav, P. P. Eds.; Apple Academic Press: USA, 2017; pp 197–258.

Patel, S.; Majumder, A.; Goyal, A. Potentials of Exopolysaccharides from Lactic Acid Bacteria. *Indian J. Microbiol.* **2012**, *52*(1), 3–12.

Perez, R. H.; Zendo, T.; Sonomoto, K. Novel Bacteriocins from Lactic Acid Bacteria (LAB): Various Structures and Applications. *Microb. Cell Fact.* **2014**, *13*(Suppl. 1). 1–13. DOI: 10.1186/1475-2859-13-S1-S3.

Reis, N. A.; Saraiva, M. A. F.; Duarte, E. A. A.; de Carvalho, E. A.; Vieira, B. B.; Evangelista-Barreto, N. S. Probiotic Properties of Lactic Acid Bacteria Isolated from Human Milk. *J. Appl. Microbiol.* **2016**, *121*, 811–820.

Rojo-Bezares, B.; Saenz, Y.; Navarro, L.; Zarazaga, M.; Ruiz-Larrea, F.; Torre, C. Coculture-Inducible Bacteriocin Activity of *Lactobacillus Plantarum* Strain J23 Isolated from Grape Must. *Food Microbiol.* **2007**, *24*, 482–491.

Salvucci, E.; LeBlanc, J. G.; Perez, G. Technological Properties of Lactic Acid Bacteria Isolated from Raw Cereal Material. *LWT–Food Sci. Technol.* **2016**, *70*, 185–191.

Saranraj, P.; Naidu, M. A.; Sivasakthivelan, P. Lactic Acid Bacteria and Its Antimicrobial Properties: a Review. *Int. J. Pharm. Biol.* **2013**, *4*(6), 1124–1133.

Sheng, G.-P.; Yu, H.-Q.; Li, X.-Y. Extracellular Polymeric Substances (EPS) of Microbial Aggregates in Biological Wastewater Treatment Systems: A Review. *Biotechnol. Adv.* **2010**, *28*(6), 882–894.

Singha, T. K. Microbial Extracellular Polymeric Substances: Production, Isolation and Applications. *IOSR J. Pharm.* **2012**, *2*(2), 276–281.

Smid, E. J.; Erkus, O.; Spus, M.; Wolkers-Rooijackers, J. C. M.; Alexeeva, S.; Kleerebezem, M. Functional Implications of the Microbial Community Structure of Undefined Mesophilic Starter Cultures. *Microb. Cell Fact.* **2014**, *13*(Suppl. 1), 1–9. DOI: 10.1186/1475-2859-13-S1-S2.

Stoyanova, L. G.; Ustyugova, E. A.; Netrusov, A. I. Antibacterial Metabolites of Lactic Acid Bacteria: Their Diversity and Properties. *Appl. Biochem. Microbiol.* **2012**, *48*(3), 229–243.

Sudağıdan, M.; Aydın, A. Lizozimve Nisinin Gıda Kaynaklı *Staphylococcus aureus* Suşlarında Gelişimve Biyofilm Oluşumu Üzerine Etkileri. *J. Fac. Vet. Med.* **2013**, *39*(2), 254–263.

Sun, Z.; Yu, J.; Dan, T.; Zhang, W.; Zhang, H. Phylogenesis and Evolution of Lactic Acid Bacteria (Chapter 1). In *Lactic Acid Bacteria Fundamentals and Practice;* Zhang, H., Cai, Y. Eds.; Springer: Netherlands, 2014; pp 1–101. ISBN 978-94-017-8840-3.

Suskovic, J.; Kos, B.; Beganovic, J.; Pavunc, A. L.; Habjanic, K.; Matosic, S. Antimicrobial Activity–the Most Important Property of Probiotic and Starter Lactic Acid Bacteria. *Food Technol. Biotechnol.* **2010,** *48*(3), 296–307.

Valenzuela, J. F.; Pinuer, L. A.; Cancino, A. G.; Yáñez, R. B. Metabolic Fluxes in Lactic Acid Bacteria—a Review. *Food Biotechnol.* **2015,** *29,* 185–217.

Van Hijum, S. A. F. T. Fructosyltransferases of *Lactobacillus reuteri*: Characterization of Genes, Enzymes, and Fructan Polymers. PhD Thesis, University of Groningen, 2004.

Van Hijum, S. A. F. T. Structure-Function Relationships of Glucansucrase and Fructansucrase Enzymes from Lactic Acid Bacteria. *Microbiol. Mol. Biol. Rev.* **2006,** *70*(1), 157–176.

Verma, D. K.; Srivastav, P. P. *Microorganisms in Sustainable Agriculture, Food and the Environment.* as Part of Book Series on *Innovations in Agricultural Microbiology* Apple Academic Press: USA, 2017.

Verma, D. K.; Mahato, D. K.; Billoria, S.; Kapri, M.; Prabhakar, P. K.; Ajesh Kumar, V. and Srivastav, P. P. Microbial Approaches in Fermentations for Production and Preservation of Different Food. In *Microorganisms in Sustainable Agriculture, Food and the Environment.* as part of book series on *Innovations in Agricultural Microbiology;* Verma, D. K., Srivastav, P. P. Eds.; Apple Academic Press: USA, 2017a; pp 105–142.

Verma, D. K.; Mahato, D. K.; Billoria, S.; Kapri, M.; Prabhakar, P. K.; Ajesh Kumar, V. and Srivastav, P. P. Microbial Spoilage in Milk and Milk Products: Potential Solution, Food Safety, and Health Issues. In *Microorganisms in Sustainable Agriculture, Food and the Environment.* as part of book series on *Innovations in Agricultural Microbiology;* Verma, D. K, Srivastav, P. P. Eds.; Apple Academic Press: USA, 2017b. pp 171–196.

Wassie, M.; Wassie, T. Isolation and Identification of Lactic Acid Bacteria from Raw Cow Milk. *Int. J. Adv. Res. Biol. Sci.* **2016,** *3*(8), 44–49.

Yılmaz, F.; Ergene, A.; Yalçın E. Süt Fabrikası Atıksuyundan _zole Edilen Mikroorganizmalarile Biyosürfektan Üretimi: Optimum Kosullarının Arastırılması. *Elektronik Mikrobiyoloji Dergisi TR (Or Lab On-Line Mikrobiyoloji Dergisi),* **2010,** *8*(1), 20–30.

Zhang, W.; Zhang, H. Genomics of Lactic Acid Bacteria (Chapter 3). In *Lactic Acid Bacteria: Fundamentals and Practice;* Zhang, H., Cai, Y. Eds.; Springer: Netherlands, 2014; pp 205–247. ISBN 978-94-017-8840-3.

CHAPTER 6

UTILIZATION OF BACTERIOPHAGES TARGETING *LISTERIA MONOCYTOGENES* IN THE DAIRY AND FOOD INDUSTRY

ANDREI SORIN BOLOCAN[1,*], LUMINIŢA CIOLACU[2], ELENA-ALEXANDRA ONICIUC[3], LORRAINE DRAPER[1,4], ANCA IOANA NICOLAU[3,5], MARTIN WAGNER[2,6], and COLIN HILL[1,7]

[1]*APC Microbiome Institute, University College Cork, College Road, Cork, Ireland, Tel.: +353214901781, Mob.: +353851529997*

[2]*Department for Farm Animal and Public Health in Veterinary Medicine, Institute of Milk Hygiene, Milk Technology and Food Science, University of Veterinary Medicine Vienna, Veterinαrplatz 1, 1210 Vienna, Austria, Tel.: +43 1 25077 3510, E-mail: luminita. Ciolacu@vetmeduni.ac.at, Tel.: 0043 1 25077 3500, Fax: 0043 1 25077 3590*

[3]*Department of Food Science and Engineering and Applied Biotechnology, Faculty of Food Science and Engineering, Dunarea de Jos University of Galati, Galati, Romania, Mob.: +40 744998122, E-mail: elena.Oniciuc@ugal.ro*

[4]*Tel.: +353214901781. E-mail: l.draper@ucc.ie*

[5]*Mob.: +40 755746227, E-mail: anca.nicolau@ugal.ro*

[6]*Tel.: 0043 1 25077 3500, Fax: 0043 1 25077 3590, E-mail: martin.wagner@vetmeduni.ac.at*

[7]*Tel.: +353214901781, E-mail: c.hill@ucc.ie*

Corresponding author. E-mail: andrei.bolocan@ucc.ie; andrei.s.bolocan@gmail.com

6.1   INTRODUCTION

The modern microbial food safety perspective is based more and more on "natural" antimicrobials that allow the control of the microbial ecology of foods (Verma et al., 2017b). Bacteriophages (phages), viruses that kill bacteria, are the oldest and most ubiquitous biological systems on the Earth (Chibani-Chennoufi et al., 2004; Verma and Srivastav, 2017) that are suitable candidates for the environment friendly biocontrol of foodborne pathogens. Their high specificity to target the bacterial host, ability to withstand food processing environmental stresses together with their low toxicity, and prolonged shelf life make phages a good alternative to desinfectants, antibiotics, and/or food preservatives. From the food safety perspective, phages are used to target foodborne pathogens at different stages of the food chain; in livestock they are administrated as therapy to prevent or reduce animal illness, thus preventing the entry of pathogens into the food processing environment; in the food processing environment they are used as biosanitation agents for cleaning and disinfecting thus reducing colonization and/or biofilm formation. They are also added to food products or applied on the food surface to inhibit and/or eradicate the growth of pathogenic and spoilage bacteria reviewed (Sillankorva et al., 2012; Sulakvelidze, 2013).

This chapter gives an overview of the foodborne pathogen *Listeria monocytogenes* as the causative agent of listeriosis and its occurrence in milk and dairy products and discusses the use of bacteriophages for the biocontrol of this pathogen in the dairy industry.

6.2   LISTERIA MONOCYTOGENES

Human listeriosis is one of the most serious foodborne diseases under European Union (EU) surveillance, causing high morbidity, hospitalization, and mortality leading to gastroenteritis, meningoencephalitis, and maternofetal infections. The risk groups are the elderly and immunocom promised persons as well as pregnant women and infants (EFSAECDPC, 2016).

L. monocytogenes is the causative agent of listeriosis, it is a Gram-positive facultative intracellular pathogen able to persist in food processing environments for longtime periods increasing the risk of (re) contamination of food products (Rychli et al., 2016).Owing to its ability to survive and grow in acidic and alkaline conditions, its tolerance for high salt concentrations, low temperatures, and low water activity (a_w) (Gandhi and Chikindas, 2007), the pathogen successfully colonizes in food and food processing environments (Ferreira et al., 2014). *L. monocytogenes* can adhere to food-contact surfaces such as stainless steel, plastic, and glass, where it can form biofilms or find suitable conditions for growing on floors, drains, the cold and wet atmosphere of refrigerated rooms within food processing facilities (Carpentier and Cerf, 2011).

Food products are the vehicle of 99% of listeriosis cases, with outbreaks mainly associated with ready-to-eat (RTE) dairy, fish, and meat products (EFSAECDPC, 2016). According to Regulation 2073/2005 concerning microbiological criteria for foodstuffs, the criterion of less or equal to 100 colony-forming units (cfu)/g in 25 g of sample is required during the entire shelf life of RTE foods on the market. Despite this, a statistically significant increasing trend in listeriosis cases was reported between 2008 and 2015 with an EU case fatality of 17.7% among the 1524 confirmed cases in 2015 (EFSAECDPC, 2016).

Different *L. monocytogenes* strains show differences in epidemic potential and virulence. The majority of human listeriosis outbreaks worldwide have been linked to lineage I serotype 4b isolates and only some outbreaks have been caused by lineage I serotype 1/2b isolates and lineage II serotype 1/2a isolates. On the other hand, lineage II (serotypes 1/2a, 1/2c, and 3a) isolates are usually more frequently recovered from foods and food plant environments as compared to lineage I (serotypes 1/2b, 3b, 3c, and 4b) isolates (Orsi et al., 2011).

In order to cause an infection through transmission of contaminated foods, L. monocytogenes needs to contaminate and survive food processing environments, be present in raw materials or finished products, and to overcome the conditions encountered during the passage through the gastrointestinal tract of the host (Lecuit, 2005). The strain, the environmental conditions, and the nature of food are the vehicles of listeriosis which have an impact on the pathogenicity of *L. monocytogenes* (Kathariou, 2002; Mahoney and Henriksson, 2003; Duodu et al., 2010). For example, milk and milk-specific characteristics, such as fat content, have an influence on the virulence of *L. monocytogenes*, a fact that has been demonstrated in-vitro (Pricope-Ciolacu et al., 2013).

Several listeriosis outbreaks have been reported in the last decade and many were related to dairy products. In Europe, the large listeriosis outbreak reported in 2009/2010 in Austria, Germany, and the Czech Republic was due to contaminated acid curd cheese with 34 cases, including eight fatalities (Schoder et al., 2014). An outbreak of listeriosis, including 12 human cases was reported in Belgium in 2011, presumably due to the consumption of hard cheese made with pasteurized milk, produced by a Belgium manufacturer (Yde et al., 2012). The most recent listeriosis outbreak was reported in the United States (2016–2017) and included six infected people from four states. The outbreak was traced back to soft raw milk cheese (CDCP, 2017).

6.2.1   L. MONOCYTOGENES IN DAIRY INDUSTRY

Owing to their high content of proteins, minerals, and vitamins, milk and dairy products are recommended as one of the main components of a healthy diet across all categories of the population (Verma et al., 2017a). However, the high content of nutrients also makes them a favorable media for the survival and growth of many different foodborne pathogens, including *L. monocytogenes*. Although, *L. monocytogenes* is efficiently inactivated during processing through thermal treatments, for example, pasteurization, many listeriosis outbreaks have nonetheless been traced back to pasteurized milk and dairy products. Thus, indicating that pasteurization alone is not enough to control milk-borne pathogens (Koch et al., 2010; Swaminathan and Gerner-Smidt, 2007). Recontamination can occur from the processing environment, equipment, or by human handling during post-pasteurization, packaging, and storage of food (Ferreira et al., 2014). Nevertheless, raw milk or raw milk dairy products are still consumed worldwide and related listeriosis outbreaks were also reported (CDCP, 2016).

Two different scenarios are likely to explain the presence of the pathogen in milk and dairy products. The first one refers to the consumption of *L. monocytogenes* present in contaminated raw milk and dairy products manufactured from contaminated raw milk without prior heating to ensure the inactivation of the pathogen. *L. monocytogenes* can cause mastitis in cows and be excreted from the udder of the infected animal. The prevalence of *L. monocytogenes* in raw milk was previously investigated and was reported to range between 0 and 20% (Waak et al., 2002; D'Amico and Donnelly, 2010; Fox et al., 2011).Silage, fecal shedding, and infected animals were also reported as sources of contamination of raw milk at livestock farms (Nightingale et al., 2004). Maintenance and on-farm transmission of L. monocytogenes appear to rely on ingestion of L. monocytogenes-contaminated feeds and amplification of L. monocytogenes by bovine hosts both with or without clinical disease, followed by fecal dispersal into the farm environment (Nightingale et al., 2004).There are different factors such as farm size and number of animals on the farm, hygiene and farm management practices, sampling and detection methods as well as geographical location, and season impact on the prevalence of *L. monocytogenes* in milk (Oliver et al., 2005). In total, 23% of human sporadic strains had pulse field gel electrophoresis patterns identical to that of farm isolate(s) (Borucki et al., 2004).

The second scenario is likely the contamination of post-pasteurization of otherwise safe milk and/or dairy products. In this case, the main source of contamination is the food processing environments, were the pathogen can survive and persist for long periods of time. Improper application of good manufacturing practices and good hygiene practices (GHP), together with improper human handling and transition between processing lines correlated with unhygienic design of the equipment may lead to the colonization and spread of *L. monocytogenes* throughout the environment of the dairy processing facilities, resulting in contamination of RTE products (Muhterem-Uyar et al., 2015; Da Silva and De Martinis, 2013) (Fig. 6.1).

GHP as well as environmental monitoring and efficient sanitation procedures are critical in order to prevent and control the spreading of *L. monocytogenes* strains in niches in dairy plants.

Van Asselt et al. (2017) analyzed the food safety hazards in the European dairy supply chain and reported that soft and semisoft cheeses are most frequently associated with *L. monocytogenes* due to a high moisture content that allows growth during refrigerated storage. The occurrence of *L. monocytogenes* in soft and semisoft cheeses made from raw or low heat-treated milk was 1.4%; only slighter higher than in soft and semisoft cheeses made from pasteurized milk (1.3%). Hard cheeses are assumed not to support the growth of *L. monocytogenes* and no counts of *L. monocytogenes* > 100 cfu/g were reported by European Food Safety Authority (EFSA) in hard cheeses made either from pasteurized or raw milk in 2015 (EFSAECDPC, 2016). This is also confirmed by many challenge tests, that in hard cheeses such as raw milk farmhouse cheddar cheeses low prevalence or absence of *L. monocytogenes* was seen (Dalmasso and Jordan, 2014).

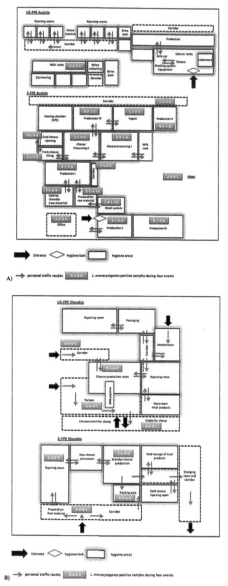

FIGURE 6.1   (See color insert.) Example of *Listeria monocytogenes* spreading in Austrian industrial dairy (A) and Slovakian farmhouse cheese making (B) food processing environments.

Reprinted with permission from Muhterem-Uyar, M.; Dalmasso, M.; Bolocan, A. S.; Hernandez, M.; Kapetanakou, A. E.; Kuchta, T.; Manios, S. G.; Melero, B. and Minarovičová, J. Environmental sampling for Listeria monocytogenes control in food processing facilities reveals three contamination scenarios. Food Control, 2015, 51, 94–107. © 2015 Elsevier.

6.2.2　L. MONOCYTOGENES IN PERSISTENCE AND BIOFILMS

The main risk posed by *L. monocytogenes* arises from the ability of the pathogen to overcome different environmental conditions and to quickly colonize new niches in the food processing environment where it is able to persist for long periods of time. Persistence, defined as repeated isolation of the same subtype over a period of time from the same food-processing plant, results in intermittent contamination of foods and leads to transmission of the pathogen to humans (Ferreira et al., 2014).Even if the exact mechanisms or specific genetic determinates for persistence are not totally understood or defined, persistent strains were reported (Ferreira et al., 2014; Schoderet al., 2014; Stesslet al., 2014) and were linked to several listeriosis outbreaks such as the human listeriosis outbreak in2000 caused by a *L.* monocytogenes strain that persisted in a food processing facility over 12 years (Orsi et al., 2008). *L. monocytogenes* of sequence type 121, known to be persistent in the food processing environment across Europe harbor the transposon, Tn*6188*, which is responsible for increased tolerance against quaternary ammonium compounds (QAC) that might contribute to survival and persistence (Müller et al., 2014). Stessl et al. (2014) investigated the occurrence and potential persistence of *L. monocytogenes* across 19 producers of traditional Irish, Czech, and Austrian cheese products and observed that *L. monocytogenes* persistence was a widespread epidemiological event. Despite the fact that contamination with *L. monocytogenes* could be traced back for more than a decade, the niche of in-house strain survival in the investigated cheese processing facilities remained unknown.

In order to persist in the food processing environment, *L. monocytogenes* needs to survive cleaning and disinfection processes. One of the strategies used is biofilm formation (Giaouris et al., 2014). Biofilms are recognized as a dominant lifestyle of microorganisms attached to a biotic or abiotic surface and embedded within self-produced extracellular polymeric substances (EPS). The ability of *L. monocytogenes* to form biofilms varies between strains and depends on different intrinsic and extrinsic factors (e.g., nutrient level and temperature) (Chihib et al., 2003; Di Bonaventura et al., 2008; Pan et al., 2010; Kadam et al., 2013; Doijad et al., 2015). L. monocytogenes can adhere to many materials frequently used in milk handling equipment, milk lines, or milk tanks such as stainless steel, rubber, or plastic (Beresford et al., 2001). Furthermore, Cocolin et al. (2009) found that 30% of the plastic and wood shelves used for

Gorgonzola cheese maturation were contaminated. Bacterial biofilms are an alarming issue in the food industry owing to their increased resistance to antimicrobials, desiccation, heat, and disinfectants.

6.3   BACTERIOPHAGES VERSUS *L. MONOCYTOGENES:* OLD FIGHT AND SAME ENEMIES

First reports on *L. monocytogenes*-specific bacteriophages date back to the 1940s and 1960s. The isolation and characterization of *L. monocytogenes* phages started with the work done by Sword and Pickett. They have isolated phages from lysogenic strains and based on their characteristics were proposed as tools for the positive generic identification of fresh *L. monocytogenes* isolates (Sword and Pickett 1961). The method was further developed and in 1986 a phage typing system for *L. monocytogenes* based on 28 phages was evaluated to be used for epidemiological studies (McLauchlin et al., 1986). Till date more than 500 bacteriophages infecting *L. monocytogenes* and other *Listeria* spp. are isolated from different sources such as sewage, plants, silage, mushrooms (Fig. 6.2) as well as knowledge of lysogenic *L. monocytogenes* strains (Schmuki et al., 2012; Vongkamjan et al., 2012; Klumpp and Loessner, 2013; Casey et al., 2015). All of them are members of the *Caudovirales*, featuring either the long noncontractile tails of the *Siphoviridae* family or the complex contractile tail machines of the *Myoviridae* family (Klumpp and Loessner, 2013). No Podoviridae infecting *Listeria* have been found and reported thus far (Hagens and Loessner, 2014). The commercially available main *L. monocytogenes* phages are listed in Table 6.1.

One of the first phages used to eliminate *L. monocytogenes* is phage P100, commercially available as Listex™ P100 (Micreos Food Safety, Wageningen, the Netherlands) and recommended at a dosage of 1×10^8pfu/ml or 1 ml/100 cm². Member of the Myoviridae family, the virulent phage P100, isolated from the wastewater of a dairy plant is able to kill the host as soon the infection has been established. However, based on genomic and toxic analysis, the phage showed no indications for any potential risk and in 2007 it was also confirmed as Generally Regarded as Safe (GRAS) by Food and Drug Administration (FDA) and EFSA. Similar to phage P100 and member of the same family is phage A551 which is able to infect about 95% of *L. monocytogenes* strains across the major serovar groups 1/2 and 4b (Guenther et al., 2009).

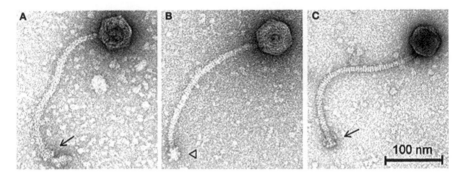

FIGURE 6.2 Transmission electron micrographs of the lytic phages vB_LmoS_188 (isolate from wild mushrooms—A) and vB_LmoS_293 (isolate from mushroom compost—B,C) negatively stained using 2% uranyl acetate. The double-discs conformation of the baseplates is indicated by the arrows (A,C). The six-pointed star conformation of the phage vB_LmoS_293 baseplate is indicated by the triangle in (B).

Sources: Reprinted from Casey et al. (2015). https://creativecommons.org/licenses/by/4.0/

TABLE 6.1 Bacteriophages Commercially Available Against *Listeria monocytogenes* in Dairy Industry

Phage name	Commercial name	Producer	Approved for use	References
P100	Listex P100	Micreos Food Safety, Wageningen, the Netherlands	In all food materials (2007)	(Carlton et al., 2005; Soniet al., 2012; Silva et al., 2014)
List-36 (ATCC# PTA-5376), LMSP-25 (ATCC# PTA-8353), LMTA-57 (ATCC# PTA8355), LMTA-94 (ATCC# PTA-8356), LMTA-148 (ATCC# PTAPTA-8357)	ListShield™	Intralytix, Inc, Baltimore, MD, USA	On food and surfaces in food production facilities (2006)	(Sulakvelidze, 2013; Ly-Chatain, 2014; Sadekuzzamanet al., 2017)

Another bacteriophage product ListShield™, Intralytix, Inc., Baltimore, MD, USA, consists of a cocktail of equal quantities of five individually purified phages such as List-36 (ATCC# PTA-5376), LMSP-25 (ATCC# PTA-8353), LMTA-57 (ATCC# PTA8355), LMTA-94 (ATCC#

PTA-8356), and LMTA-148 (ATCC# PTAPTA-8357). Each of these monophages is effective against a specific *L. monocytogenes* serotype. ListShield™ is a liquid preparation, with a minimum phage concentration of 10.0 ± 0.3 log·pfu/ml and was first approved by the USFDA and US Department of Agriculture's Food Safety and Inspection Service for applications in foods in 2006 and reapproved as GRAS status by FDA in 2014.

6.4 APPLICATION OF BACTERIOPHAGES TARGETING *L. MONOCYTOGENES*

6.4.1 ON NONFOOD-CONTACT SURFACES

As post-processing contamination of dairy products is well-documented, the control of the processing environments is critical. One of the first directions to use phages against *L. monocytogenes* was as prevention method for spreading of the pathogen or its possible latter persistence on food and nonfood-contact surfaces. Persistence and biofilms are very often related to improper cleaning and sanitation processes or resistance of strains to the disinfectants routinely used in the processing environments. Therefore, there is always the need to develop new formulas or find other methods to eliminate the pathogen from food processing surfaces. One of the proposed methods was the use of bacteriophages.

AISI 304 stainless steel is the most commonly used equipment in the dairy industry. Therefore, many studies focused on removal of *L. monocytogenes* biofilms established on this type of surface. The use of phageP100 treatment proved to be efficient in significantly reducing *L. monocytogenes* cell populations under biofilm conditions irrespective of the serotype, growth conditions, or biofilm levels on stainless steel surfaces (Soni and Nannapaneni 2010; Iacumin et al., 2016). The same bacteriophage, P100, was used to remove 14 days old biofilms on stainless steel AISI type 304 plates with grooves (depths between 0.2 and 5 mm). When compared to chemical disinfectants (sodium dichloroisocyanurate and QAC), bacteriophages showed the best antimicrobial effect in most cases in shallow grooves, but not in the deep grooves (Chaitiemwong et al., 2014). ListShield™ preparation, applied onto the surfaces by spraying, or with a cloth or sponge, to cover the targeted surface reduced the biofilms more than 2 log· cfu/cm^2 on stainless steel surface and more than 1 log·cfu/cm^2 on rubber surfaces at all temperatures tested (Sadekuzzaman et al., 2017).

Investigating the susceptibility of biofilms to bacteriophage attack, Hughes et al. (1998) revealed that the biofilm EPS initially needs to be degraded by phage-borne polysaccharide depolymerase, to allow the phage to gain access to bacterial surface receptors. Therefore, the specificity of the enzyme attacking the biofilm, the number of phages and the different phenotypic states of bacteria at different levels in a biofilm are major factors that impact the efficacy of the phage application (Hughes et al., 1998). Indeed, Montañez-Izquierdo et al. (2012) confirmed, by using fluorescent microscopy (Fig. 6.3), the need of combining other hygienization measures to phage treatment in order to increase efficacy. Furthermore, Ganegama Arachchi et al. (2013) observed that phages LiMN4L, LiMN4p, and LiMN17 were more effective on biofilm cells dislodged from the surface compared with undisturbed biofilm cells.

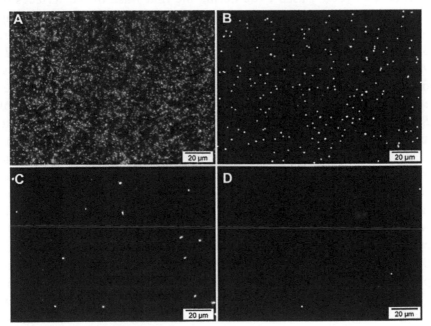

FIGURE 6.3 (See color insert.) Vital status of *L. monocytogenes* in biofilms on stainless steel surfaces treated with (A) 5 log·pfu/ml phage P100, (B) 6 log·pfu/ml phage P100, (C) 7 log·pfu/ml phage P100, and (D) 8 log·pfu/ml of phage P100 at 48 h. Epifluorescence digital images of the stain with LIVE/DEAD.

Reprinted from with permission from Montañez-Izquierdo, V. Y.; Salas-Vázquez, D. I.; Rodríguez-Jerez, J. J. Use of Epifluorescence Microscopy to Assess the Effectiveness of Phage P100 in Controlling *Listeria monocytogenes* Biofilms on Stainless Steel Surfaces. *Food Control.* **2012**, *23*, 470–477. © 2012 Elsevier.

6.4.2   IN DAIRY PRODUCTS

Some *L. monocytogenes* phages have already been approved by EFSA and FDA as SAFE to be used in dairy products due to their potential to inhibit pathogenic and spoilage bacteria. Their efficiency, however, needs to be optimized for each product and its specific properties in order to convince producers to add them in combination or instead of the already used antibacterial ingredients.

The majority of studies performed to date investigate *L. monocytogenes* survival using different cheese models and by applying the phage treatment on the surface of the cheese. The reported results are consistent with and underline the importance of phage dosage, level of *L. monocytogenes* contamination, and time point of treatment application. Carltonet al. (2005) used P100 during the rind wash of the surfaces of red smear soft cheese (type "munster") contaminated at the beginning of the ripening period with a low concentration of *L. monocytogenes*. Significant reduction (at least 3.5 logs) or complete eradication of *L. monocytogenes* viable counts was reached depending on the time points, frequency, and dose of phage applications (Carlton et al., 2005).Guenther et al. (2011)evaluated the efficacy of phage A511 to control the growth of *L. monocytogenes* during the production and ripening of Camembert-like and red smear soft cheeses. The singlehigher dose of phage application (1×10^9 PFU/cm^2) was found to be more effective compared to multiple low-dose treatments. The highest efficacy of the phage was seen at target cell concentrations of 100 cfu/cm^{2}or below, while at higher contamination levels a significant reduction of the *L .monocytogenes* viable counts was reported. Therefore, application of the phage treatment needs to be done at an early stage as both cheeses are supporting the growth of the pathogen (Guenther and Loessner, 2011). The inhibitory effect of phage P100 against *L. monocytogenes* on soft cheese (Minas Frescal and Coalho cheeses) was investigated by Silva et al. (2014).The efficacy of the treatment depends of the initial concentration of *L. monocytogenes* and needs high concentration of phages per unit area (Silva et al., 2014). Guenther et al. (2009) showed that the efficiency of phage treatment is not affected by storage time and temperature but depends on the host to target ratio. Furthermore, the food matrix itself can prevent, due to specific characteristics (e.g. ionic strength, pH), the binding of phages to their receptors on the bacterial surfaces. Indeed both P100 and A511 phages suppressed or prevented the growth of *L. monocytogenes* strains (serovar 1/2c and 4b) at higher levels in chocolate milk and mozzarella cheese brine compared to solid foods (Guenther et al., 2009). Even if phages show extreme host specificity their effect on starter culture and/ or on the natural microflora of milk and dairy products needs to be also considered. They might negatively impact on the organoleptic proper-ties or reduce beneficial microflora (Sulakvelidze, 2013). Even if it was reported that phage A511 did not affect the "normal" microbiota of white and red smear soft cheeses (Guenther and Loessner, 2011) other studies are needed.

Endolysins or bacteriophage-derived cell wall hydrolytic enzymes could be a good alternative to prevent listerial contamination during the production of fresh cheeses. Van Tassell et al. (2017) recombinantly expressed endolysin PlyP100 and tested its efficacy against a cocktail of *L. monocytogenes* in a fresh cheese model. PlyP100 demonstrates optimal activity under pH and salt concentrations consistent with a low-acid food matrix such as fresh cheese and displays considerable inhibitory activity against *L. monocytogenes* in fresh cheese for at least 4 week under refrig-erated storage (Van Tassell et al., 2017).

The combination of two phages (FWLLm1 and FWLL3) and one bacteriocin (coagulin C23) was also tested as a possible alternative for the efficient eradication of *L. monocytogenes* in milk due to an induced synergistic effect. Only one combination (FWLLm1 and coagulin C23) showed an enhanced inhibitory effect compared to each antimicrobial alone. The failure of the second combination was due to a higher rate of resistance development and the presence of cross-resistance (Rodríguez-Rubio et al., 2015).

In a new approach to control *L. monocytogenes* in queso fresco cheese, Soni et al. (2012) evaluated the combined effects of GRAS antimicrobials (lauricarginate and potassium lactate–sodium diacetate mixture) with bacteriophage P100 (phage P100). Between 2 and 3.5–4 log·cfu/cm² reduction of *L. monocytogenes* counts were detected by using 7.8 log·pfu/ cm² of phage P100 or 200 ppm of lauricarginate, but a subsequent regrowth of *L. monocytogenes* at 4°C was seen. However, the combination treatment of phage P100 or lauricarginate with potassium lactate–sodium diacetate mixture reduced the initial *L. monocytogenes* counts by 2–4 log· cfu/cm² and kept the *L. monocytogenes* counts at that reduced level for 28 days at 4°C (Soni et al., 2012).

6.5   LIMITATION OF PHAGE APPLICATION

The effectiveness of phage applications in eradicating foodborne pathogens depends on several factors such as the bacteriophages/target bacteria ratio, the mode and moment of treatment, environmental conditions, the neutralization of phage, and accessibility to target bacteria, amongst others (Ly-Chatain, 2014). Therefore, all specific food and food-processing environments factors need to be considered when designing the right application procedures. Fister and coworker showed that phage P100 is stable under most conditions typically encountered in dairy production environments, but phage replication, which is essential for all active control strategies, is reduced by many important factors such as the extreme pH values, high salt concentrations, and/or presence of detergents (Fister et al., 2016a, b).

The natural microbiota of food or other food ingredients might also impact the efficacy of the phages as it might impair the accessibility of the host to the specific bacterial receptors. For example, it has been shown that for phages used against *Staphylococcus aureus*, whey proteins and raw milk are inhibiting phage attachment (Gill et al., 2006) and phage proliferation, respectively (O'Flaherty et al., 2005). Contradictory results were also published. For example, bacteriophages were effective against *Escherichiacoli* strains in ultrahigh temperature and raw bovine milk at various storage temperatures (O'Flaherty et al., 2005). Therefore, all this data need to be considered carefully and individual studies are essential for each case.

Another concern of phage application is the possibility that *L. monocytogenes* might be insensitive or might develop resistance to bacteriophages. The sensitivity of 486 *L. monocytogenes* strains isolated from 59 dairy plants in Austria over 15 years against phage P100 was investigated and only 2.7% of the isolates were found to be resistant. The occurrence of resistance was related to phage P100 application (Listex™) and could be due to bacterial cell wall modifications that hinder entry or attachment of the phage to the bacteria (Fister et al., 2016b). Conversely, Carlton and coworkers found no evidence for phage resistance in the *Listeria* isolates recovered from cheese samples treated with low concentrations of P100 (Carlton et al., 2005). It is, however, recognized that phages have the ability to adapt to new receptors and to overcome bacterial resistance (Labrie et al., 2010). This should be further investigated insubsequent studies, especially in food processing environments where phages are used intensively.

6.6 PERSPECTIVES IN UTILIZATION OF BACTERIOPHAGES IN FOOD INDUSTRY

In recent years, there is an increasing trend in consumer attitude to avoid chemicals and favor environment-friendly treatments and new "green" strategies to control foodborne pathogens. The use of bacteriophages as novel and natural antibacterials are supported by numerous studies that prove their efficacy.

Even if some phages or phage mixtures are already approved to be used in food and food processing environments, their use should be still looked carefully. Successful phage infection and killing of host cells is strongly dependent on the environmental conditions, food matrix, and phage–host ratio. Therefore, all protocols developed by the food safety authorities and/or food producers to apply phages against *L. monocytogenes* should be individually optimized not only with respect to the phages and target organisms but also by considering specific properties of the food matrix, for example, pH and storage temperature. Furthermore, the concentration of the host and phages is essential for efficient inhibitory effect and should be adapted considering the contamination point along the food chain, type of surface, and other cleaning and disinfection processes used. Combined application or alternative solutions, as the use of endolysins have been also considered with the main advantage of an easier approval process compared to viruses.

Although bacteriophages have their own limitation and much information is still lacking they arise as a good alternative to increase or optimize food safety.

6.7 CONCLUSIONS AND SUMMARY

L. monocytogenes is a major concern within the dairy industry as it is the causative agent of human listeriosis. There are numerous publications detailing the spread of *L. monocytogenes* from the dairy processing environment to the final dairy products. Owing to the ability of this pathogen to persist within the food processing environment, despite stringent hygiene practices, bacteriophages or bacterial viruses have been proposed as an alternative for the biocontrol of *L. monocytogenes*. The potential use of bacteriophages within food processing facilities has been previously demonstrated through research studies. Currently, two commercial products, that is, Listex P100 and ListShield™, are available for use in the food industry. As a novel alternative to current methods, the lysins produced by bacteriophages were explored as a possiblebiosanitizer for *L. monocytogenes* and the results are promising.

Therefore, this book chapter focuses on the utilization of bacteriophage as abiocontrol agent to target *L. monocytogenes* within the dairy industry; specifically by the application of bacteriophages for the elimination of biofilms and the control of growth in individual dairy products. In addition, this chapter discussesthe limitations and the perspectives of the bacteriophages application in the food industry. The food matrices, environmental conditions, and other preservative agents present in the food product and the ability of the bacteria to become resistant to bacteriophages will always raise troubles inthe designofa new bacteriophage commercial product intended for use in food industry.

6.8   AUTHOR'S CONTRIBUTION

Authors, Andrei Sorin Bolocan of University College Cork and Luminița Ciolacu of University of Veterinary Medicine, Vienna, both contributed equally to this chapter.

KEYWORDS

- antibacterial ingredients
- antimicrobials
- biofilms
- bacteriophages
- biocontrol
- biosanitation
- cheeses
- endolysin
- Generally Regarded as Safe
- listeriosis

- *Listeria monocytogenes*
- maternofetal infections
- meningoencephalitis
- phage cocktail
- phages
- podoviridae
- siphoviridae
- virulent phage

REFERENCES

Beresford, M. R.; Andrew, P. W.; Shama, G. *Listeria monocytogenes* Adheres to Many Materials Found in Food-Processing Environments. *J. Appl. Microbiol.* **2001**, *90*, 1000–1005.

Borucki, M. K.; Reynolds, J.; Gay, C. C.; McElwain, K. L.; Kim, S. H.; Knowles, D. P.; Hu, J. Dairy Farm Reservoir of *Listeria monocytogenes* Sporadic and Epidemic Strains. *J. Food Prot.* **2004**, *67*, 2496–2499.

Carlton, R. M.; Noordman, W. H.; Biswas, B.; De Meester, E. D.; Loessner, M. J. Bacteriophage P100 for Control of *Listeria monocytogenes* in Foods: Genome Sequence, Bioinformatic Analyses, Oral Toxicity Study, and Application. *Regul. Toxicol. Pharmacol.* **2005**,*43*, 301–312.

Carpentier, B.; Cerf, O. Review—Persistence of *Listeria monocytogenes* in Food Industry Equipment and Premises. *Int. J. Food Microbiol.* **2011**, *145*, 1–8.

Casey, A.; Jordan, K.; Neve, H.; Coffey, A.; McAuliffe, O. A Tail of Two Phages: Genomic and Functional Analysis of *Listeria monocytogenes* Phages vB_LmoS_188 and vB_LmoS_293 Reveal the Receptor-Binding Proteins Involved in Host Specificity. *Front Microbiol.* **2015**, *6*, 1–14. DOI: 10.3389/fmicb.2015.01107.

CDCP (Centers for Disease Control and Prevention) (2016). Multistate Outbreak of Listeriosis Linked to Raw Milk Produced by Miller's Organic Farm in Pennsylvania (Final Update).https://www.cdc.gov/listeria/outbreaks/raw-milk-03-16/(accessed May 05, 2017).

CDCP (Centers for Disease Control and Prevention) (2017). Multistate Outbreak of Listeriosis Linked to Soft Raw Milk Cheese Made by Vulto Creamery (Final Update). https://www.cdc.gov/listeria/outbreaks/soft-cheese-03-17/index.html (accessed May 05, 2017).

Chaitiemwong, N.; Hazeleger, W. C.; Beumer, R. R. Inactivation of *Listeria monocytogenes* by Disinfectants and Bacteriophages in Suspension and Stainless Steel Carrier Tests. *J. Food Prot.* **2014**, *77*, 2012–2020.

Chibani-Chennoufi, S.; Bruttin, A.; Dillmann, M. L.; Brüssow, H. Phage-Host Interaction: An Ecological Perspective. *J. Bacteriol.* **2004**, *186*, 3677–3686.

Chihib, N.; Ribeiro, M.; Delattre, G.; Laroche, M.; Federighi, M. Different Cellular Fatty Acid Pattern Behaviours of Two Strains of *Listeria monocytogenes* Scott A and CNL 895807 Under Different Temperature and Salinity Conditions. *FEMS Microbiol. Lett.* **2003**, *218*, 155–160.

Cocolin, L.; Nucera, D.; Alessandria, V.; Rantsiou, K.; Dolci, P.; Grassi, M. A.; Lomonaco, S.; Civera, T. Microbial Ecology of Gorgonzola Rinds and Occurrence of Different Biotypes of *Listeria monocytogenes*. *Int. J. Food Microbiol.* **2009**, *133*, 200–205.

D'Amico, D. J.; Donnelly, C. W. Microbiological Quality of Raw Milk Used for Small-Scale Artisan Cheese Production in Vermont: Effect of Farm Characteristics and Practices. *J. Dairy Sci.* **2010**, *93*, 134–147.

Da Silva, E. P.; De Martinis, E. C. P. Current Knowledge and Perspectives on Biofilm Formation: the Case of *Listeria monocytogenes*. *Appl. Microbiol. Biotechnol.* **2013**, *97*, 957–968.

Dalmasso, M.; Jordan, K. Absence of Growth of *Listeria monocytogenes* in Naturally Contaminated Cheddar Cheese. *J. Dairy Res.* **2014**, *81*, 46–53.

Di Bonaventura, G.; Piccolomini, R.; Paludi, D.; D'Orio, V.; Vergara, A.; Conter, M.; Ianieri, A. Influence of Temperature on Biofilm Formation by *Listeria monocytogenes* on Various Food-Contact Surfaces: Relationship with Motility and Cell Surface Hydrophobicity. *J. Appl. Microbiol.* **2008**, *104*, 1552–1561.

Doijad, S. P.; Barbuddhe, S. B.; Garg, S.; Poharkar, K. V.; Kalorey, D. R.; Kurkure, N. V.; Rawool, D. B.; Chakraborty, T. Biofilm-Forming Abilities of *Listeria monocytogenes* Serotypes Isolated from Different Sources. *PLoS One* **2015**, *10*, e0137046.

Duodu, S.; Holst-Jensen, A.; Skjerdal, T.; Cappelier, J. M.; Pilet, M. F.; Loncarevic, S. Influence of Storage Temperature on Gene Expression and Virulence Potential of *Listeria monocytogenes* Strains Grown in a Salmon Matrix. *Food Microbiol.* **2010**, *27*, 795–801.

EFSAECDPC (European Food Safety Authority and European Centre for Disease Prevention and Control).The European Union Summary Report on Trends and Sources of Zoonoses, Zoonotic Agents and Food-Borne Outbreaks in 2015. *EFSA J.* **2016**, *14*(12), 4634. DOI: 10.2903/j.efsa.2016.4634.

Ferreira, V.; Wiedmann, M.; Teixeira, P.; Stasiewicz, M. J. *Listeria monocytogenes* Persistence in Food-Associated Environments: Epidemiology, Strain Characteristics, and Implications for Public Health. *J. Food Prot.* **2014**, *77*, 150–170.

Fister, S.; Fuchs, S.; Stessl, B.; Schoder, D.; Wagner, M.; Rossmanith, P. Screening and Characterisation of Bacteriophage P100 Insensitive *Listeria monocytogenes* Isolates in Austrian Dairy Plants. *Food Control* **2016b**, *59*, 108–117.

Fister, S.; Robben, C.; Witte, A. K.; Schoder, D.; Wagner, M.; Rossmanith, P. Influence of Environmental Factors on Phage-Bacteria Interaction and on the Efficacy and Infectivity of Phage P100. *Front. Microbiol.* **2016a**, *7*(1152), 1–13. DOI: 10.3389/fmicb.2016.01152.

Fox, E.; Hunt, K.; O'Brien, M.; Jordan, K. *Listeria monocytogenes* in Irish Farmhouse Cheese Processing Environments. *Int. J. Food Microbiol.* **2011**, *145*(Suppl 1), S39–S45.

Gandhi, M.; Chikindas, M. L. Listeria: A Foodborne Pathogen that Knows How to Survive. *Int. J. Food Microbiol.* **2007**, *113*, 1–15.

GanegamaArachchi, G. J.; Cridge, A. G.; Dias-Wanigasekera, B. M.; Cruz, C. D.; McIntyre, L.; Liu, R.; Flint, S. H.; Mutukumira, A. N. Effectiveness of Phages in the Decontamination of *Listeria monocytogenes* Adhered to Clean Stainless Steel, Stainless Steel Coated with Fish Protein, and as a Biofilm. *J. Ind. Microbiol. Biotechnol.* **2013**, *40*, 1105–1116.

Giaouris, E.; Heir, E.; Hébraud, M.; Chorianopoulos, N.; Langsrud, S.; Møretrø, T.; Habimana, O.; Desvaux, M.; Renier, S.; Nychas, G. J. Attachment and Biofilm Formation by Foodborne Bacteria in Meat Processing Environments: Causes, Implications, Role of Bacterial Interactions and Control by Alternative Novel Methods. *Meat Sci.* **2014**, *97*, 289–309.

Gill, J. J.; Sabour, P. M.; Leslie, K. E.; Griffiths, M. W. Bovine Whey Proteins Inhibit the Interaction of *Staphylococcus aureus* and Bacteriophage K. *J. Appl. Microbiol.* **2006**, *101*, 377–386.

Guenther, S.; Loessner, M. J. Bacteriophage Biocontrol of *Listeria monocytogenes* on Soft Ripened white Mold and Red-Smear Cheeses. *Bacteriophage* **2011**, *1*, 94–100.

Guenther, S.; Huwyler, D.; Richard, S.; Loessner, M. J. Virulent Bacteriophage for Efficient Biocontrol of *Listeria monocytogenes* in Ready-to-Eat Foods. *Appl. Environ. Microbiol.* **2009**, *75*, 93–100.

Hagens, S.; Loessner, M. J. Phages of *Listeria* offer Novel Tools for Diagnostics and Biocontrol. *Front. Microbiol.* **2014**, *5*, 1–6.

Hughes, K. A.; Sutherland, I. W.; Jones, M. V. Biofilm Susceptibility to Bacteriophage Attack: The Role of Phage-Borne Polysaccharide Depolymerase. *Microbiology* **1998**, *144*, 3039–3047.

Iacumin, L.; Manzano, M.; Comi, G. Phage Inactivation of *Listeria monocytogenes* on San Daniele Dry-Cured Ham and Elimination of Biofilms from Equipment and Working Environments. *Microorganisms* **2016**, *4*(1), 4. DOI: 10.3390/microorganisms4010004.

Kadam, S. R.; Den Besten, H. M. W.; Van der Veen, S.; Zwietering, M. H.; Moezelaar, R.; Abee, T. Diversity Assessment of *Listeria monocytogenes* Biofilm Formation: Impact of Growth Condition, Serotype and Strain Origin. *Int. J. Food Microbiol.* **2013**, *165*, 259–264.

Kathariou, S. *Listeria monocytogenes* Virulence and Pathogenicity, a Food Safety Perspective. *J. Food Prot.* **2002**, *65*, 1811–1829.

Klumpp, J.; Loessner, M. J. Listeria Phages: Genomes, Evolution, and Application. *Bacteriophage* **2013**, *3*(1), e26861. DOI: 10.4161/bact.26861.

Koch, J.; Dworak, R.; Prager, R.; Becker, B.; Brockmann, S.; Wicke, A.; Wichmann-Schauer, H.; Hof, H.; Werber, D.; Stark, K. Large listeriosis Outbreak linked to Cheese Made from Pasteurized Milk, Germany, 2006–2007. *Foodborne Pathog. Dis.* **2010**, *7*, 1581–1584.

Labrie, S. J.; Samson, J. E.; Moineau, S. Bacteriophage Resistance Mechanisms. *Nat. Rev. Micro.* **2010**, *8*, 317–327.

Lecuit, M. UnderstandingHow *Listeria monocytogenes* Targets and Crosses Host Barriers. *Clin. Microbiol. Infect.* **2005**, *11*, 430–436.

Ly-Chatain, M. H. The Factors Affecting Effectiveness of Treatment in Phages Therapy. *Front. Microbiol.* **2014**, *5*(51), 1–7. DOI: 10.3389/fmicb.2014.00051.

Mahoney, M.; Henriksson, A. The Effect of Processed Meat and Meat Starter Cultures on Gastrointestinal Colonization and Virulence of *Listeria monocytogenes* in Mice. *Int. J. Food Microbiol.* **2003**, *84*, 255–261.

McLauchlin, J.; Audurier, A.; Taylor, A. G. The Evaluation of a Phage-Typing System for *Listeria monocytogenes* for use in Epidemiological Studies. *J. Med. Microbiol.* **1986**, *22*, 357–365.

Montañez-Izquierdo, V. Y.; Salas-Vázquez, D. I.; Rodríguez-Jerez, J. J. Use of Epifluorescence Microscopy to Assess the Effectiveness of Phage P100 in Controlling *Listeria monocytogenes* Biofilms on Stainless Steel Surfaces. *Food Control.* **2012**, *23*, 470–477.

Muhterem-Uyar, M.; Dalmasso, M.; Bolocan, A. S.; Hernandez, M.; Kapetanakou, A. E.; Kuchta, T.; Manios, S. G.; Melero, B.; Minarovičová, J. Environmental Sampling for *Listeria monocytogenes* Control in Food Processing Facilities Reveals Three Contamination Scenarios. *Food Control* **2015**, *51*, 94–107.

Müller, A.; Rychli, K.; Zaiser, A.; Wieser, C.; Wagner, M.; Schmitz-Esser, S. The *Listeria monocytogenes* Transposon Tn6188 Provides Increased Tolerance to Various Quaternary Ammonium Compounds and Ethidium Bromide. *FEMS Microbiol. Lett.* **2014**, *361*, 166–173.

Nightingale, K. K.; Schukken, Y. H.; Nightingale, C. R.; Fortes, E. D.; Ho, A. J.; Her, Z.; Grohn, Y. T.; McDonough, P. L.; Wiedmann, M. Ecology and Transmission of *Listeria monocytogenes* Infecting Ruminants and in the Farm Environment. *Appl. Environ. Microbiol.* **2004**, *70*, 4458–4467.

O'Flaherty, S.; Coffey, A.; Meaney, W. J.; Fitzgerald, G. F.; Ross, R. P. Inhibition of Bacteriophage K Proliferation on *Staphylococcus aureus* in Raw Bovine Milk. *Lett. Appl. Microbiol.* **2005**, *41*, 274–279.

Oliver, S. P.; Jayarao, B. M.; Almeida, R. A. Foodborne Pathogens in Milk and the Dairy Farm Environment: Food Safety and Public Health Implications. *Foodborne Pathog. Dis.* **2005**, *2*, 115–129.

Orsi, R. H.; Bakker, H. C. den; Wiedmann, M. *Listeria monocytogenes* lineages: Genomics, Evolution, Ecology, and Phenotypic Characteristics. *Int. J. Med. Microbiol.* **2011**, *301*, 79–96.

Orsi, R. H.; Borowsky, M. L.; Lauer, P.; Young, S. K.; Nusbaum, C.; Galagan, J. E.; Birren, B. W.; Ivy, R. A.; Sun, Q.; Graves, L. M.; Swaminathan, B.; Wiedmann, M. Short-Term Genome Evolution of *Listeria monocytogenes* in a Non-Controlled Environment. *BMC Genomics* **2008**, *9*(539), 1–17. DOI: 10.1186/1471-2164-9-539.

Pan, Y.; Breidt, F.; Gorski, L. Synergistic Effects of Sodium Chloride, Glucose, and Temperature on Biofilm Formation by *Listeria monocytogenes* Serotype 1/2a and 4b Strains. *Appl. Environ. Microbiol.* **2010**, *76*, 1433–1441.

Pricope-Ciolacu, L.; Nicolau, A. I.; Wagner, M.; Rychli, K. The Effect of Milk Components and Storage Conditions on the Virulence of *Listeria monocytogenes* as Determined by a Caco-2 Cell Assay. *Int. J. Food Microbiol.* **2013**, *166*, 59–64.

Rodríguez-Rubio, L.; García, P.; Rodríguez, A.; Billington, C.; Hudson, J. A.; Martínez, B. Listeriaphages and CoagulinC23 Act Synergistically to Kill *Listeria monocytogenes* in Milk Under Refrigeration Conditions. *Int. J. Food Microbiol.* **2015**, *205*, 68–72.

Rychli, K.; Grunert, T.; Ciolacu, L.; Zaiser, A.; Razzazi-Fazeli, E.; Schmitz-Esser, S.; Ehling-Schulz, M.; Wagner, M. Exoproteome Analysis Reveals Higher Abundance of Proteins Linked to Alkaline Stress in Persistent *Listeria monocytogenes* Strains. *Int. J. Food Microbiol.* **2016**, *218*, 17–26.

Sadekuzzaman, M.; Yang, S.; Rahaman Mizan, M. F.; Kim, H.-S.; Ha, S.-D. Effectiveness of a Phage Cocktail as a Biocontrol Agent Against *L. monocytogens* Biofilms. *Food Control* **2017**, *78*, 256–263.

Schmuki, M. M.; Erne, D.; Loessner, M. J.; Klumpp, J. Bacteriophage P70: Unique Morphology and Unrelatedness to Other Listeria Bacteriophages. *J. Virol.* **2012**, *86*, 13099–13102.

Schoder, D.; Stessl, B.; Szakmary-Brändle, K.; Rossmanith, P.; Wagner, M. Population Diversity of *Listeria monocytogenes* in Quargel (acid curd cheese) Lots Recalled During the Multinational Listeriosis Outbreak2009/2010. *Food Microbiol.* **2014**, 39, 68–73.

Sillankorva, S. M.; Oliveira, H.; Azeredo, J. Bacteriophages and Their Role in Food Safety. *Int. J. Microbiol.* **2012**, *2012*, 1–13. http://dx.doi.org/10.1155/2012/863945.

Silva, E. N. G.; Figueiredo, A. C. L.; Miranda, F. A.; de Castro Almeida, R. C. Control of *Listeria monocytogenes* Growth in Soft Cheeses by Bacteriophage P100.*Brazilian J. Microbiol.* **2014**, *45*, 11–16.

Soni, K. A.; Nannapaneni, R. Removal of *Listeria monocytogenes* Biofilms with Bacteriophage P100. *J. Food Prot.* **2010**, *73*(8), 1519–1524.

Soni, K. A.; Desai, M.; Oladunjoye, A.; Skrobot, F.; Nannapaneni, R. Reduction of *Listeria monocytogenes* in Queso Fresco Cheese by a Combination of Listericidal and Listeriostatic GRAS Antimicrobials. *Int. J. Food Microbiol.* **2012**, *155*, 82–88.

Stessl, B.; Fricker, M.; Fox, E.; Karpiskova, R.; Demnerova, K.; Jordan, K.; Ehling-Schulz, M.; Wagner, M. Collaborative Survey on the Colonization of Different Types of Cheese-Processing Facilities with *Listeria monocytogenes*. *Foodborne Pathog. Dis.* **2014**, *11*, 8–14.

Sulakvelidze, A. Using lytic Bacteriophages to Eliminate or Significantly Reduce Contamination of Food by Foodborne Bacterial Pathogens. *J. Sci. Food Agric.* **2013**, *93*, 3137–3146.

Swaminathan, B.; Gerner-Smidt, P. The Epidemiology of Human Listeriosis. *Microbes Infect.* **2007**, *9*, 1236–1243.

Sword, C. P.; Pickett, M. J. The Isolation and Characterization of Bacteriophages from *Listeria monocytogenes*. *J. Gen. Microbial.* **1961**, *25*, 241–248.

Van Asselt, E. D.; Van der Fels-Klerx, H. J.; Marvin, H. J. P.; Van Bokhorst-van de Veen, H.; Groot, M. N. Overview of Food Safety Hazards in the European Dairy Supply Chain. *Compr. Rev. Food Sci. Food Saf.* **2017**, *16*, 59–75.

Van Tassell, M. L.; Ibarra-Sánchez, L. A.; Hoepker, G. P.; Miller, M. J. Hot Topic: Antilisterial Activity by EndolysinPlyP100 in Fresh Cheese. *J. Dairy Sci.* **2017**, *100*, 2482–2487.

Verma, D. K.; Srivastav, P. P. Microorganisms in Sustainable Agriculture, Food and the Environment. In *as part of Book Series on Innovations in Agricultural Microbiology*; Verma, D. K., Srivastav, P. P. Eds.; Apple Academic Press: USA; 2017.

Verma, D. K.; Mahato, D. K.; Billoria, S.; Kapri, M.; Prabhakar, P. K.; Ajesh Kumar, V.; Srivastav, P. P. Microbial Approaches in Fermentations for Production and Preservation of Different Food. In *Microorganisms in Sustainable Agriculture, Food and the Environment. as part of book series on Innovations in Agricultural Microbiology*; Verma, D. K., Srivastav, P. P. Eds.; Apple Academic Press: USA; 2017a.

Verma, D. K.; Mahato, D. K.; Billoria, S.; Kapri, M.; Prabhakar, P. K.; Ajesh Kumar, V.; Srivastav, P. P. Microbial Spoilage in Milk Products, Potential Solution, Food Safety and

Health Issues. In *Microorganisms in Sustainable Agriculture, Food and the Environment. as part of book series on Innovations in Agricultural Microbiology*; Verma, D. K.; Srivastav, P. P. Eds.; Apple Academic Press, USA, 2017b.

Vongkamjan, K.; Switt, A. M.; den Bakker, H. C.; Fortes, E. D.; Wiedmann, M. Silage Collected from Dairy Farms Harbors an Abundance of Listeriaphages with Considerable Host Range and Genome Size Diversity. *Appl. Environ. Microbiol.* **2012**, *78*, 8666–8675.

Waak, E.; Tham, W.; Danielsson-Tham, M. L. Prevalence and Fingerprinting of *Listeria monocytogenes* Strains Isolated from Raw Whole Milk in Farm Bulk Tanks and in Dairy Plant Receiving Tanks. *Appl. Environ. Microbiol.* **2002**, *68*, 3366–3370.

Yde, M.; Naranjo, M.; Mattheus, W.; Stragier, P.; Pochet, B.; Beulens, K.; De Schrijver, K.; Van den Branden, D.; Laisnez, V.; Flipse, W.; Leclercq, A.; Lecuit, M.; Dierick, K.; Bertrand, S. Usefulness of the European Epidemic Intelligence Information System in the management of an Outbreak of Listeriosis, Belgium, 2011. *Euro Surveill.* **2012**, *17*(38), 1–5.

PART III

Emerging Scope and Potential Applications in the Dairy and Food Industry

SIGNIFICANCE OF FOOD-GRADE MICROORGANISMS IN BIODEGRADATION AND BIOTRANSFORMATION OF HARMFUL TOXIC COMPOUNDS

AMI PATEL[1,*], POOJA THAKKAR[2], NIHIR SHAH[1,3], DEEPAK KUMAR VERMA[4], and PREM PRAKASH SRIVASTAV[4,5]

[1]*Division of Dairy and Food Microbiology, Mansinhbhai Institute of Dairy and Food Technology-MIDFT, Dudhsagar Dairy Campus, Mehsana, Gujarat 384002, India, Mob.: +00–91–9825067311, Tel.: +00–91–2762243777 (O), Fax: +91–02762–253422*

[2]*Department of Life Sciences, Gujarat University, Ahmedabad, India, Mob.: +00–91–9909804647, E-mail: poojathakkar16603@gmail.com*

[3]*Mobile: +00–91–9925605480, E-mail: nihirshah13@yahoo.co.in*

[4]*Agricultural and Food Engineering Department, Indian Institute of Technology, Kharagpur, West Bengal 721302, India, Tel.: +91 3222281673, Mob.: +91 7407170259, Fax: +91 3222282224, E-mail: deepak.verma@agfe.iitkgp.ernet.in, rajadkv@rediffmail.com*

[5]*E-mail: pps@agfe.iitkgp.ernet.in*

Corresponding author. E-mail: amiamipatel@yahoo.co.in

7.1   INTRODUCTION

In the biosphere, microorganisms are the only entities with an incomparable aptitude to utilize diverse organic and inorganic substances for growth and convert it to simple form which is no longer harmful to human health and environment (Verma and Srivastav, 2017). Their potential to alter natural and synthetic hazardous chemical compounds into simple energy sources and/or raw materials for own growth offer a viable alternative to expensive chemical as well as physical remediation processes; further, biological processes are environment-friendly and economic (Sasikumar and Papinazath, 2003). Many microorganisms including bacteria, yeast, and molds have shown their biodegradation and bioremediation significance. However, little is known about the application of food-grade microorganisms in this aspect. Among various bacteria, lactic acid bacteria (LAB) and related genera such as *Bifidobacterium, Propionibacteria,* as well as several yeast species are considered as food-grade microorganisms, inherently associated with fermentation of many food products (Patel et al., 2017; Verma et al., 2017a). Some of these microorganisms have gained probiotic status due to their noteworthy contributions in health stimulatory activities as a natural gut microflora in animals and humans (Patel, 2012; Patel et al., 2017).

Apart from playing essential role in food fermentation processes, recent research activities have suggested new possibilities of employing these Generally Regarded as Safe status microorganisms for environment-friendly approaches such as recycling waste; biotransformation and/or biodegradation of mycotoxins, heavy metals, polycyclic aromatic hydrocarbons (PAHs), pesticides, azo dyes, and other toxic metabolites (Verma et al., 2017b). The strains of Lactobacillus, present in the various fermented food products and in human mouth, gut, and vagina, have been proven to have an ability to bind and detoxify some of the toxic metabolites into harmless form (Hernandez-Mendoza et al., 2009). LABs have been investigated to evaluate their capability to degrade some petroleum hydrocarbons and to decolorize several synthetic dyes in order to reveal their potential application in bioremediation (Abou-Arab et al., 2010; Elbanna et al. 2017). Furthermore, recent reports on the biotechnological production of lactic acid represent an appropriate solution to the environmental pollution caused by the petrochemical industries (Hamdan and Sonomoto, 2011).

In context to this, we had given an overview of diverse studies dealing with several toxic compounds, discussed their outcomes, summarized the current understanding of biodegradation and/or detoxification mechanisms of food-grade bacteria, and future application of these magic bugs in the prevention and control of food and environmental contamination with harmful toxic compounds in this chapter.

7.2 GENERAL MECHANISM OF ACTION OF FOOD-GRADE MICROORGANISMS

LAB and related food-grade bacteria generally produce organic acids and decrease the pH of specific environment. Among these, lactic acid has a strong sterilizing effect that suppresses spoilage types as well as harmful microorganisms and also augments decomposition of organic compounds. LAB promote the decomposition of complex materials such as lignin, cellulose, hemicellulose, and fermentation of these substances, thereby eliminating undesirable effects of these nondecomposed organic materials. Higa and Chinen (1998) proposed that effective microorganisms contain a variety of organic acids (such as lactate, acetate, propionate) aid by the existing LAB, which also secrete enzymes, antioxidants, and metal sequesters/chelates that help in detoxifying harmful substances by deionization of toxic substances. Further, the chelation of heavy metals like iron consequently induces microorganisms to secrete decomposing enzymes such as lignin peroxidase (Higa, 1999).

In many cases, bacterial cell wall components (proteins, polysaccharide, and peptidoglycan) have been observed to be responsible for reducing the bioavailability of toxic metabolites such as mycotoxins or heavy metals (Fig. 7.1). It involves simple adsorption or ion-exchange mechanism. Further, researchers had mentioned that several food-grade strains produce enzymes (such as hydrolases, reductases, dehydrogenases) that can modify or degrade aromatic compounds such as azo dyes and polycyclic hydrocarbons. For example, *Lactobacillus casei* TISTR 1500 was found to possess cytoplasmic azoreductase which could break down azo bonds under microaerophilic conditions (Seesuriyachan et al., 2011). Earlier *Lactobacillus* strains were assumed to synthesize chromate-reducing enzyme; following the partial purification process (two- to five-fold), an enzyme chromate reductase have been characterized in these strains (Mishra et al., 2012).

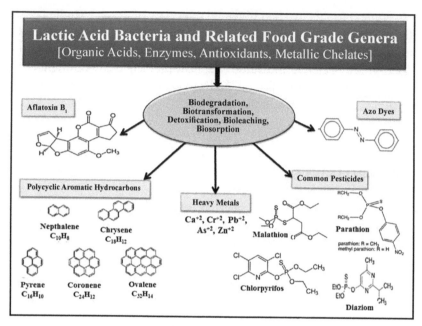

FIGURE 7.1   (See color insert.) Schematic representation of various mechanisms of food-grade bacteria for biological control of toxic compounds.

7.3   BIOLOGICAL CONTROL OF VARIOUS TOXIC COMPOUNDS

7.3.1   MYCOTOXINS

Mycotoxins are synthesized by molds like *Aspergillus, Penicillium,* and *Fusarium,* which are generally found as natural contaminants in foods. Mycotoxins can cause toxicities to humans and animals but several bacteria and fungi species have been revealed to degrade mycotoxins enzymatically (Reddy et al., 2010). Among molds, several species of *Aspergillus* such as *A. flavus, A. parasiticus, A. ochraceoroseus, A. nomius, A. bombycis, A. tamari,* and *A. pseudotamarii* are known to produce aflatoxins (AFs), a group of highly toxic secondary metabolites. It is found in improperly stored milk, cereals, tree nuts, oilseeds, and spices and when such contaminated foods are processed, AFs enter the general food supply (Cousin et al., 2005).

AF is a well-known carcinogen; International Agency for Research in Cancer and classified AF as a class 1 human carcinogen. There are about

18 different kinds of AFs such as B_1, B_2, B_{2a}, G_1, G_2, M_1, M_2, P, Q, and so forth, identified so far. The B_1, B_2, G_1, and G_2 AF types are the most frequently occurring ones in fungi cultures, while M_1 and M_2 types are usually observed in milk. Once AFs enter in the body by any means, in the liver, it may be metabolized to a reactive epoxide intermediate or hydroxylated to be converted into the less dangerous AF M_1 (Boonen et al., 2012). According to another research, 8,9-epoxide—an AFB_1 metabolite—forms DNA adducts primary with guanine on N7 (Qinghua et al., 2009). AFB_1 is the most potent toxin among various AFs that can permeate through the skin. AFB_1 exerts its noxious effects upon the liver cells, further responsible for imposing damages on DNA, mutation, stimulation of cancer and abortion, birth deformity, immune system suppression, and phytotoxic reactions (Hamidi et al., 2013). Toxic effects of AFs, in particular, carcinogenic effects have been reported in humans and diverse animals. A number of aflatoxicosis outbreaks have taken place in Western Europe, Asia and Africa countries; primarily in adults with poor nutritional status plus maize as a staple food. An outbreak of aflatoxicosis was resulted in about 125 deaths among 317 cases of poisoning in Kenya during May 2004 (Probst et al., 2007). Numerous animal studies have shown that the liver is the main target organ and therefore the main symptoms of AF exposure in domestic and laboratory animals are hepatic injuries (Silvia, 2007).

Several species of the LAB have been investigated for their prospective to either degrade or reduce the bioavailability of mycotoxins (Table 7.1). They have been found to protect against toxic compounds contained in foods such as amino acids, pyrolysates, heterocyclic aromatic amines, PAHs, and mycotoxins (Topcu et al., 2010). Some strains of LAB had shown great ability to bind AF in contaminated medium. It is proposed that a physical union, an adhesion to microbial cell wall constituents (polysaccharide and peptidoglycan), might be responsible for reducing the bioavailability of mycotoxins instead of covalent binding or degradation (Elsanhoty, 2014) though the exact mechanism/mode of action has not been clarified yet. Previously, El-Nezami et al. (1998) showed that probiotic lactobacilli have the ability to remove AFs from contaminated liquid media. Further, metabolically inactivated cells (heat and acid treated) performed AF removal. Haskard et al. (2001) reported that viable cells of *Lactobacillus rhamnosus* strain GG (ATCC 53103) and *L. rhamnosus* strain LC-705 (DSM 7061) removed AFB_1 from aqueous solution. Abdella et al. (2005) observed that *Lactobacillus* strains could remove more AF in comparison to the strains of *Pediococcus* and *Leuconostoc*.

TABLE 7.1 Investigations Showing Effect of Food-Grade Microorganisms Against Mycotoxins.

Probiotic strain(s)	Food/medium	Major effects	References
Lactobacillus delbrueckii UFV H2b20, *Saccharomyces cerevisiae* var. *boulardii*, and *S. cerevisiae* UFMG 905	Peanut grains	All probiotics either in live or inactivated state notably reduced *A. parasiticus* sporulation, though best results were achieved with live cells only; combination of *S. boulardii* and *L. delbrueckii* (96.1%) observed the best reduction followed by *Saccharomyces boulardii* with *S. cerevisiae* and *L. delbrueckii* with *S. cerevisiae*, that is, 71.1 and 66.7%, respectively	Da Silva et al. (2015)
Lactic acid bacteria (LAB) and bifidobacteria against aflatoxin M_1 (AFM_1)	Yoghurt	Observed significant differences ($P < 0.05$) between the strains in their ability to decrease AFM_1 in De Man, Rogosa and Sharpe (MRS) broth in the viable state, heated state, and acid treatment; *Lactobacillus plantarum* was the highest capable strain of removing AFM_1; at the end of storage period yoghurt fermented by 50% yoghurt culture (*Streptococcus thermophilus* and *Lactobacillus bulgaricus*) and 50% *L. plantarum* recorded the highest reduction in the AFM_1 concentration	Elsanhoty et al. (2014)
LAB and bifidobacteria against aflatoxin M_1 (AFM_1)	Yoghurt	AFM_1 levels decreased significant ($P < 0.05$) in the probiotic yoghurts as compared with those initially added to milk; AFM_1 had no remarkable effect on the viability of tested bacteria	Montaseri et al. (2014)
Bacteriocinogenic *L. plantarum*, *Lactococcus lactis* and aflatoxin B_1 (AFB1)	–	*L. plantarum* had a superior detoxification rate (46%) in comparison to *L. lactis* (27%); the most successful detoxification rate (81%) was achieved when strains were incubated together in a single broth culture	Sezer et al. (2013)
Lactobacillus. pentosus and *Lactobacillus. brevis* against AFB_1	–	Both the strains demonstrated the capability to absorb and isolate AFB_1 by absorbing and discharging 17.4% and 34.7%, respectively	Hamidi et al. (2013)

TABLE 7.1 *(Continued)*

Probiotic strain(s)	Food/medium	Major effects	References
Enterococcus faecium M74 and *E. faecium* EF031 against AFB$_1$ and patulin	—	Strain M74 removed 19.3–30.5% of AFB$_1$ and 15.8–41.6% of patulin, while strain EF031 removed 23.4–37.5% of AFB$_1$ and 19.5–45.3% of patulin all over incubation period of (48 h)	Topcu et al. (2010)
Lactobacillus acidophilus, L. brevis, Lactobacillus casei, Lactobacillus delbruekii, and *L. plantarum* against AFB$_1$	Maize	Treatments with all LAB showed significant reductions ($P<0.05$) in AFB$_1$ concentration; *L. plantarum* appeared as the most competent strain for AFB$_1$ degradation	Oluwafemi et al. (2010)
L. plantarum PTCC 1058 against AFB$_1$	—	Within 1 h, 45%, while within 90 h, 100% removal of AFB$_1$ was reported from solution by the strain	Khanafari et al. (2007)
Probiotic LAB and bifidobacteria against AFB$_1$	—	For the six different strains of *Lactobacillus* and *Bifidobacterium,* the aflatoxin-binding capacity was ranged from 5.8 to 31.3%	Peltonen et al. (2001)

In one recent study, the amount of AFB_1 was decreased by 55% after 6 h of fermentation with the probiotic microorganisms (*Lactobacillus paracasei* LOCK 0920, *Lactobacillus plantarum* LOCK 0945, *Lactobacillus brevis* LOCK 0944, *Saccharomyces cerevisiae* LOCK 0140) with a low level of AFB_1 (1 mg/kg) in a feed mixture (Sliżewskaand and Smulikowska, 2011). In 2009, Kabak et al. first reported the ability of six probiotic LAB on the bioaccessibility of ochratoxin A and AFB_1 through an in vitro digestion model under fed conditions. Depending on the types of toxin and their food and contamination levels, all the strains observed varying binding ability to both the mycotoxins. Overall, for all the tested probiotic strains, a maximum of 37% reduction and 73% reduction was detected for both AFB_1 and Ochratoxin A (OTA), respectively.

The toxin binding level differs according to several environmental conditions such as temperature, pH, cell density, toxin concentration, and physical state (live/dead) of microbial cells (Zinedine et al., 2005; Fuchs et al., 2008). Furthermore, the application of heat and acid affects the microbial cell structure which ultimately affects the stability of toxin–bacteria complex (Haskard et al., 2001). A recent experiment suggested that feed additive (0.15%) containing *L. casei, Bacillus subtilis,* and yeast *Pichia anomala* in broiler diets could considerably relieve the harmful effect of AFB_1 on the production performance of chicken and nutrient metabolic rates ($P<0.05$). Moreover, it significantly improved AFB_1 metabolism, antioxidant activity, hepatic cell structure, and various hepatic enzyme gene expressions involved in oxidoreductase, cell growth, apoptosis, metabolic process, and immune system ($P<0.05$) (Zuo et al., 2013).

7.3.2   PESTICIDES

There are several studies showing the efficiency of food-grade bacteria on the pesticide biodegradation on various substrates such as milk, yoghurt, and different silages. Earlier, in the presence of 200 mg/l of chlorpyrifos (CP), four different CP-degrading LAB, identified as *Leuconostoc mesenteroides* WCP907, *L. brevis* WPC902, *Lactobacillus sakei* WCP904, and *L. plantarum* WPC 931 were isolated from kimchi fermentation by Cho et al. (2009). The authors observed that these strains were able to utilize CP as the sole source of carbon and phosphorous. These strains were also found to degrade diazinon (DZ), coumaphos (CM), methyl parathion (MPT), and parathion (PT) as a sole source of carbon and phosphorous.

In the concomitant experiment, it was established that among all the isolates, *L. brevis* WCP902 was able to degrade organophosphorus pesticides (OPPs) by synthesizing organophosphorus hydrolase (OpdB) enzyme (Islam et al., 2010; Azizi, 2011). In that, optimum OpdB enzyme activity appeared at 35°C and at pH of 6.0 during the degradation of DZ, CM, CP, PT, and MPT. Researchers mentioned that OpdB contains the same motif "Gly–X–Ser–X–Gly" found in majority of bacterial and eukaryotic enzymes such as serine hydrolases, lipases, and esterase; and the conserved serine residue "Ser82," considerably involved with enzyme activity, is supposed to be linked with the removal of several pesticides (Islam et al., 2010).

Ayana and coworkers investigated the effect of fungicides (Anadol and Tasolen), herbicides (Roundup and Saturn), and insecticides (Lannate and Reldan) on acid, acetaldehyde, and diacetyl production from yoghurt culture and ABT cultures such as *Streptococcus salivarius* ssp. *thermophilus* H, *Lactobacillus delbrueckii* ssp. *bulgaricus*, *Lactobacillus acidophilus* (Type 145), *L. casei* ssp. *casei*, and *Bifidobacterium* spp. (Ayana et al., 2011). Most lactobacilli exhibited sensitivity toward insecticides, whereas *Bifidobacterium* spp. appeared excessive sensitive. Few more such investigations are comprised in Table 7.2.

7.3.3 POLYCYCLIC AROMATIC HYDROCARBONS

PAHs may be described as a group of chemicals comprising two or more fused aromatic rings that are synthesized from the incomplete combustion or high-temperature pyrolysis of wood, coal, fossil fuel, oil, gas, garbage, or other organic compounds, like tobacco smoke, charbroiled meat, and also exhaust from automobile and trucks (Mottier et al., 2000). These PAHs enter in the environment through volcano, forest fire, domestic combustion, cigarette smoke, and many such related sources and release to air, water, soil, and ultimately in food. PAHs have proven for toxic and mutagenic effects in animals and carcinogenic effects on humans. Some PAHs including benzo(a)pyrene and dibenzo(a,h)anthracene can damage genetic materials, in turn, initiate cancer development (Schneider et al., 2014).

Several PAHs are found to be present in foodstuffs from diverse environmental sources, and therefore, development of some technological processes to reduce the contents of these chemical pollutants to enhance the food safety and quality would serve as a better viable alternative

TABLE 7.2 Investigations Showing Effect of Food-Grade Microorganisms Against Pesticides.

Pesticide(s) and substrate type	Employed strain(s)	Major findings	References
Chlorantraniliprole and chlorpyrifos (CP); silage Alfalfa (*Medicago sativa* L.)	*L. plantarum*	Fermentation with strain increased the contents of lactate and other short-chain fatty acids of silage treated with pesticides ($P<0.05$) with simultaneous reduction in butyric acid content and pH; treatment also slowed down the dissipation of CP	Zhang et al., (2017)
CP; rice straw silage	*L. casei* WYS3	Viable cells of strain were shown to bind 33.3–42% of spiked CP in silage; as determined in gas chromatography–mass spectrometry (GC-MS) analysis strain detoxified CP through P-O-C cleavage and the gene expression of organophosphorus hydrolase tripled after the pesticide addition to lactic culture, in comparison to the control unit (without CP)	Wang et al. (2016)
CP and phorate; whole corn silage	LAB strains	Combination of strains has shown greater organophosphorus pesticides (OPPs) degradation (decreased OPPs levels from 24.9 to 33.4%) than a distinct strain or the wild microbiota. Further, CP had lower degradation rate constants (0.0274–0.0381 vs. 0.0295–0.0355/week) in comparison to phorate	Zhang et al. (2016)
Skimmed milk and CP, diazinon (DZ), fenitrothion (FNT), malathion (MT), and methyl parathion (MPT)	10 LAB strains	Degradation of pesticides was enhanced with the inoculated lactic strains, which resulted in 0.8–225.4% increase in the rate constants. DZ and MPT were more stable, while CP, FNT, and MT were more labile	Zhang et al. (2014)
Skimmed milk and nine OPPs	Five strains of the LAB and two commercial yoghurt starters	In the milk, OPP dissipation was enhanced by the inoculated strains and starters, resulting in decreased OPP concentration by 7.0–64.6% and 7.4–19.2%, respectively. All nine OPPs were more susceptible to *Lactobacillus bulgaricus*, which enhanced their degradation rate constants by 18.3–133.3%	Zhou and Zhoa (2015)

(Abou-Arab et al., 2010). Food-grade bacteria are commonly employed in dairy and food industries for various purposes; it is possible that incorporated starters microorganisms would degrade such pollutants and would be of great interest and value in context to a public health concern. A very limited information is available in this aspect. Three different bacteria, namely *Bifidobacterium bifidum, S. thermophilus,* and *Lactobacillus bulgaricus* were investigated for biodegradation of PAHs in liquid medium and fermented milk (yoghurt) by Abou-Arab et al. (2010). The reductions were 46.6, 87.7, and 91.5% with *B. bifidum, S. thermophiles,* and *L. bulgaricus,* respectively, after 72 h of incubation at 37°C. The authors suggested that the bacterial cell might adsorb PAHs and the persistence of PAHs was dependent on numerous factors including the composition of the medium, kind of microorganism and their concentration, the interaction between PAHs and target microbial cell, and the growth conditions such as temperature and pH.

In a recent investigation, Bartkiene and coworkers evaluated the effect of LAB treatment on the formation of PAHs as well as biogenic amines in cold-smoked pork meat sausages (Bartkiene et al., 2017). The authors noticed that surface treatment of sausages with selected lactic strains (such as *L. sakei, Pediococcus acidilactici, Pediococcus pentosaceus*) before and after smoking, significantly reduced both chrysene and benzo(*a*) pyrene contents ($P<0.05$). The LAB treatment before smoking decreased spermidine and cadaverine content, while after smoking, it either diminished putrescine content or totally eliminated from outer layers and center of sausages depending on the strain used.

7.3.4   AZO DYES

In comparison to natural dyes, azo dyes are the most widely employed synthetic colorants owing to several advantages, such as cost-effectiveness, ease of synthesis, stability, and availability in different colors. Azo dyes are utilized in food, brewing, pharmaceutical, and cosmetic industries. Conversely, a number of studies suggest that some of the azo dyes have toxic, mutagenic, and/or carcinogenic effects (Spadaro et al. 1992; Lu et al., 2010). Azoreductase produced by several microorganisms can break azo bonds (–N=N–) and form nitro-aromatic compounds. The biodegradation process can be aerobic, anaerobic, or a combination of the two. In azo dyes, reductive cleavage of azo bond (–N=N–) generates colorless aromatic amines (Mohammed, 2009). Several non-lactic bacteria and fungus species have been reported for azo dyes degradation and mineralization ability; however, information on the capability of food-grade microorganisms is scant.

In one of the earlier experiments, *L. casei* was capable to decolorize azo dye from textile wastewater and it was proven that LAB have a capability to decolorize dye (Seesuriyachana et al., 2007). In another study, strain *L. casei* TISTR 1500 had shown good prospective of azo decolorization by altering the dye to N, N-dimethyl-p-phenylenediamine and 4-aminobenzenesulfonic acid (Seesuriyachan et al., 2007). Further, the strain had specific decolorization rate of 14.2 mg/gCell/h (Seesuriyachan et al., 2009). The authors also noticed that growing cells were more tolerant to a high initial concentration of dye than the freely suspended cells. In a subsequent experiment, it was observed that the composition of medium plays a significant role in decolorization of methyl orange and production of biomass (Seesuriyachan et al., 2011). *L. paracasei* had shown a complete reduction of food-grade dyes, that is, *amaranth, tartrazine,* and *allura* red dye within a period of 48 h (Hassan et al., 2014). Several such studies showing percent dye decolorization using food-grade bacteria have been comprised in Table 7.3.

TABLE 7.3 Investigations Showing Percent Dye Decolorization Using Food Grade Bacteria.

Food-grade bacteria	Azo dye	Percentage decolonization	References
Four LAB strains (LAB13, LAB100, Ent2, and Eco5)	Food-grade dyes: tartrazine (E102), sunset yellow (E110), carmoisine (E122), and ponceau 4R (E124)	96–98%	Elbanna et al. (2017)
Lactobacillus sp.	Remazol golden yellow	81%	Palani et al. 2012
L. delbrueckii	Reactive orange 16	57%	Zuraida et al., 2013a
	Reactive black 5	51%	
L. delbrueckii	Batik wastewater	51%	Zuraida et al. 2013b
Oenococcus oeni ML34	Fast red	93%	El Ahwany, 2008
Subspecies of *L. lactis* and *L. casei*	Reactive black 5	99%	You and Teng, 2009

Earlier, Pérez-Díaz et al. (2007) observed that microaerophilic *L. para-casei* LA 0471 degraded 80% of tartrazine with a growth rate 0.023 h^{-1} in less than 13 days, whereas *L. casei* LA 1133 degraded 35% of initial tartrazine with a growth rate at 0.052 h^{-1} in 17 days. In another study, *L. acidophilus* ATCC 4356 showed a complete reduction of methyl red, orange G, Sudan III, and Sudan IV. On the other hand, *Lactobacillus fermentum* ATCC 23271 showed complete degradation of only Sudan III and Sudan IV, but not of methyl red (Chen et al., 2009). Moreover, when *Oenococcus oeni* ML34 was supplied with 5 g/l of glucose, bacterium showed up to 93% decolorization of fast red dye (El Ahwany, 2008).

Even at a higher concentration of 1000 mg/l (RO16) dye, a textile soil isolate of *Lactobacillus* exhibited notable color removal capability (95%) within 24 h (Shah, 2014). In another approach, a total of 80 LAB isolates were shown to decolorize the textile azo dyes, namely eriochrome red B, reactive Lanasol Black B, and 1:2 metal complexes I Yellow (SGL) from 75 to 100% within 4 h. The highly efficient strains were belonging to *L. casei, L. rhamnosus,* and *L. paracasei.* The genes required for the azo dyes degradation were found to be located on a 3 kb plasmid (Elbanna et al., 2010). In a recent in vivo experiment in rats, Elbanna et al. (2017) observed significant decreases in inflammation and a noticeable improvement in the liver, kidney, spleen, and small intestine of rats treated with LAB and food azo dyes simultaneously were observed.

Most of the bacterial genera from the azo dye-Remazol golden yellow contaminated soil belonged to *Bacillus* spp. (47.91%), subsequently followed by *Enterobacteriaceae* spp. (17.70%), and *Lactobacillus* spp. (4.16%); *Lactobacillus* spp. exhibited 81% dye decolonization (Palani et al., 2012). Xu et al. (2010) mentioned that majority of colonic gut bacteria including *L. rhamnosus, Bifidobacterium infantis, Enterococcus faecalis, Clostridium indolis,* and *Ruminococcus obeum* were capable of reducing all four tested Sudan dyes and para red completely.

7.3.5 HEAVY METALS

The interactions of the LAB with metal ions and possible applications have received less attention. Binding of heavy metals on the LAB could be a promising solution for the removal or reducing toxic heavy metal ions from water, liquid food, and from the body (Schut et al., 2011), while leaching of trace elements could enhance the value of fermented foods. From water, probiotic also LAB had proven to bind cadmium ions (Halttunen et al., 2003). Though very limited information is available in this direction (as comprised in Table 7.4), the outcomes of different investigations done so far are encouraging and indicative of prospective use of LAB as metal bioquencher/biosequesters, and thus detoxifying agent.

Mishra et al. (2012) revealed that sequential subculturing of *Lactobacillus* strains in hexavalent chromium Cr(VI) resulted in the development of chromium resistance. Resistant bacteria showed better tolerance to metal and Cr(VI) reduction through soluble enzyme biosynthesis. It did not build up chromium inside bacterial cell, which consequently played vital role in the chromium detoxification in the gastrointestinal tract. Furthermore, lactobacilli are the group of harmless food-grade bacteria and thus, in comparison to other non-food-grade chromium-resistant bacteria, chromium-resistant lactobacilli can be a better choice for the removal of chromium from the contaminated environment.

According to Shrivastava et al. (2003) gut microflora provide the first line of defense to the body by transforming toxic Cr(VI) to a less toxic Cr(III) form. In a recent study, exopolysaccharide obtained from *Lactobacillus* isolate showed 70% decrease in the concentration of Cr^{+2} after 20-day incubation period with the heavy metal from wastewater (Khan and Dona, 2015). In another experiment, Zhai and group investigated the protective role of *L. plantarum* CCFM8610 strain against acute cadmium (Cd) toxicity in mice (Zhai et al., 2013). Bacterial treatment could successfully reduce metal absorption in intestine, decreased accumulation of Cd in tissues, alleviated renal and hepatic oxidative stress, and improve hepatic histopathological changes as compared to the animals that received Cd only. Therefore, it is suggested that the administration of living bacterial cells after Cd exposure offered the most considerable protection.

7.4 CONCLUSION AND FUTURE PROSPECTS

Food-grade bacteria play a vital role in food fermentation processes since ages. However, these technologically important microorganisms did not receive much attention for their significance as biodegradation/biotransformation agent of harmful toxic substances. Several strains of LAB and

TABLE 7.4 Investigations Showing the Effectiveness of Food-Grade Microorganisms Against Different Heavy Metals.

Heavy metal(s)	Employed bacterial strain(s)	Major findings	References
Cadmium and lead	L. rhamnosus GG (ATCC 53103), L. casei Shirota, Lactobacillus fermentum ME3, Bifidobacterium longum 2C, B. longum 46, and B. lactis Bb12, and two commercial starter cultures: FVDVS XT-303-eXact and YO-MIX 401	Food-grade strains shown significant metal removal ability which was found to be more specific to strain and metal, both. The observed removal was rapid and metabolic independent surface process; also it was highly influenced with pH	Halttunen et al. (2007)
Lead, cadmium and copper in fish farming system	probiotic strains of L. rhamnosus GG, L. fermentum ME3, L. bulgaricus (a commercial strain), and L. acidophilus X37	The percentage of heavy metals removal was found to be temperature dependent; highest net removal was observed for L. acidophilus X37 (97.6%), followed by L. rhamnosus GG (74.8%), L. fermentum ME3 (71.16%), and L. bulgaricus (61.00%). The optimum pH was 6.0 for all LABs except for L. bulgaricus, it was 5.0	Rayes (2012)
Copper, chromium, and arsenic in the chromated copper arsenate (CCA)-treated wood	L. acidophilus NBRC13951, L. bulgaricus NBRC13953, and L. plantarum NBRC15891, S. thermophilus NBRC13957	Strains were able to extract the heavy metals through organic acid production, in particular, pyruvic acid and lactic acid	Chang and Kikuchi (2012)
Zinc	Leuconostoc mesenteroides, L. brevis, and L. plantarum	L. mesenteroides was the most effective in zinc binding with 27 mg/g as compared to L. brevis and L. plantarum (20 and 10 mg/g, respectively)	Mrvcic et al. (2009a, 2009b)
Zinc and iron	L. delbrueckii ssp. bulgaricus Lb-12, S. thermophilus STM-7	Biomass of all tested strains observed 90–100% Fe^{+2} and 70–90% Zn^{+2} removal under optimal value of process parameters	Sofu et al. (2015)
Cadmium, lead, and arsenic as well as aflatoxin B_1 (AFB_1) from contaminated water	Strains of LAB	L. acidophilus and Bifidobacterium anglulatum were detected as the most effective heavy metal removers among all tested LAB	Elsanhoty et al. (2016)

related genera have been proven in vitro and in vivo to bind mycotoxins, PAHs, pesticides, heavy metals, and azo dyes. Either bacterial cell–cell components or their metabolic products like enzymes bind or transform the toxic compounds into nontoxic form. The removal of mycotoxins and other discussed compounds from contaminated food and feed is urgently needed to improve the safety and use of biological agents such as food-grade bacteria or yeast cells appears as a vital approach. Nevertheless, additional investigations are needed to investigate the exact mechanisms involved in the biodegradation/biotransformation process by food-grade microorganisms aiming their application in dairy and food industry.

KEYWORDS

- ABT cultures
- acetate
- adsorption
- AFB$_1$
- AFB$_1$ metabolism
- aflatoxicosis
- aflatoxicosis outbreaks
- aflatoxin
- aflatoxin B$_1$
- aflatoxin M$_1$
- amino acid
- antimicrobial effect
- antioxidant activity
- antioxidants
- apoptosis
- aromatic compounds
- *Aspergillus*
- *Aspergillus flavus*
- azo dye
- azoreductase
- *Bacillus* spp.
- *Bacillus subtilis*
- bacterial cell
- bacterial cell wall
- bifidobacteria
- *bifidobacterium*
- *Bifidobacterium* spp.
- bioavailability
- biodegradation
- biodegradation process
- biogenic amines
- biological processes
- biomass
- bioquenching
- bioremediation
- biosphere
- biotechnological production
- biotransformation
- broiler diets
- carcinogen

- reductases
- remediation processes
- saccharomyces
- secondary metabolites
- serine hydrolases
- spoilage
- staple food
- sterilizing effect
- *Streptococcus*

- *Streptococcus salivarius*
- synthetic dyes
- toxic compounds
- toxin concentration
- toxin–bacteria complex
- viable cells
- yeast
- yoghurt
- yoghurt culture

REFERENCES

Abdella, Z.; Mohamed, F.; Mohamed, B. In vitro Reduction of Aflatoxin B1 by Strains of Lactic Acid Bacteria Isolated from Moroccan Sourdough Bread. *Int. J. Agric. Biol.* **2005,** *7*(1), 67–70.

Abou-Arab, A. A. K.; Abou-Bakr, S.; Maher, R. A.; El-Hendawy, H. H.; Awad, A. A. Degradation of Polycyclic Aromatic Hydrocarbons as Affected by Some Lactic Acid Bacteria. *J. Am. Sci.* **2010,** *6*(10), 1237–1246

Ayana, I. A. A. A.; Gamal El Deen, A. A.; El-Metwally, M. A. Behavior of Certain Lactic Acid Bacteria in the Presence of Pesticides Residues. *Int. J. Dairy Sci.* **2011,** *6,* 44–57.

Azizi, A. Bacterial-Degradation of Pesticides Residue in Vegetables During Fermentation. In *Pesticides—Formulations, Effects, Fate;* Stoytcheva, M., Ed.; InTech: UK, 2011; 651–660. ISBN: 978-953-307-532-7, http://www.intechopen.com/books/pesticides-formulations-effects-fate/bacterial-degradation-ofpesticides-residuc-in-vegetables-during-fermentation.

Bartkiene, E.; Bartkevics, V.; Mozuriene, E.; Krungleviciute, V.; Novoslavskij, A.; Santini, A.; Cizeikiene, D. The Impact of Lactic Acid Bacteria with Antimicrobial Properties on Biodegradation of Polycyclic Aromatic Hydrocarbons and Biogenic Amines in Cold Smoked Pork Sausages. *Food Control.* **2017,** *71,* 285–292.

Boonen, J.; Malysheva, S. V.; Taevernier, L.; Di Mavungu, J. D.; De Saeger, S.; De Spiegeleer, B. Human Skin Penetration of Selected Model Mycotoxins. *Toxicology* **2012,** *301*(1), 21–32.

Chang, Y.-C.; Kikuchi, S. Extraction of Heavy Metals from CCA Preservative-Treated Wood Waste Using Lactic Acid Bacteria: An Environment-Friendly Technology. *J. Biorem. Biodegrad.* **2012,** *3*(12), 1–2. DOI: 10.4172/2155-6199.1000e126.

Chen, H.; Xu, H; Heinze, T. M.; Cerniglia, C. E. Decolorization of Water and Oil-Soluble Azo Dyes by *Lactobacillus acidophilus* and *Lactobacillus fermentum. J. Ind. Microbiol. Biotechnol.* **2009,** *36,* 1459–1466.

Cho, K. M.; Math, R. K.; Islam, S. M. A.; Lim, W. J.; Hong, S. Y.; Kim, J. M.; Yun, M. G.; Cho, J. J.; Yun, H. D. Biodegradation of Chlorpyrifos by Lactic Acid Bacteria During Kimchi Fermentation. *J. Agric. Food Chem.* **2009,** *57*(5), 1882–1889.

Cousin, M. A.; Riley, R. T.; Pestka, J. Foodborne Mycotoxins: Chemistry, Biology, Ecology, and Toxicology. In Foodborne Pathogens: Microbiology and Molecular Biology; Fratamico, P. M., Bhunia, A. K., Smith, J. L., Eds.; Caister Academic Press: UK, 2005.

Da Silva, J. F. M.; Peluzio, J. M.; Prado, G.; Madeira, J. E. G. C.; Silva, M. O.; de Morais, P. B.; Rosa, C. A.; Pimenta R. S.; Nicoli, J. R. Use of Probiotics to Control Aflatoxin Production in Peanut Grains. *Sci. World J.* **2015,** *2015,* 1–8. http://dx.doi.org/10.1155/2015/959138.

El Ahwany, A. M. D. Decolorization of Fast Red by Metabolizing Cells of *Oenococcus oeni* ML34. *World J. Microbiol. Biotechnol.* **2008,** *24,* 1521–1527.

Elbanna, K.; Hassan, G.; Khider, M.; Mandour, R. Safe Biodegradation of Textile Azo Dyes by Newly Isolated Lactic Acid Bacteria and Detection Of Plasmids Associated with Degradation. *J. Biorem. Biodegrad.* **2010,** *1*(3), 1–6.

Elbanna, K.; Sarhan, O. M.; Khider, M.; Elmogy, M.; Abulreesh, H. H.; Shaaban, M. R. Microbiological, Histological, and Biochemical Evidence for the Adverse Effects of food Azo Dyes on Rats. *J. Food Drug Anal.* **2017.** http://dx.doi.org/10.1016/j.jfda.2017.01.005.

El-Nezami, H.; Kankaanpaa, P.; Salminen, S.; Ahokas, J. Ability of Dairy Strains of Lactic Acid Bacteria to Bind a Common Food Carcinogen, Aflatoxin B1. *Food Chem. Toxicol.* **1998,** *36*(4), 321–326.

Elsanhoty, R. M.; Salam, S. A.; Ramadan, M. F.; Badr, F. H. Detoxification of Aflatoxin M1 in Yoghurt Using Probiotics and Lactic Acid Bacteria. *Food Control* **2014,** *43,* 129–134.

Elsanhoty, R. M.; Al-Turki, I. A.; Ramadan, M. F. Application of Lactic Acid Bacteria in Removing Heavy Metals and Aflatoxin B1 from Contaminated Water. *Water Sci. Technol.* **2016,** *74*(3), 625–638. DOI: 10.2166/wst.2016.255.

Fuchs, S.; Sontag, G.; Stidl, R.; Ehrlich, V.; Kundi, M.; Knasmüller, S. Detoxification of Patulin and Ochratoxin A, Two Abundant Mycotoxins, by Lactic Acid Bacteria. *Food Chem. Toxicol.* **2008,** *46,* 1398–1407.

Halttunen, T.; Kankaanpaa, P.; Tahvonen, R.; Salminen, S.; Ouwehand, A. C. Cadmium Removal by Lactic Acid Bacteria. *Biosci. Microflora* **2003,** *22*(3), 93–97.

Halttunen, T.; Salminen, S.; Tahvonen, R. Rapid Removal of Lead and Cadmium from Water by Specific Lactic Acid Bacteria. *Int. J. Food Microbiol.* **2007,** *114*(1), 30–35.

Hamdan, A. M.; Sonomoto, K. Production of Optically Pure Lactic Acid For Bioplastics. In *Lactic Acid Bacteria and Bifidobacteria: Current Progress in Advanced Research;* Sonomoto, K., Yokota, A., Eds.; Caister Academic Press: Portland, USA, 2011.

Hamidi, A.; Mirnejad, R.; Yahaghi, E.; Behnod, V.; Mirhosseini, A.; Amani, S.; Sattari, S.; Khodaverdi Darian, E. The Aflatoxin B1 Isolating Potential of Two Lactic Acid Bacteria. Asian Pac. J. *Trop. Biomed.* **2013,** *3*(9), 732–736.

Haskard, C. A.; El-Nezami, H. S.; Kankaanpaa, P. E.; Salminen, S.; Ahokas, J. T. Surface Binding of Aflatoxin B1 by Lactic Acid Bacteria. *Appl. Environ. Microbiol.* **2001,** *67*(7), 3086–3091.

Hassan, M. A.; Fakhry, S. S.; Moslah, A. H.; Jabur, Z. A. Biodegradation of Food Dyes by *Lactobacillus Paracasei. Indian J. Appl. Res.* **2014,** *4*(7), 486–488

Hernandez-Mendoza, A.; Garcia, H. S.; Steele, J. L. Screening of *Lactobacillus casei* Strains for their Ability to Bind Aflatoxin B1. *Food Chem. Toxicol.* **2009,** *47,* 1064–1068.

Higa T. Application of Effective Microorganisms for Sustainable Crop Production. College of Agriculture, University of Ryukyus, Okinawa, Japan, 1999. http://www.eminfo.nl/wp-content/uploads/2013/08/Application-of-EM-for-Sustainable-Crop-Production-paper-1999.pdf (accessed April 19, 2017).

Higa, T.; Chinen, N. EM Treatment of Odor, Wastewater, and Environmental Problems Okinawa, College of Agriculture, University of Ryukyus, Okinawa, Japan, 1998.

Islam, S. M. A.; Math, R. K.; Cho, K. M.; Lim, W. J.; Hong, S. Y.; Kim, J. M.; Yun, M. G.; Cho, J. J.; Yun, H. D. Organophosphorus Hydrolase (OpdB) of *Lactobacillus* Brevis WCP902 from Kimchi is able to Degrade Organophosphorus Pesticides. *J. Agric. Food chem.* **2010,** *58*(9), 5380–5386.

Kabak, B.; Brandon, E. F.; Var, I.; Blokland, M.; Sips, A. J. Effects of Probiotic Bacteria on the Bioaccessibility of Aflatoxin B1 and Ochratoxin A Using an in vitro Digestion Model Under Fed Conditions. *J. Environ. Sci. Health, Part B* **2009,** *44*(5), 472–480.

Khan, R.; Dona, J. Extraction and Optimization of Exopolysaccharide Production from Lactic Acid Bacteria and Its Application in Biosorption of Chromium from Waste Water. *Eur. Acad. Res. III* **2015,** *3*(4), 4576–4588.

Khanafari, A.; Soudi, H.; Miraboulfathi, M.; Osboo, R. K. An in vitro Investigation of Aflatoxin B1 Biological Control by *Lactobacillus plantarum. Pak. J. Biol. Sci.* **2007,** *10*(15), 2553–2556.

Lu, K.; Zhang, X. L.; Zhao, Y. L.; Wu, Z. L. Removal of Color from Textile Dyeing Wastewater by Foam Separation. *J. Hazard. Mater.* **2010,** *182*(1–3), 928–932.

Mishra, R.; Sinha, V.; Kannan, A.; Upreti, R. K. Reduction of Chromium-VI by Chromium Resistant Lactobacilli: A Prospective Bacterium for Bioremediation. *Toxicol. Int.* **2012,** *19*(1), 25–30.

Mohammed, S. A. S. Physiological Effect of a Food Additive on some Hematological and Biochemical Parameters of Male Albino Rats. *Egypt. Acad. J. Biol. Sci.* **2009,** *2*(1), 143–151.

Montaseri, H.; Arjmandtalab, S.; Dehghanzadeh, G.; Karami, S.; Razmjoo, M. M.; Sayadi, M.; Oryan, A. Effect of Production and Storage of Probiotic Yogurt on Aflatoxin M1 Residue. *J. Food Qual. Hazards Control,* **2014,** *1*(1), 7–14

Mottier, P.; Parisod, V.; Turesky, R. J. Quantitative Determination of Polycyclic Aromatic Hydrocarbons in Barbecued Meat Sausages by Gas Chromatography Coupled to Mass Spectrometry. *J. Agric. Food Chem.* **2000,** *48,* 1160–1166

Mrvcic, J.; Prebeg, T.; Barisic, L.; Stanzer, D.; Bacun-Druzina, V.; Stehlik-Tomas, V. Zinc Binding by Lactic Acid Bacteria. *Food Technol. Biotechnol.* **2009a,** *47*(4) 381–388.

Mrvcic, J.; Stanzer, D.; Bacun-Druzina, V.; Stehlik-Tomas, V. Copper Binding by Lactic Acid Bacteria (LAB). *Biosci. Microflora* **2009b,** *28*(1), 1–6.

Oluwafemi, F.; Kumar, M.; Bandyopadhyay, R.; Ogunbanwo, T.; Kayode, B. Biodetoxification of Aflatoxin B1 in Artificially Contaminated Maize Grains Using Lactic Acid Bacteria. *Toxin Rev.* **2010,** *29*(3–4), 115–122.

Palani, V. R.; Rajakumar, S.; Ayyasamy, P. M. Exploration of Promising Dye Decolorizing Bacterial Strains Obtained from Erode and Tiruppur Textile Wastes. *Int. J. Environ. Sci.* **2012,** *2*(4), 2470–2481.

Patel, A. Characterization of Potential Probiotic Lactic Acid Bacteria and Quantification of their Exopolysaccharides. Ph.D. Desecration, Department of Dairy Microbiology, SMC College of Dairy Science, AAU, Anand, Gujarat (India), 2012.

Patel, A.; Shah, N.; Verma, D. K. Lactic Acid Bacteria (LAB) Bacteriocins: An Ecological and Sustainable Biopreservative Approach to Improve the Safety and Shelf-life of Foods. In *Microorganisms in Sustainable Agriculture, Food and the Environment;* Verma, D. K., Srivastav, P. P., Eds.; (as part of book series on *Innovations in Agricultural Microbiology)*; Apple Academic Press: USA, 2017; pp 197–258.

Peltonen, K.; El-Nezami, H.; Haskard, C.; Ahokas, J.; Salminen, S. Aflatoxin B1 Binding by Dairy Strains of Lactic Acid Bacteria and *bifidobacteria. J. Dairy Sci.* **2001,** *84,* 2152–2156.

Pérez-Díaz, I. M.; Kelling, R. E.; Hale, S.; Breidt, F.; McFeeters, R. F. *Lactobacilli* and Tartrazine as Causative Agents of Red-Color Spoilage in Cucumber Pickle Products. *J. Food Sci.* **2007,** *72,* 240–245.

Probst, C.; Njapau, H.; Cotty, P. J. Outbreak of an Acute Aflatoxicosis in Kenya in 2004: Identification of the Causal Agent. *Appl. Environ. Microbiol.* **2007,** *73*(8), 2762–2764.

Qinghua, W.; Jezkova, A.; Yuan, Z.; Pavlikova, L.; Dohnal, V.; Kuca, K. Biological Degradation of Aflatoxins. *Drug Metabol. Rev.* **2009,** *41*(1), 1–7.

Rayes, A. A. H. Field Studies on the Removal of Lead, Cadmium and Copper by the Use of Probiotic Lactic Acid Bacteria from the Water for Culturing Marine Tilapia *T. spilurus. N. Y. Sci. J.* **2012,** *5*(11), 74–82.

Reddy, K. R. N.; Salleh, B.; Saad, B.; Abbas, H. K.; Abel, C. A.; Sheir, W. T. An Overview of Mycotoxin Contamination in Foods and Its Implications for Human Health. *Toxin Rev.* **2010,** *29,* 3–26.

Sasikumar, C. S.; Taniya Papinazath T. Environmental Management: Bioremediation of Polluted Environment. *Proceedings of the Third International Conference on Environment and Health,* Chennai, India, 2003, pp 15–17.

Schneider, K.; Shoemaker, U. S.; Olthmanns, J.; Kalberlah, F.; Roller, M. PAK (Polyzyklische aromatische Kohlenwasserstoffe). In: *Gefährdungsabschätzung von Umweltschadstoffen Ergänzbares Handbuch toxikologischer Basisdaten und ihre Bewertung;* Eikmann, T., Heinrich, U., Heinzow, B., Konietzka, R. Eds.; Erich Schmidt Verlag GmbH and Co. KG: Berlin, 2014.

Schut, S.; Zauner, S.; Hampel, G.; König, H.; Claus, H. Biosorption of Copper by Wine-Relevant *Lactobacilli. Int. J. Food Microbiol.* **2011,** *145*(1), 126–131.

Seesuriyachan, P.; Takenaka, S.; Kuntiya, A.; Klayraung, S.; Murakami, S.; Aoki, K. Metabolism of Azo Dyes by *Lactobacillus casei* TISTR 1500 and Effects of Various Factors on Decolourisation. *Water Res.* **2007,** *41,* 985–992.

Seesuriyachan, P.; Kuntiya, A.; Sasaki, K.; Techapun, C. Comparative Study on Methyl Orange Removal by Growing Cells and Washed Cell Suspensions of *Lactobacillus casei* TISTR 1500. *World J. Microbiol. Biotechnol.* **2009,** *25,* 973–979.

Seesuriyachan, P.; Kuntiya, A.; Techapun, C.; Chaiyaso, T.; Hanmuangjai, P.; Leksawasdi, N. Nutritional Requirements for Methyl Orange Decolourisation by Freely Suspended Cells and Growing Cells of *Lactobacillus casei* TISTR 1500. *Maejo Int. J. Sci. Technol.* **2011,** *5*(1), 32–46.

Sezer, C.; Guven, A.; Oral, N. B.; Vatansever, L. Detoxification of Aflatoxin B1 by Bacteriocins and Bacteriocinogenic Lactic Acid Bacteria. *Turk. J. Vet. Anim. Sci.* **2013,** *37,* 594–601.

Shah, M. P. Exploited Application of *Lactobacillus* in Microbial Degradation and Decolorization of Acid Orange. *Int. J. Environ. Biorem. Biodegrad.* **2014,** *2*(4), 160–166.

Shrivastava, R.; Upreti, R. K.; Chaturvedi, U. C. Various Cells of the Immune System and Intestine Differ in their Capacity to Reduce Hexavalent Chromium. *FEMS Immunol. Med. Microbiol.* **2003,** *38*(1), 65–70.

Silvia, G. Aflatoxin Binding by Probiotics: Experimental Studies on Intestinal Aflatoxin Transport, Metabolism and Toxicity. Ph.D. Dissertation, Faculty of Medicine, School of Public Health and Clinical Nutrition, Clinical Nutrition and Food and Health Research Centre, University of Kuopio, Finland, 2007.

Sliżewskaand, K.; Smulikowska, S. Detoxification of Aflatoxin B1 and Change in Microflora Pattern by Probiotic in vitro Fermentation of Broiler Feed. J. Anim. Feed Sci. **2011,** *20,* 300–309.

Sofu, A.; Sayilgan, E.; Guney, G. Experimental Design for Removal of Fe (II) and Zn (II) Ions by Different Lactic Acid Bacteria Biomasses. *Int. J. Environ. Res.* **2015,** *9*(1), 93–100.

Spadaro, J. T.; Gold, M. H.; Renganathan, V. Degradation of Azo Dyes by the Lignin Degrading Fungus Phanerochaete Chrysosporium. *Appl. Environ. Microbiol.* **1992,** *58*(8), 2397–2401.

Topcu, A.; Bulat, T.; Wishah, R.; Boyacı, I. H. Detoxification of Aflatoxin B 1 and Patulin by *Enterococcus faecium* Strains. *Int. J. Food Microbiol.* **2010,** *139*(3), 202–205.

Verma, D. K.; Srivastav, P. P. *Microorganisms in Sustainable Agriculture, Food and the Environment* (as part of book series on *Innovations in Agricultural Microbiology); Apple Academic Press: USA, 2017.

Verma, D. K.; Mahato, D. K.; Billoria, S.; Kapri, M.; Prabhakar, P. K.; Ajesh Kumar, V.; Srivastav, P. P. Microbial Approaches in Fermentations for Production and Preservation of Different Food. In *Microorganisms in Sustainable Agriculture, Food and the Environment;* Verma, D. K., Srivastav, P. P., Eds.; (as part of book series on *Innovations in Agricultural Microbiology);* Apple Academic Press: USA, 2017a; pp 105–142.

Verma, D. K.; Mahato, D. K.; Billoria, S.; Kapri, M.; Prabhakar, P. K.; Ajesh Kumar, V. Srivastav, P. P. Microbial Spoilage in Milk Products, Potential Solution, Food Safety and Health Issues. In *Microorganisms in Sustainable Agriculture, Food and the Environment;* Verma, D. K., Srivastav, P. P., Eds.; (as part of book series on *Innovations in Agricultural Microbiology)*; Apple Academic Press: USA, 2017b; pp 171–196.

Wang, Y. S.; Wu, T. H.; Yang, Y.; Zhu, C. L.; Ding, C. L.; Dai, C. C. Binding and Detoxification of Chlorpyrifos by Lactic Acid Bacteria on Rice Straw Silage Fermentation. *J. Environ. Sci. Health, Part B* **2016,** *51*(5), 316–325.

Xu, H.; Heinze, T. M.; Paine, D. D.; Cerniglia, C. E.; Chen, H. Sudan Azo Dyes and Para Red Degradation by Prevalent Bacteria of the Human Gastrointestinal Tract. *Anaerobe* **2010,** *16*(2),114–119.

You, S. J.; Teng, J. U. Anaerobic Decolorization Bacteria for the Treatment of Azo Dye in a Sequential Anaerobic and Aerobic Membrane Bioreactor. *J. Taiwan Inst. Chem. Eng.* **2009,** *4*(5), 500–504.

Zhai, Q.; Wang, G.; Zhao, J.; Liu, X.; Tian, F.; Zhang, H.; Chen, W. Protective Effects of *Lactobacillus* Plantarum CCFM8610 Against Acute Cadmium Toxicity in Mice. *Appl. Environ. Microbiol.* **2013,** *79*(5), 1508–1515.

Zhang, Y. H.; Xu, D.; Liu, J. Q.; Zhao, X. H. Enhanced Degradation of Five Organophosphorus Pesticides in Skimmed Milk by Lactic Acid Bacteria and Its Potential Relationship with Phosphatase Production. *Food Chem.* **2014,** *164,* 173–178.

Zhang, Y. H.; Xu, D.; Zhao, X. H.; Song, Y.; Liu, Y. L.; Li, H. N. Biodegradation of Two Organophosphorus Pesticides in Whole Corn Silage as Affected by the *Cultured Lactobacillus Plantarum. 3 Biotech.* **2016,** *6*(73), 1–6.

Zhang, Q.; Yu, Z.; Wang, X.; Na, R. Effects of Chlorpyrifos and Chlorantraniliprole on Fermentation Quality of Alfalfa (*Medicago sativa* L.) Silage Inoculated with or without *Lactobacillus plantarum* LP. *Anim. Sci. J.* **2017,** *88*(3),456–462.

Zhou, X. W.; Zhao, X. H. Susceptibility of Nine Organophosphorus Pesticides in Skimmed Milk Towards Inoculated Lactic Acid Bacteria and Yogurt Starters. *J. Sci. Food Agric.* **2015,** *95*(2), 260–266.

Zinedine, A.; Faid, M.; Benlemlih, M. In vitro Reduction of Aflatoxin B1 by Strains of Lactic Acid Bacteria Isolated from Moroccan Sourdough Bread. *Int. J. Agric. Biol.* **2005;** *7,* 67–70.

Zuo, R. Y.; Chang, J.; Yin, Q. Q.; Wang, P.; Yang, Y. R.; Wang, X.; Wang G. Q.; Zheng, Q. H. Effect of the Combined Probiotics with Aflatoxin B1-Degrading Enzyme on Aflatoxin Detoxification, Broiler Production Performance and Hepatic Enzyme Gene Expression. *Food Chem. Toxicol.* **2013,** *59,* 470–475

Zuraida, M. S.; Nurhaslina, C. R.; Ku Halim, K. H. Influence of Agitation, pH and Temperature on Growth and Decolorization of Batik Wastewater by Bacteria, *Lactobacillus delbrueckii. Int. J. Res. Rev. Appl. Sci.* **2013a,** *14*(2), 269–275.

Zuraida, M. S.; Nurhaslina, C. R.; Ku Halim, K. H. Removal of Synthetic Dyes from Wastewater by Using Bacteria, *Lactobacillus delbrueckii. Int. Refereed J. Eng. Sci.* **2013b,** *2*(5), 1–7.

CHAPTER 8

TAILORING THE FUNCTIONAL BENEFITS OF WHEY PROTEINS BY ENCAPSULATION: A BOTTOM-UP APPROACH

NICOLETA STĂNCIUC[1,*], GABRIELA RBPEANU[1,2], and
IULIANA APRODU[1,3]

[1]Faculty of Food Science and Engineering, "Dunărea de Jos"
University of Galati, Domnească Street 111, 800201, Galati,
Romania, Tel.: +40336130183, Mob.: +40729270954,
Fax: +40236460165

[2]E-mail: Gabriela.Rapeanu@ugal.ro

[3]E-mail: Iuliana.Aprodu@ugal.ro

*Corresponding author. E-mail: Nicoleta.Stanciuc@ugal.ro

8.1 INTRODUCTION

Being the most widely distributed secondary metabolites in plants, phenolic and polyphenolic compounds in food and nutraceuticals exert their beneficial effect as free radical scavengers and metal chelators, thereby preventing oxidation of low-density lipoproteins and DNA strand scission and improve immune function (Shahidi and Naczk, 2004). Due to their properties, these food components have strong chemopreventive properties against the most common contemporary human diseases, such as cardiovascular diseases, cancer, and neurodegenerative pathologies (Latruffe et al., 2014). Polyphenols contribute to the organoleptic properties of plant-based foods, especially due to their astringency, bitter taste, color, and by their participation in haze formation (Le Bourvellec and

Renard, 2012). A wide range of phenolic compounds, such as phenols, benzoic acid derivatives, phenylpropanoids, flavonoids, stilbenes, tannins, lignans, and lignins can be found in plants and foods (Lättia et al. 2011). Probably, the most studied compounds from the perspective of functionality in human diet are flavonoids. The structure of flavonoids (2-phenyl benzopyran skeleton) is characterized by the presence of a basic skeleton C_6–C_3–C_6, with two aromatic rings and a heterocyclic ring, the later containing one oxygen atom. Flavonoids are known for their wide range of biological effects including antioxidant, antibacterial, antiviral, anti-inflammatory, antiallergic, antiestrogenic, and anticarcinogenic capacities (Cook and Samman, 1996; Middleton et al., 2000; Heim et al., 2002; Nandave et al., 2005). The use of flavonoids in different matrices is limited due to their low bioavailability caused by the poor water solubility (Manach et al., 2005). It has been suggested that the bioavailability of these compounds can be increased through the use of proteins able to complex flavonoids (Gholami and Bordbar, 2014).

The red-colored anthocyanins containing extracts gained increasing popularity in the food industry as the natural alternatives to synthetic dyes. The bioactivity of anthocyanins is attributed to their high antioxidant capacity, confirmed by in vitro and in vivo experiments by numerous studies (Sellappan et al., 2002; Lätti et al., 2008; Schantz et al., 2010). Due to their antioxidant capacity, the anthocyanins gained increased importance in the development of functional foods. However, their stability is highly influenced by pH value, ionic strength, and concentration, presence of other solutes, UV light, oxygen, and temperature. Additionally, it has been reported by Xiong et al. (2006) that the antioxidant activity significantly decreases with progressive running of oxidation reactions.

Another important class of biologically active compounds (BACs) widely distributed in nature consists of carotenoids. Plants and microorganisms are able to synthesize over 600 yellow and orange-red pigments which belong to the carotenoids (Vachali et al., 2012). Carotenes are carotenoid compounds consisting of hydrocarbons or xanthophylls with one or more oxygen atoms. They are omnipresent dyes with antioxidant and provitamin A activities. In order to accomplish their physiological function, carotenes must be transported to target tissues for absorption and stabilization after being absorbed from dietary sources (Bhosale and Bernstein, 2007). However, xanthophylls are involved in non-provitamin A activities, being effective in preventing the development of cancers or cardiovascular diseases (Bhosale and Bernstein, 2007; Knockaert et al., 2012). In this respect, xanthophylls act as antioxidants through quenching singlet and reactive oxygen species, as well as some other free radicals resulting as by-products of the cellular metabolic process. Regarding the biological functions of these compounds, it has been suggested that carotenoid intake decreases the risk of neovascular age-related macular degeneration by up to 43% (Møller et al., 2000).

The β-carotene (BC) along with lutein, lycopene, zeaxanthin, α-carotene, and β-cryptoxanthin account for 90% of circulating carotenoids (Parker, 1989). The typical structure of BC consists of a polyene chain having β-rings at the two extremities of the chain and 11 conjugated double bonds. Due to the presence of conjugated bonds, BC is susceptible to isomerization. In nature, BC is mostly present as all-*trans*-BC, which is thermodynamically the most stable form (Knockaert et al., 2012). Industrial processing of food leads to isomerization of *trans*- to *cis*-forms, which are relatively thermodynamically stable. The occurrences of *cis* isomer lead to a significant decrease in the biological properties of BC, accompanied by a reduction of bioavailability (Deming et al., 2002).

One of the most used and promising methods to improve the stability of polyphenols and carotenoids is microencapsulation, which allows solving some problems related to different limitations of low stability in food matrices. Teng et al. (2015) have highlighted the main purpose of microencapsulation, namely protecting the sensitive compounds from the action of environmental factors, such as oxygen, acidity, light, and so forth; controlling the bioavailability and bioavailability of encapsulated compounds, and promoting the controlled release in target sorption areas. In addition, microencapsulation offers significant advantages associated with the transport and storage, masking undesirable flavor, concentration of the BACs, and compartmentalization of two or more molecules species (Munin and Edwards-Lévy, 2011). The microencapsulation of polyphenols is possible due to the ability of these compounds to interact with different molecules, such as proteins. Different methods were reported for microencapsulation, including emulsification, ionotropic gelation (or coacervation), spray drying, supercritical fluids, and thermal gelation.

Food-grade proteins are biopolymers often used as ingredients in different food formulations (Chen et al., 2006). Many data from the literature describe different strategies developed for the protection and delivery of BACs. Therefore, a significant number of reviews support the hypothesis that designing new delivery systems is an effective approach for improving the stability and bioavailability of bioactive compounds (Livney, 2010; Joye et al., 2015; McClements, 2015; McClements et al., 2015). However, these techniques widely vary depending on the nature of the bioactive compound and the coating material, and the molecular particularities of the resulting microparticles (Tavares et al., 2014). Advanced techniques like enzyme inhibition, front analysis capillary electrophoresis, microcalorimetry, nephelometry, nuclear magnetic resonance (NMR) spectroscopy, protein precipitation, and turbidity are widely used for investigating complexation reactions between proteins and polyphenols (Fickel et al., 1999; Horne et al., 2002; De Freitas et al., 2003; Frazier et al., 2003; Papadopoulou et al., 2004; Carvalho et al., 2006). Many previous studies have demonstrated the interactions between different polyphenols and food-grade proteins. Valuable details on the interactions between flavonols or flavonoids and different proteins, such as α-lactalbumin (α-LA), β-lactoglobulin (β-LG), β-casein, ovalbumin, lysozyme, phosvitin, and gelatin, are available in the literature (Jöbstl et al., 2006; Prigent et al., 2003; Yan et al., 2009; Kanakis et al., 2011; Zorilla et al., 2011).

The objective of this chapter is to discuss the encapsulation of different BACs, such as polyphenols and carotenoids, using whey proteins as wall material. Our focus in this chapter is to provide a general overview of the fundamental principles of fluorescence and to highlight the employment of steady-state techniques for the detection of molecular interactions and determination of binding constants. The discussion was mainly based on an evaluation of binding parameters and atomic events revealed through molecular docking experiments.

8.2 STRUCTURAL AND FUNCTIONAL FEATURES OF WHEY PROTEINS: A BINDING PERSPECTIVE

Milk proteins act like natural vehicles, which are responsible for delivering from mother to the newly born the important compounds like essential micronutrients (e.g., calcium and phosphate), building blocks (e.g., amino acids), and immune system components (e.g., immunoglobulins and lactoferrin [LF]) (Livney, 2010). In addition, proteins such as β-LG and bovine serum albumin (BSA) were reported to bind several small hydrophobic molecules. Taking into account that whey proteins are Generally

Recognized as Safe, and have very good biological, chemical, and physical functionalities, they are widely used in food and pharmaceutical industries (Verma and Srivastav, 2017; Verma et al., 2017). Besides their recognized nutritional value, the whey proteins are suitable for carrying different compounds, allowing the possibility to fabricate advanced vectors for loading the bioactive compounds (Livney, 2010; Tavares et al., 2014, Zhu et al., 2017).

8.2.1 BOVINE β-LACTOGLOBULIN

Bovine β-LG, the main whey protein of cow's milk at a concentration of 0.3 g/100 ml, has 162 amino acid residues, and is a member of the lipocalin protein superfamily. Provisionally, it has been suggested that the physiological function of β-LG consists in binding and transport of small hydrophobic molecules, such as retinol and fatty acids (Pérez and Calvo, 1995). Distinctive for the members of this family is the presence of the eight-stranded β-barrel structure, with a three-turn α-helix on its outer surface. The central cavity defined by this region binds a variety of hydrophobic molecules. Three (Cys^{106}, Cys^{119}, and Cys^{121}) of the five-Cys residues are placed in the hydrophobic pocket of the protein, being located within the gap defined by the G and H strands and the protein helix (Brownlow et al., 1997) (Fig. 8.1a). The phenyl ring of Phe^{136} is located between the Cys^{121} residue and the disulfide bond connecting Cys^{106} and Cys^{119}. Molecular dynamics simulations performed on the 3NPO.pdb model from RCSB protein data bank indicated that the abovementioned secondary and tertiary structural elements are involved in defining the thermal-dependent behavior of the protein. According to Stănciuc et al. (2012a), partial β-LG secondary structure loss occurs when heating the proteins up to 90°C. The strand content decreased from 29.7% at 30°C to 15.9%, suggesting important protein unfolding events up to 80°C. The further increase of the number of amino acids involved in defining strands with temperature (26.6% at 90°C) was ascribed to the partial protein refolding, following a pattern most probably different to the one specific to the native proteins (Stănciuc et al., 2012a). New native-like strands organized as antiparallel β-sheet or β-turns were reported for the protein structure equilibrated at high temperature. The hydrophobic core of the protein appears rather stable at thermal treatment. Only slight increase of the total length of the disulfide bridges involved in stabilizing the β-LG hydrophobic core was observed at high temperature (Cys[66]–Cys[160] increases from 2.04 to 2.09 Å, and Cys[106]–Cy[119] increases from 1.98 to 2.01 Å). Therefore, the heat-induced molecular events involved slight enlargement of the cavity between CD loop and the strands G and H, coupled with significant torsion of the C–S–S–C angles in the disulfide bridges connecting the mentioned secondary structure motifs (Loch et al., 2011). It is, therefore, possible to tailor the β-LG binding properties for different small molecules in the central cavity of the molecule.

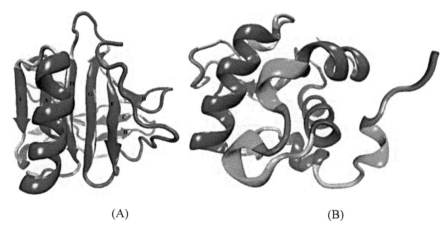

(A)   (B)

FIGURE 8.1   (See color insert.) Details on the structure of (a) bovine β-LG (model 3NPO.pdb from RCSB protein data bank) and (b) α-LA (1F6S.pdb). (The proteins are represented in new cartoon style using VMD software [Humphrey et al., 1996]. The secondary structure elements are represented as follows: α-helices in purple; 3–10 helices in green; extended β-sheets in violet; turns in red; and coils in pink.)
Source: (a) Adapted from Loch et al. (2011) and (b) adapted from Chrysina et al. (2000).

β-LG is an exceptionally acid-stable protein; it can be found in dimer state at pH around neutrality, and dissociates reversibly into monomers at pH 2.0, without changes of the native structure. This contributes to making β-LG very resistant to pepsin degradation at low pH (Marengo et al., 2016). These conformational transitions can be exploited in terms of binding different compounds with specific functionality. The major associative forces to ligand binding are considered the hydrogen bonding, hydrophobic interaction, and van der Waal interaction (Van de Weert and Stella, 2011).

The use of β-LG as a carrier matrix is based on two unique properties, as suggestively described by Teng et al. (2015), related to its lower susceptibility to enzymatic hydrolysis in the gastrointestinal tract. The resistance of β-LG to pepsin proteolysis is due to the specificity of the enzyme, which is known to cleave peptide bonds at the hydrophobic patch of protein, whereas the protein has a high content of charged and polar amino acids. The exposure of the side chains of charged and polar amino acids on protein surface was reported to change with the temperature, possibly affecting the stability of the protein (Simion et al., 2015). Although the protein surface available to the solvent increased with the temperature from 25 to 85°C, burial of some initially exposed polar groups was found. Additionally, the enzyme–substrate binding is limited due to the content of rigid β-sheet structure, which is higher at high temperatures (Simion et al., 2015), and is therefore responsible for the reduced molecular flexibility. Moreover, the molecular flexibility is affected by the presence of the two disulfide bridges (Cys^{160}–Cys^{66} and Cys^{106}–Cys^{119}), which prevents dissociation of molecules. Additionally, β-LG can be slowly digested by trypsin in the small intestine. This behavior at digestion confers attractive properties as encapsulating material for the controlled release of bioactive compounds in the gastrointestinal tract.

8.2.2 α-LACTALBUMIN

The monomeric α-LA is an acidic Ca^{2+}-binding protein consisting of 123 amino acids. The α-helical conformation prevails in the native structure of α-LA molecule, which is highly stable because of the existence of four disulfide bonds (Fig. 8.1b). In respect to the total serum proteins, α-LA represents about 20% (w/w), being the second most abundant protein in whey. It has been reported that this small protein (14.2 kDa) has diverse biological and pharmaceutical important functions, such as biosynthesis of lactose in the mammary gland, whereas some specific forms of the protein were found to be responsible for the induction of apoptosis in tumor cells. Chaudhuri et al. (2010) suggested that α-LA molecule can be successfully used as model for investigating the folding pattern of proteins because of the existence of several partially folded intermediate states and the molten globule form under various conditions of this protein. It has been reported that human α-LA has therapeutic value since its *apo* state can bind oleic acid to form the so-called human α-LA made lethal to tumor cells (HAMLET) complex, a folded variant that induces tumor cell apoptosis (Svensson et al., 2000). This folded variant can effectively cure cutaneous warts caused by human papillomavirus (Gustafsson et al., 2004), suggesting that might be a potential antitumor drug (Zhang et al., 2010).

α-LA contains four tryptophan residues, located either in the aromatic cluster I (Trp[118]) or in the aromatic cluster II (Trp[26], Trp[60], Trp[104], and Trp[118]) (Chrysina et al., 2000). The native tertiary structure of α-LA consists of two different domains divided by a cleft: a large domain (α-domain) and a small domain (β-domain) (Pike et al., 1996). The tertiary structure of the protein was characterized in detail: the α-domain (1–34 and 86–123) contains three pH-stable α-helices (helix H1, 5–11; H2, 23–34; H3, 86–98), a pH-dependent flexible loop/helix (H4, 105–110), and two short 3_{10} helices (h1, 18–20; h3, 115–18). Between pH 6.5 and 8.0, the flexible 105–110 loop region adopts a helical conformation (H4) (Pike et al., 1996). The 35–85 β-domain is composed of a small three-stranded antiparallel β-pleated sheet (strands S1 consisting of 41–44, S2 of 47–50, and S3 of 55–56) and of a short 3_{10} helix (h2, 77–80) (Pike et al., 1996; Chrysina et al., 2000). Regarding the thermal-dependent behavior of these structural elements, it has been shown that the α-helical content of the native α-LA increases by 22.84% and the number of β-turns decreased from 21 to 15 when heating up to 80°C (Stănciuc et al., 2012b). Both the length of the α-LA helices and the interaction within pairs of helices vary with the temperature, suggesting thermal-dependent stability and availability to interact with other molecules.

8.2.3   BOVINE SERUM ALBUMIN

BSA contains three similar lobes, and consists of 583 amino acid residues, 17 disulfide bridges that ensure high stability of protein conformation, and one free cysteine residue (Cys[34]) available to establish intermolecular bonds responsible for aggregation behavior as it is located on the surface of protein (He and Carter, 1992; Giancola et al., 1997). Serum albumin is among the most abundant carrier proteins involved in the transport and delivery of endogenous and exogenous compounds present in the blood (Sarkar et al., 2013). Among serum albumins, the ones from human and bovine origins (human serum albumin [HAS] and BSA, respectively) are probably the most studied proteins. A detailed analysis performed on BSA (model 4F5S.pdb from RCSB Protein Data Bank) after running molecular dynamics simulations indicated the high content of helical structures (over 70% on total amino acids) with high stability at thermal treatment (Ursache et al., 2016). The most important changes in the protein structure at high temperature (75°C) were related to shift of H8 helix (Pro[199]–Phe[126]) and melting of H28 helix (Leu[480]-Asn[482]). Most of the helices are over 10 residues long and their length tends to increase with the temperature. It has been suggested that the high entropic cost associated to protein folding is well balanced by the increased stability of the BSA molecule.

Although the two proteins are 76% homologous, the main difference consists in the number and position of Trp residues (HSA has one Trp residue on position 214, while BSA contains two Trp residues on positions 134 and 212, respectively) (Togashi and Ryder, 2008). Therefore, in case of the BSA, the intrinsic fluorescence arises from Trp[134] and Trp[212] located in subdomains IA and IIA, respectively (He and Carter, 1992). Trp[212] is located in a cleft or in the hydrophobic fold, whereas Trp[134] is located on the surface of the molecule (Sugio et al., 1999).

8.2.4   LACTOFERRIN

LF is a minor whey protein (concentration of 20 and 200 mg/ml, depending on the lactation period), with particular nutritional and functional properties. LF is a glycoprotein found in milk and other secretions such as mucins, saliva, seminal fluids, tears, and the secondary granules of neutrophils of different mammals, which is (Farnaud and Evans, 2003). Bovine LF has a molecular weight of about 80 kDa, is bilobal, and made of 689 amino acids; each lobe (C- and N-lobe) has two domains hosting one metal-binding site able to bind a ferric ion (Fe^{3+}) each (Baker et al., 2002; Dupont et al., 2006). The analysis performed on 1BLF.pdb model at single molecule level indicated that the secondary structure of the protein is dominated by the helical motifs; about 23.5% of the amino acids are involved in α-helices and 3.8% in 3–10 helices whose stability slightly vary with the temperature (Stănciuc et al., 2013). The strands content of LF secondary structure (about 17.3%) is given by the three parallel, three antiparallel, and one mixed β-sheet. In addition to the changes in hydrogen-bonding network, known to be involved in stabilizing the protein motifs, the significant decrease of the reverse turns connecting the secondary structure elements with the temperature increase (Stănciuc et al., 2013) suggests the occurrence of important denaturation events even at 80°C, which might result in loss of protein bioactivity. Some differences were reported between the two lobes of the LF molecule. Due to the more compact structure, the C-lobe appears more heat stable with respect to the N-lobe (Sánchez et al., 1992). On the other hand, the closure of the domains in the C-lobe appears milder, suggesting the possibility of easier release of the iron ion. The in silico observation of the relative distance between amino acids, which are responsible for the iron-binding activity (Asp[60], Tyr[92], Tyr[192], and His[253] in N-lobe and Asp[395], Tyr[433], Tyr[526], and His[595] in C-lobe) decreases at high temperature (Stănciuc et al., 2013), explains the changes in the iron-binding ability of the LF molecule at thermal treatment (Sui et al., 2010). The physiological role of the protein mainly consists of inhibiting the growth of Gram-positive and Gram-negative bacteria as well as yeasts by the iron-chelating effect (Steijns and van Hooijdonk, 2000). It also possesses some interesting biological properties, including anti-inflammatory and immunomodulating properties (Farnaud and Evans, 2003; Legrand et al., 2004). Fang et al. (2014) demonstrated that, similar to α-LA, LF has an ability to bind oleic acid to form an antitumor complex similar to HAMLET.

8.3 METHODS FOR INVESTIGATING THE BINDING MECHANISMS

Knowing the system particularities at the molecular level is crucial for efficient probing the interaction between ligands and proteins and understanding the nature of the complex formation involving the detailed characterization of the molecular events leading to molecules binding (Bandyopadhyay et al., 2012). Significant work has been performed in order to characterize the interactions between different ligands and proteins, relying on the use of different techniques such as fluorescence spectroscopy, circular dichroism spectroscopy, dynamic light scattering (small-angle X-ray scattering and small-angle neutron scattering), Fourier-transform infrared spectroscopy, isothermal titration calorimetry, and NMR, mass spectroscopy, computational methods, and microscopy. Ulrih (2015) classified these methods into three classes as follows: screening methods (in situ methods), methods for studying the mechanisms behind the ligand–protein interactions, and methods for investigating protein aggregation and precipitation.

Due to the high sensitivity, as well as relative simplicity, the fluorescence techniques are very convenient and allow efficient and reliable investigations on the environment-dependent behavior of different molecules. Mocz and Ross (2013) gave a detailed overview on the basic concepts of fluorescence, focusing on the interaction between different molecules. The basics of these techniques involve the inherent fluorescence property of protein molecules given by the presence amino acid residues having fluorescent properties (tryptophan, tyrosine, and phenylalanine). In an aqueous environment, the emission peaks specific to Phe, Tyr, and Trp residues are located at 280, 305, and 348 nm, respectively. The Phe residues are not the significant contributors to the fluorescence spectra of most of the proteins. On the other hand, the emission of Tyr residues is not usually detected in undenatured proteins. Therefore, the emission of proteins is dominated by Trp, which absorbs at the longest wavelength (Lakowicz, 1999). Knowledge of these properties is important when studying the binding of different ligands to the protein molecules.

Ligands are able to bind to the backbone of the protein molecules, causing the unfolding, therefore enabling the exposure of previously buried hydrophobic amino acids. The unfolding can cause the exposure of tryptophan residue to solvent, resulting in fluorescence quenching.

The fluorescence intensity recorded at maximum emission wavelength can be used to estimate the binding parameters using the Stern–Volmer Equation 8.1 (Lakowicz, 1999):

$$\frac{F_0}{F} = 1 + K_q x \ \sigma_0 x [Q] = 1 + K_{SV}[Q] \tag{8.1}$$

where F_0 and F are the fluorescence intensities of the protein before and after the addition of quencher, respectively, K_q is the biomolecular quenching rate constant, σ_0 is the lifetime of fluorophore in the absence of the quencher (10^{-8} s), K_{sv} is the Stern–Volmer quenching constant, and $[Q]$ is the concentration of the quencher. The K_{sv} is estimated as the slope of the linear relationship between F_0/F and $[Q]$.

Nonlinear Stern–Volmer plot can also be observed in a multi-tryptophan protein with different exposure degree of the Trp residues varying from fully buried to fully exposed. In these cases, the fraction of total fluorophore accessible to the quencher can be calculated from the modified Stern–Volmer Equation 8.2:

$$\frac{F_0}{\Delta F} = \frac{1}{K_Q.f_a[Q]} + \frac{1}{f_a} \tag{8.2}$$

where ΔF is the change in the fluorescence intensity due to quenching, F_0 and $[Q]$ have the same meaning as in Equation 8.1, K_Q is the Stern–Volmer quenching constant of the exposed Trp residues, and f_a is the fraction of the initial fluorescence, which is accessible to the quencher.

For a static quenching mechanism, the binding constant (K_a) and the number of the binding sites (n) for the ligand–protein complex can be estimated using Equation 8.3:

$$Log\frac{(F_0 - F)}{F} = LogK_a + nlog[Q] \tag{8.3}$$

Although widely used for studying the ligand–protein interactions, fluorescence quenching has the limitation like difficulties in providing any direct information on the events governing the molecular behavior of the proteins or ligands (Bandyopadhyay et al., 2012). The fluorescence data can be expanded with observations from computational methods. Important atomic level details on molecules behavior can be obtained by performing molecular mechanics and molecular dynamics simulations for computing all bonded and nonbonded interaction at appropriate timescale. Different molecular docking software succeeded in predicting the molecular recognition and binding activities of different molecules. The receptor–ligand recognition is usually decided based on shape complementarity criteria (Aprodu, 2016). Among the output docking solutions, the best fit is chosen factoring the score of the geometric match of the molecules within the complex, as well as different components of the binding energy.

8.4   EVIDENCE ON BINDING OF POLYPHENOLS AND CAROTENOIDS TO WHEY PROTEINS

It has been demonstrated that β-LG has the ability to bind anthocyanins from different fruits and vegetables in model systems. Oancea et al. (2017) studied the interactions between anthocyanins from sour cherries extract and β-LG preliminarily heat treated at temperatures ranging from 25 to 100°C for 15 min by using fluorescence spectroscopy and molecular modeling techniques. The fluorescence intensity of the protein was significantly influenced by increasing the concentration of ligand, tested at all temperatures. The fluorescence emission spectra for all samples were similar to that measured in the absence of anthocyanins extract, which suggested that the protein dominated the overall fluorescence behavior. A typical fluorescence quenching phenomenon at 25°C and respectively 90°C is shown in Figure 8.2.

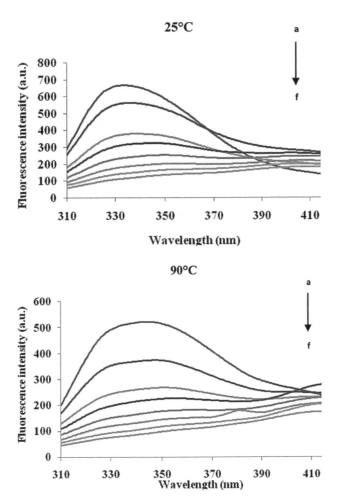

FIGURE 8.2 (See color insert.) The emission spectra of the β-LG and sour cherries extract (the anthocyanins concentration in the extract [from a–f] varied from 0 to 0.093 μm cyanidin glycoside equivalents).
Source: Adapted from Oancea et al. (2017) (with permission from Taylor & Francis).

When 0.093-µm cyanidin glycoside equivalents were added to untreated β-LG, a 75 nm redshift in maximum wavelength λ_{max} was found, whereas heating in the temperature range of 50–90°C caused significant 78–76 nm redshifts. These results indicate that the interactions between β-LG and sour cherries extract modified the polarity of the Trp microenvironment (Lakowicz, 1999). Moreover, the significant redshifts in λ_{max} suggest that Trp residues in the protein are more exposed to the hydrophilic solvent due to the interaction of protein molecules with polyphenolic compounds from the extract.

The data were fitted to the Stern–Volmer equation and the quenching constants are given in Table 8.1. The linearity of the plots typically indicates the existence of a single type of fluorophore in the protein molecule; the fluorophores display similar exposure, therefore being equally accessible for binding (Acharya et al., 2013; Chung et al., 2015). The quenching of the protein fluorophore by anthocyanins may occur through either *collisional quenching* (dynamic/nonbinding quenching) or binding-related (static) quenching (Lakowicz, 2006). The lowest K_{SV} value was measured at 100°C (0.23 ± 0.03 × 106/M), whereas the highest at 80°C (0.59 ± 0.15 × 106/M), suggesting a higher exposure of Trp exposure at lower temperatures (Table 8.1).

TABLE 8.1 The Binding Parameters of Heat-Treated β-lactoglobulin and Sour Cherry Extract.

T (°C)	K_{SV} (106 l/mol)	K_a (1013/mol/s)	K_a (l/mol)	n
25	0.46 ± 0.09[a]	15.55 ± 3.00	1.25 ± 0.22	0.60 ± 0.11
50	0.50 ± 0.07	16.85 ± 2.64	1.19 ± 0.18	0.50 ± 0.02
60	0.46 ± 0.07	15.37 ± 2.41	1.18 ± 0.16	0.57 ± 0.04
70	0.46 ± 0.04	15.43 ± 1.49	1.17 ± 0.16	0.55 ± 0.03
80	0.59 ± 0.15	19.56 ± 5.02	1.19 ± 0.17	0.34 ± 0.01
90	0.36 ± 0.03	12.04 ± 1.26	1.18 ± 0.17	0.55 ± 0.08
100	0.23 ± 0.03	7.63 ± 1.19	1.18 ± 0.16	0.88 ± 0.03

Sources: Adapted from Oancea et al. (2017) (with permission from Taylor & Francis).
[a]Standard deviation

The F_0/F versus [Q] plots can provide useful information on the nature of the quenching mechanism. In this regard, Chung et al. (2015) suggested that when dealing with linear F_0/F versus [Q] plots, one cannot distinguish between dynamic and static quenching. In most of the cases, the dynamic quenching mechanism is indicated by the increase in K_{SV} constant with increasing temperature. On the other hand, Joye et al. (2015) suggested that an increase or decrease of K_{SV} constant is normally obtained in static quenching, in direct relation to the nature of the interaction, for example, hydrophobic, hydrogen bonding, or electrostatic interactions. In the study of Oancea et al. (2017), a combination of dynamic, at lower temperatures, and static mechanisms, at higher temperatures, was highlighted (Table 8.1).

Further, in order to distinguish the interaction between the β-LG and anthocyanins from sour cherries extract, the bimolecular quenching constant (K_q) was calculated by using Equation 8.1 and compared to the maximum value for diffusion-limited quenching in water (1010/M/s) (Lakowicz, 2006). The interaction between heat-treated β-LG and sour cherry extract was specific allowing static quenching rather than dynamic quenching, as indicated by the significantly higher K_q values (Table 8.1) in respect to the maximum value specific for the diffusion-limited quenching in water (Chung et al., 2015).

Gholami and Bordbar (2014) explained that high temperature can cause the rapid dissociation of weakly formed complexes, and accordingly, the Stern–Volmer constant values decrease. These authors suggested K_{SV} value for naringenin binding to β-LG of 0.1370×106/M at 25°C, whereas Li et al. (2013) reported values of 6.0×10^4/M for curcumin binding to β-LG at pH 7.0. Arroyo-Maya et al. (2016) also suggested a decrease in K_{SV} values in a lower temperature range (25–45°C).

In order to determine the equilibrium between the bound and free molecules at different temperatures, the apparent binding constant (K_a) and the number of binding sites (n) were derived from the slope and intercept of the log $[(F_o - F)/F]$ vs. log $[Q]$ plot (Table 8.1). The K_a values are not significantly different in the temperature range studied ($p > 0.05$), varying between $1.25 \pm 0.22 \times 1014$/M at 25°C and $1.18 \pm 0.16 \times 1014$/M at 100°C. Oancea et al. (2017) also indicated that at each tested temperature, the n values are lower than one, suggesting that the weak binding of anthocyanins to β-LG might be due to the multitude of compounds present in the extract, which can compete for the binding sites. However, an increase from 0.60 ± 0.11 to 0.88 ± 0.03 when increasing the temperature up to 100°C was reported, thus indicating an improved affinity for anthocyanins. Mohammadi et al. (2009) suggested K_a and n of 2.49×1012/M/s and 0.85, respectively, whereas Li et al. (2013) reported higher but still comparable values of $6.0 \times 1012/M/s$ and 1.1. The K_a and n values obtained by Gholami and Bordbar (2014) were $0.5685 \times 106/M$ and 1.11.

Chung et al. (2015) studied the potential of native and denatured whey protein to limit the degradation of anthocyanin in the model beverages containing anthocyanin (0.025%), ascorbic acid (0 or 0.05%), and calcium salt (0 or 0.01%). When using heat-denatured whey proteins, the stability of the anthocyanin during storage was significantly improved. At 25°C, these authors calculated a K_{SV} value of 1217.3/M, whereas an increase was observed with increasing temperature up to 40°C. These authors suggested n values around 0.5–0.7 for anthocyanin molecules bound per protein molecule, which are similar to the data reported by Oancea et al. (2017).

Wiese et al. (2009) studied the affinity of binding between cyanidin-3-glucoside (Cy3glc) and different proteins, such as HSA and β-LG, by the quenching of protein Trp fluorescence. For all investigated proteins, the highest emission was collected between 330 and 350 nm, when excited at wavelength of 290 nm. A higher association constant of $73 \pm 1 \cdot 10^3$ 1/mol was calculated for human serum albumin at pH 7.0, whereas β-LG showed the lowest value of $27 \pm 2 \cdot 10^3$ 1/mol at pH 4.5 and 37°C. Cy3glc molecule is highly stable under acidic conditions (pH < 1.5) and tends to decompose at higher pH values. Wiese et al. (2009) suggested the possibility of the involvement of HSA in transportation and distribution of the phenolic compounds in the organism, given the high affinity of the protein for these compounds at pH 7.0.

The interactions between HSA and different B-ring-hydroxyl-groups-substituted anthocyanins, namely cyanidin-3-O-glucoside (C3G), delphinidin-3-O-glucoside (D3G), and pelargonidin-3-O-glucoside (P3G), and under physiological pH conditions have been investigated by Tang et al. (2014) by means of fluorescence spectroscopy, UV-vis absorbance, and circular dichroism. In fluorescence analysis, these authors found that fluorescence intensity of HSA decreased with the gradual addition of P3G, C3G, or D3G, which indicated that these compounds might all interact with HSA. As indicated by the redshifts registered in UV-vis maximum peak positions, P3G, C3G, or D3G binding to the HSA resulted in important changes in the protein conformation, whereas the hydrophobicity in the vicinity of aromatic amino acid residues increased. Tang et al. (2014) suggested a typical static quenching mechanism between P3G/C3G and HSA, whereas D3G interacted with HSA through a combination of the static and dynamic quenching mechanisms. The binding constants and number of binding sites of ligand–protein were 1.066×105 l/mol for P3G, 1.180×105 l/mol for C3G, and 1.271×105 l/mol for D3G, respectively. Moreover, for all tested polyphenols, the number of binding sites was close to one, suggesting the presence of the one class of binding sites for P3G/C3G/ D3G and HSA. The affinity of anthocyanins for HSA decreased in the following order: D3G > C3G > P3G (Tang et al., 2014).

Arroyo-Maya et al. (2016) studied the interactions between the pelargonidin and different proteins, such as β-LG using fluorescence spectroscopy. The pelargonidin quenching experiments were performed at pH 3.0 and 45°C. The results showed that β-LG fluorescence intensity was quenched with increased concentration of anthocyanin, whereas the only slight wavelength shift (about 2 nm) of the spectral maximum suggested no significant changes in the secondary structure. The estimated K_{SV} was 2.4×104/M, whereas the bimolecular rate constant was significantly larger (8×1012/M/s) compared to the value expected for a bimolecular rate constant. It is therefore believed that β-LG quenching most likely occurred through a static mechanism relying on the ligand binding to the protein (Lakowicz, 2006). These authors suggested that at 25°C, the binding affinity of β-LG for pelargonidin was higher at pH 3.0 (K_b of approximately 1.5×105/M) than at pH 7.0 (7.6×104/M). It is obvious that the binding ability of β-LG changes with pH because the protein is found predominantly in a monomeric form around pH 3.0 and in a dimer form around pH 7.0 (Bello et al., 2008). Protein monomers associations into dimers are highly probable at pH 7.0, causing burial of potential binding sites with an affinity for different small molecules. Moreover, Arroyo-Maya et al. (2016) revealed the importance of electrical characteristics of the protein with pH, with significant changes from positive at pH 3.0 to negative at pH 7.0, which may also have influenced the nature of any binding interactions. Based on the negative value of ΔG, the interaction between pelargonidin and β-LG is thermodynamically favorable, whereas the positive value of ΔS indicates that molecules binding relies to a high extent on the hydrophobic interactions.

Another study of Soares et al. (2007) based on measuring the quenching of BSA intrinsic fluorescence provided evidence on the binding of several phenolic compounds such as (+)-catechin, (−)-epicatechin, (−)-epicatechin gallate, malvidin-3-glucoside, tannic acid, procyanidin B4, procyanidin

B2 gallate, and procyanidin oligomers to the protein. These authors suggested that the binding/quenching process was significantly affected by the structure of polyphenols: a direct relationship between binding affinity increased and the molecular weight of polyphenol compounds was reported, as well as the positive effect of the presence of galloyl groups. Different values for K_{SV} values were reported, as follows: 14,100 and 13,800/M for catechin monomer and procyanidin dimer B4, respectively; 19,500 and 21,900/M for galloyl derivatives; 8700/M for (+)-catechin; and 100,548/M for tannic acid. When quenching BSA with different flavonoids, the fluorescence intensity decreased, but with no significant shift in λ_{max}, both at pH 5.0 and 4.0, highlighted no change in the immediate environment of the Trp residues other than the fact that the polyphenols were situated at close proximity to the Trp residue. Soares et al. (2007) suggested that tannic acid may lead to total quenching, while (+)-catechin and malvidin-3-glucoside quenched only 20–30% of the total BSA fluorescence. A static quenching mechanism was suggested for the interaction between BSA and (+)-catechin and malvidin-3-glucoside, whereas for the tannic acid, these authors suggested a complex mechanism described by "sphere of action model." A higher affinity of BSA for tannic acid when compared with (+)-catechin and malvidin-3-glucoside was suggested.

The thermodynamic parameters were used by Oancea et al. (2017) to describe the molecular forces involved in the complex formation. Two temperature regions were obtained when plotting ln K versus T^{-1} with linear correlations, between 25 and 70°C, and between 80 and 100°C, respectively. The ΔG values are positive and negative, respectively, in the two defined temperature ranges, suggesting that the binding process is nonspontaneous and spontaneous, respectively. Between 25 and 70°C, the positive and negative values of ΔH and ΔS indicate that the interaction between the anthocyanins from the sour cherry extract and β-LG is an endothermic process accompanied by entropy decrease. At higher temperatures, the reaction is endothermic with a positive value for ΔS. This type of reaction is reactant-favored at low temperatures and product-favored at high temperatures. Based on the sign and magnitude of the thermodynamic parameters, Ross and Subramanian (1981) classified the interactions mainly occurring between two molecules, as follows: hydrophobic interactions when $\Delta H > 0$ and $\Delta S > 0$, van der Waals' and hydrogen bonds interactions when $\Delta H < 0$ and $\Delta S < 0$, and electrostatic interactions when $\Delta H < 0$ and $\Delta S > 0$. The hydrophobic forces were involved in the formation of the β-LG-sour cherry extract complex at higher temperatures. The magnitude of the free energy decreased at higher temperature, suggesting that the binding interaction between β-LG and anthocyanins from sour cherries extract was weaker at higher temperatures, which may be an indicative of hydrogen bonding (Joye et al., 2015).

According to Chung et al. (2015), the interaction between denatured whey proteins and anthocyanin is exothermic, as indicated by the negative ΔH value (-98.85 kJ/M), the complex being mainly stabilized through hydrogen bonding. Given the negative ΔS values (-292.89 J/mol/K), they also suggested that the system migrates toward a less disordered configuration.

Similar observations are provided by Arroyo-Maya et al. (2016) who studied the pelargonidin binding to β-LG and showed that the process is thermodynamically favorable and complex formation is spontaneous (negative ΔG values) under acidic and neutral pH solvent conditions. The binding enthalpy (ΔH) was negligible at pH 3.0 (~ 0 kJ/mol) but seems to be positive at pH 7.0. (86 ± 5 kJ/mol). The positive ΔS values observed at both pH values are a measure of the hydrophobic contacts mainly involved in the binding interaction. Moreover, Arroyo-Maya et al. (2016) emphasized that in case of all molecular mechanisms standing behind the binding thermodynamics and regardless of the pH conditions, the binding of the molecules was entropically driven. Therefore, the favorable entropic contributions compensate for the unfavorable enthalpy contributions, resulting in a net negative free energy change.

There are mainly three sites on β-LG molecule proposed to be engaged in binding different small molecules: one internal cavity partially hosted by the β-barrel and two exposed sites consisting on a hydrophobic pocket located on the protein surface in a hollow surrounded by α-helix and β-barrel motifs, and one patch located nearby Trp[19]–Arg[124] residues (Roufik et al., 2006). Oancea et al. (2017) provided atomic level details on the interaction between β-LG and cyanidin-3-rutinoside (CYR), one of the major components of anthocyanin extract from sour cherry. The results of in silico analysis on β-LG-CYR complex generated through molecular docking technique resemble the fluorescence spectroscopy observations, suggesting better recognition of the thermally treated protein for the anthocyanin molecule compared to the native structure (Fig. 8.3).

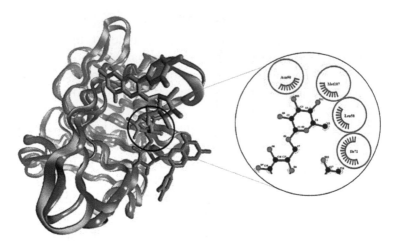

FIGURE 8.3   (See color insert.) Superposition of the β-LG-CYR docking models at different temperatures. The image was prepared using the VMD software. The protein at 25 and 90°C is represented in new ribbons style in gray and blue, respectively, whereas, the ligand is represented in orange and red, respectively, in licorice style. In detail are marked the amino acids located in the hydrophobic binding site which are in direct contact with atoms of the CYR at both temperatures. The image presented in the inset was prepared using LigPlot software.

Protein models equilibrated at 25 and 90°C hosted the CYR molecule in a large exposed hydrophobic cavity, and the binding sites share the following amino acids Leu[58], Asn[90], Ile[71], and Met[107] (Oancea et al., unpublished results). The higher stability of the complex obtained after the preliminary treatment of the protein at 90°C was explained through the rearrangements of the residues located in the CYR binding site, leading to the exposure of new functional groups or of some initially buried amino acids, therefore favoring new hydrophobic interactions and hydrogen bonds between the receptor and ligand molecules. Diamino acids and nonpolar neutral amino acids are mainly involved in the interaction with the ligand. After heating protein, four additional residues (Lys[60], Asn[88], Asn[109], and Gln[115]) become available for providing free amino groups which might stabilize the interface within the assembly. The hydrogen bonds connecting the two molecules of the complex were identified only at high temperature, and exclusively involved the newly exposed residues Lys[60], Asn[109], and Ser[116]. As in case of quercetin, quercitrin, and rutin used as ligands for the β-LG in the study of Sahihi et al. (2012), the phenolic groups in the CYR structure play key role in the binding process. The number of hydrogen bonds connecting the ligand by the receptor protein increased with the OH groups in the flavonoids structure (Sahihi et al., 2012). Regarding the protein surface in contact with the CYR atoms, only a slight increase (from 619.60 to 623.20 Å²) was reported with the temperature. Moreover, given the significantly higher binding energy found for the complex including β-LG preliminary treated at 90°C with respect to the native structure, one can easily understand the improvement of the affinity between molecules within the complex.

In another study, Aprodu et al. (2017) investigated the interaction and thermal stability of the complex formed between β-LG and carotenoid extract from sea buckthorn (*Hippophae rhamnoides L.*) (CSE) in correlation with protein structural changes monitored by fluorescence spectroscopy. The used methods were extrinsic and intrinsic fluorescence, phase diagram, synchronous spectra, 3D spectra, and quenching experiments with acrylamide and KI. Binding sites for β-LG–CSE complex and its effect on conformational stability and the secondary structure of β-LG were evaluated using quenching experiments with β-carotene and molecular docking and molecular dynamics simulations. The study considered the exceptional functional and nutritional properties of sea buckthorn. Sea buckthorn contains more than 190 compounds distributed in various parts of the plant, with different effects in promoting health, such as antioxidant, anti-inflammatory, immunomodulatory, hypolipidemic, hypoglycemic, and antiatherogenic (Xue et al., 2015). The fruits contain high amounts of nutrients such as vitamin C, flavonoids and tocopherols, tocotrienols, carotenoids, sterols, and triacylglycerols. Concerning the carotenoids content, Pop et al. (2014) suggested BC content between 1.9 and 7.4 mg/100 g dm, β-cryptoxanthin of 1.3–1.6 mg/100 g dm, lycopene of 1.4–2.3 mg/100 g·dm and zeaxanthin of 1.8–2.5 mg/100 g·dm, in the six Romanian varieties of sea buckthorn.

The BC binding ability by β-LG was first tested by Aprodu et al. (2017) by following the influence of increasing BC concentration on the fluorescence intensity spectra of β-LG preliminary heated at temperatures ranging from 25 to 100°C for 15 min. Significantly, when quenching the protein preliminary treated at temperatures ranging from 80 to 100°C, 7 and 3–4 nm redshifts were observed, indicating that BC addition caused the loss of the protein compact structure, exposing the hydrophobic subdomain where Trp is located. A static quenching mechanism was highlighted due to the increase of K_{SV} values in the temperature range of

25–90 (from $3.48 \pm 0.55 \times 10^{-10}$ l/mol to $6.83 \pm 0.24 \times 10^{-10}$ l/mol), followed by a decrease at higher temperature (Aprodu et al., 2017). The apparent binding constant increased up to 70°C, followed by a decrease at even higher temperatures, whereas the number of binding sites varied from 1.56 ± 0.04 at 25 to 0.96 ± 0.04 at 90°C, leading to the assumption that, regardless of the thermal treatment applied, the β-LG molecule has at least one binding site with high affinity for BC. Further details on the thermal-dependent behavior of the β-LG – BC complex were provided by Aprodu et al. (2017) after running molecular dynamics steps at 25 and 90°C. The affinity between β-LG and BC was reported to vary with the temperature, the assembly involving the protein model heated at high temperature being more thermodynamically stable. Only about 77% of the secondary struc-ture of the native β-LG was preserved at high temperature, and because of the changes in the exposure of the amino acids at the interface with BC, the total surface of the protein, which gets buried upon of the assembly formation increased from 304.7–386.1 $Å^2$. The following events were associated to the temperature increase: increased exposure to the solvent of the Ala[80] and Val[81] residues and intensification of their hydrophobic interactions with the BC molecule; burial of Lys[8] residue; and additional exposure of the Glu[89], Lys[91], and Leu[95] (Aprodu et al., 2017). After treat-ment at high temperature, the contacts between different β-LG molecules forming complex with BC are limited due to burial of the Ala[142], Leu[143], and Pro[144] (Adams et al., 2006; Aprodu et al., 2017).

In order to obtain a more complete picture of the influence of thermal treatment on the intrinsic fluorescence of the complex formed between CSE and β-LG, Aprodu et al. (2017) used three excitation wavelengths (274, 280, and 292 nm). By excitation at 274, 280, and 292 nm at 25°C, the emission peak of the corresponding β-LG hydrophobic residues were located at 336, 340, and 327 nm, respectively. Moreover, the fluorescence intensity increased with 118 and 147% when the samples were excited at 280 and 292 nm, respectively. When excited at 280 and 25 nm blueshifts have been observed in the temperature range of 50–80°C, followed by 3–5 nm redshifts at higher temperatures. The decrease in fluorescence intensity at 100°C may be explained by the effect of heating on the unfolding properties of the protein (Borkar et al., 2012), which may favor aggregation through intermolecular bonds (Rodrigues et al. 2015).

Redshifts were registered in the whole temperature range, varying from 7 nm at 60°C to 17 nm at 100°C when the complex was excited at 292 nm.

Aprodu et al. (2017) concluded that in the untreated protein molecules, the Trp and Tyr residues are located inside of protein core when λ_{max} is equal or lower than 330 nm, and are exposed to the solvent when λ_{max} is higher than 330 nm. The changes observed in the emissive properties of β-LG–CSE complex are associated with the structural changes occurring in the tertiary structure of β-LG.

In the quenching experiments with acrylamide, the K_{SV} values slightly increased in the temperature range of 25–70°C, followed by a significant increase at higher temperature, which can be attributed to the accessibility of Trp to acrylamide. The maximum value ($5.88 \pm 0.007 \times 10^{-3}$ l/mol) was calculated at 100°C, whereas the minimum value ($2.29 \pm 0.25 \times 10^{-3}$ l/mol) was estimated at 60°C. In case of quenching with KI, the minimum K_{SV} value was found at 25°C ($2.27 \pm 0.26 \cdot 10^{-3}$ l/mol) and the highest at 50°C ($4.18 \pm 0.44 \times 10^{-3}$ l/mol) (Aprodu et al., 2017).

Another method suitable for investigating the structural and conformational changes of proteins due to the binding of different ligands uses and 8-anilinonaphthalene-sulfonic acid (ANS), which has both hydrophobic and hydrophilic properties. The changes in ANS fluorescence indicate alteration of protein conformation due to ligand–protein interaction (Divsalar et al., 2006). The ANS can bind to two different places on β-LG molecules: one binding site is located on protein surface in the vicinity of the hydrophobic region of the protein and the second binding site is located within the protein core (Vetri and Militello, 2005). The site located in the surface of the protein appears to be responsible for nonspecific interaction with the ANS, whereas the internal one involves the Cys[106]–Cys[119] disulfide bridge (Collini et al., 2000). In this study, ANS complex intensity was assessed by excitation at 365 nm and collecting emission between 400 and 600 nm. An increase in fluorescence intensity of protein by heating at 90°C and a blueshift from 507 to 495 nm when increasing temperature from 25 to 70°C was observed by Aprodu et al. (2017) when heating the β-LG–CSB complex. Increasing the temperature up to 100°C resulted in 10 nm blueshift, indicating a change in the Trp microenvironment from hydrophilic to hydrophobic.

The same approach has been used by Dumitraşcu et al. (2016) to study the interaction between α-LA and CSE. The quenching process was independent of temperature, demonstrating that regardless of the thermal treatment applied, the interaction between ligand and protein occurred. The λ_{max} values of α-LA in the absence of the ligand was shifted from

326 nm at 25°C to 329 nm at 100°C, suggesting the exposure of the Trp residues at temperatures higher than 80°C. Under these conditions, the protein loses its compact subdomain structure containing hydrophobic residues. The redshift is an indicator of the Trp microenvironment nature, the increase in temperature resulting in increased hydrophilicity in the vicinity of Trp. The same behavior was observed when ligands were used, suggesting hydrophilic interaction between the two components.

The intrinsic fluorescence of α-LA is mainly due to the Trp residues: Trp[118] belongs to the rest of the cluster aromatic I, while the other three residues, in positions 26, 60, and 104, are located in the aromatic cluster II. In the native state, only Trp[60] is exposed to the solvent and contributes with 7% to the total fluorescence of the protein (Chrysina et al., 2000). It should be mentioned that the Trp residues within native α-LA molecules are surrounded by amino acids that are able to quench its fluorescence. Stănciuc et al. (2012b) suggested what in native state, Trp[26] is found in vicinity of Lys[16] and His[107], which can quench its fluorescence. Moreover, Trp[26] is located near Trp[104], Trp[60], and Trp[118]. Moreover, Trp[104] is located in vicinity of Tyr[103] and Phe[53], whereas Trp[60] is found close to Tyr[10] and Trp[118] by Phe[31].

K_{SV} values for CSE quenching were lower at 50°C ($1.54 \pm 0.12 \times 10^{-7}$ l/mol), and highest at 100°C ($2.29 \pm 0.12 \times 10^{-7}$ l/mol), highlighting that the exposure degree of Trp residues increases with increasing temperature. The increases in quenching constant values at high temperatures suggest an alteration of the tertiary structure of α-LA, with higher exposure of Trp residues. The calculated K_q values are higher than the maximum collision constant describing the interaction with different polymers, suggesting that the quenching process is initiated through a static mechanism (Table 8.2).

Regardless of the applied temperature, the K_{SV} values for CSE binding by α-LA were lower when compared with BC. For example, in case of BC quenching, an increase in K_{SV} values from $3.86 \pm 0.16 \times 10^{-7}$ l/mol to $4.25 \pm 0.06 \times 10^{-7}$ l/mol was found when increasing the temperature from 25 to 60°C, whereas lower value was measured at 100°C ($4.91 \pm 0.02 \times 10^{-7}$ l/mol) (Dumitraşcu et al., 2016). The increase in the quenching constant values at suggested temperatures indicates the alteration of tertiary structure of α-LA, with an exposure of Trp residues. The K_q presented in Table 8.2 are higher than the maximum collision constant values for the interaction with various polymers, suggesting that the quenching process is initiated by a static mechanism. These authors also suggested that more hydrophobic interactions are involved in the formation of α-LA–CSE complex when compared with α-LA–BC. However, the lower binding of CSE by α-LA may be due to the multitude of compounds in the extract, competing for the binding sites. The affinity of the protein for the CSE is decreased with increasing temperature (n decreased from 1.55 ± 0.13 at 25°C to 0.93 ± 0.02 at 100°C).

TABLE 8.2 Binding Parameters Between α-lactalbumin and carotenoid extract from sea buckthorn at Different Temperatures.

T (°C)	$K_{SV}(10^{-7}$ l/mol)	$K_q (10^{-15}$ mol/s)	$K_b (10^{-7}$ l/mol)	n
25	1.69 ± 0.02	1.69 ± 0.02	1.60 ± 0.27	1.55 ± 0.13
50	1.54 ± 0.12	1.54 ± 0.12	1.04 ± 0.08	1.13 ± 0.07
60	1.57 ± 0.02	1.57 ± 0.02	1.05 ± 0.19	1.09 ± 0.12
70	1.91 ± 0.02	1.91 ± 0.02	1.02 ± 0.02	1.13 ± 0.05
80	1.95 ± 0.16	1.95 ± 0.16	0.88 ± 0.08	1.02 ± 0.007
90	1.67 ± 0.07	1.67 ± 0.07	0.70 ± 0.08	0.93 ± 0.008
100	2.29 ± 0.12	2.29 ± 0.12	0.99 ± 0.03	0.93 ± 0.02

Source: Reprinted//adapted from Dumitraşcu et al. (2016) (with permission from Elsevier).
[a]Standard errors

Negative values for ΔH and ΔS suggested that both the hydrogen bonds and van der Waals interactions are involved in the formation of the complexes. Negative values for ΔG and ΔH indicated that the formation of the α-LA–CSE is an exothermic process. Molecular modeling investigation on the behavior of α-LA–BC assembly indicated that enthalpy- and entropy-driven events like direct interaction between ligand and protein, or conformational changes of the complex components are responsible for complex formation. The complex involving α-LA molecule preliminarily heat-treated at 90°C is more stable from the thermodynamical point of view (Dumitraşcu et al., 2016).

Ursache et al. (2017) started from the premise that the addition of bioactive ingredients in food systems are simple ways of developing functional foods that can provide physiological benefits and reduce the risk of some diseases. Therefore, based on the previously discussed studied, these authors tested the thermostability of α-LA–CSE complex by heating in the temperature range of 25–100°C in order to investigate the possibility of using it as a potential functional food ingredient.

The intrinsic fluorescence was used to investigate the heat-induced structural changes in α-LA–CSE complex. The abovementioned Trp residues from α-LA molecule cause heterogeneity in terms of accessibility in various conditions. A detailed investigation of the patterns equilibrated at 25°C showed that Trp[26] and Trp[60] are locked inside the protein molecule, while residues Trp[104] and Trp[118] are exposed on the protein surface, having a solvent available area of 1.65 and 20.19 Å^2 (Ursache et al., 2017). The association of carotenoids with α-LA before heat treatment started from the premise that the protein has the ability to act as a solubilizing agent. During the heat treatment, protein unfolding might allow carotenoids binding to regions that are not accessible in its native state. Furthermore, the increase of the energy resulting from the interaction between α-LA and BC at single molecule level, coupled with a significant decrease of the interaction surface when treating the solvated complex at increasing temperature suggests a lower affinity between the two molecules of the assembly.

The λ_{max} for Trp residues in untreated form was 326 nm. Stănciuc et al. (2012b) studied the effect of pH and heat treatment on the conformation of α-LA and suggested that at neutral pH the λ_{max} was 331 nm. The observed 5 nm blueshift may be correlated with the addition of CSE that resulted in blockage of the Trp residues in the protein core. The heat treatment resulted in a 2-nm redshift, suggesting an increase of hydrophilicity in the vicinity of Trp residues. Ursache et al. (2017) suggested that the carotenoid binding sites are not located in the neighborhood of Trp residues. Analysis performed on α-LA–BC models indicated that regardless of the simulated temperature, the binding pocket is located on the protein surface and involves the Lys[5]–Leu[12] and Asp[87]–Asp[97] helical structures (Ursache et al., 2017). Although Glu[11], Leu[12], and Ile[89] residues were reported to interact with the BC in the whole studied temperature range, the contacts established at the interface were found to be temperature dependent. For instance, Glu[11] and Leu[12] residues, which at 25°C are involved in hydrophobic contacts with the C atoms bridging the two retinyl groups, move toward each of the two equivalent β-rings of BC molecule, therefore favoring additional interaction of Ile[89] with the hydrocarbons chain (Fig. 8.4) (Ursache et al., 2017). Ursache et al. (2017) also performed quenching experiments with acrylamide and KI and suggested a K_{SV} value of $4.53 \pm 0.08 \times 10^{-2}$ l/mol at 80°C, whereas the minimum value was obtained at 60°C ($3.52 \pm 0.20 \times 10^{-2}$ l/mol). When quenching with KI, the lowest value was obtained at 25°C ($1.42 \pm 0.05 \times 10^{-2}$ l/mol) and the highest at 70°C ($2.14 \pm 0.01 \times 10^{-2}$ l/mol). The highest accessibility of the complex at 70°C is associated with an increased exposure degree of Trp residues, followed by a decrease at higher temperatures, due to the folding of polypeptide chains. It can be appreciated that the presence of CSE affected to some extent the fluorescence intensity of the Trp[118] and Trp[104] residues.

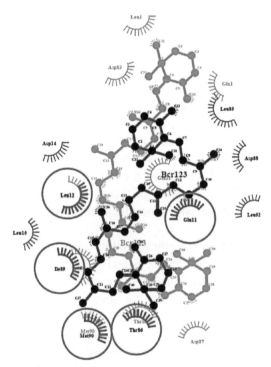

FIGURE 8.4 (See color insert.) Superposition of the α-LA–β-carotene (BC) models equilibrated at 25°C (gray) and at 90°C (colored). The hydrophobic contacts directly involving α-LA amino acids are represented by arcs radiating toward the ligand atoms they are in contact with. The amino acids interacting with the BC molecule at both temperatures are circled. The representation was prepared using LigPlot+.

Source: Adapted from Laskowski and Swindells (2011).

8.5 SUMMARY AND CONCLUSIONS

Nowadays, there is a great need for food products that provide health benefits. Therefore, advancing the fundamental knowledge for designing new functional products with antidiabetic, anti-inflammatory, antioxidative properties, and so forth, is highly desired. These products could be successfully used as an alternative to dietary supplements, offering some advantages concerning safety, consumption, and in vivo effectiveness of the delivered bioactive compounds (Šaponjac et al. 2016). In recent years, a special attention has been paid to the encapsulation of BACs, given their important biological functions and instability to various environmental factors. For example, the consumption of high amounts of polyphenolic compounds ensured important antioxidant activity in different lipid systems and in particular against oxidation of low-density lipoproteins, and was significantly correlated with the decrease of the incidence of coronary heart disease (Zhang et al., 2007). Microencapsulation is a technique for coating BACs, providing protection and controlled release of the entrapped ingredients (Esfanjani et al., 2015). Different proteins, such as whey proteins, can be used as a reasonable alternative matrix material due to their low viscosity at highly concentrated solutions (Betz and Kulozika, 2011). Whey proteins can be considered an ideal entrapping wall material because of the structural particularities which allow ligands binding and are generally considered as safe and possess high nutritional value.

The presented data are significantly important for characterization of the structural basis for whey proteins and polyphenol and/or carotenoids interactions. The detailed knowledge of the mechanisms of interaction and binding parameters are key priority for development of micro- and nano-encapsulation techniques. Future research needs to address the gap related to the knowledge on the interactions of a range of particular proteins and individual polyphenols and carotenoids in order to attempt the elucidation of structure–process–function relationships. This kind of studies would allow further developments on new functional products or ingredients with enhanced bioavailability of different compounds.

ACKNOWLEDGMENTS

This work was supported by a grant of the Romanian National Authority for Scientific Research and Innovation, CNCS-UEFISCDI, project number PN-II-RU-TE-2014-4-0115.

KEYWORDS

- anthocyanins
- carotenoids
- fluorescence spectroscopy
- functional food ingredients
- microencapsulation
- molecular modeling
- polyphenols
- whey proteins
- ANS fluorescence
- anthocyanins
- apoptosis
- biopolymers
- bovine serum albumin
- β-lactoglobulin
- immunomodulatory
- lutein
- α-lactalbumin

REFERENCES

Acharya, D. P.; Sanguansri, L.; Augustin, M. A. Binding of Resveratrol with Sodium Caseinate in Aqueous Solutions. *Food Chem.* **2013,** *141,* 1050–1054.

Adams, J. J.; Anderson, B. F.; Norris, G. E.; Creamer, L. K.; Jameson, G. B. Structure of Bovine β-Lactoglobulin (variant A) at Very Low Ionic Strength. *J. Struct. Biol.* **2006,** *154,* 246–254.

Aprodu, I. Computational Methods for Protein Analysis. In *Biomedical Engineering: Introduction to Current Approaches;* Ioniță, M., Ed.; Printech: Bucuresti, 2016; pp 101–128.

Aprodu, I.; Ursache, F. M.; Turturică, M.; Râpeanu, G.; Stănciuc, N. Thermal Stability of the Complex Formed Between Carotenoids from Sea Buckthorn (*Hippophaerhamnoides* L.) and Bovine β-Lactoglobulin. *Spectrochim. Acta Part A* **2017,** *173,* 562–571.

Arroyo-Maya, I. J.; Campos-Terán, J.; Hernández-Arana, A.; McClements, D. J. Characterization of Flavonoid-Protein Interactions Using Fluorescence Spectroscopy: Binding of Pelargonidin to Dairy Proteins. *Food Chem.* **2016,** *213,* 431–439.

Baker, E. N.; Baker, H. M.; Kidd, R. D. Lactoferrin and Transferrin: Functional Variations on a Common Structural Framework. *Biochem. Cell Biol.* **2002**, *80*, 27–34.

Bandyopadhyay, P.; Ghosh, A.; Ghosh, C. Recent Developments on Polyphenol–Protein Interactions: Effects on Tea and Coffee Taste, Antioxidant Properties and the Digestive System. *Food Funct.* **2012**, *3*, 592–605.

Bello, M.; Pérez-Hernández, G.; Fernández-Velasco; D. A. Arreguín-Espinosa, R.; García-Hernández, E. Energetics of Protein Homodimerization: Effects of Water Sequestering on the Formation of β-Lactoglobulin Dimer. *Proteins* **2008**, *70*, 1475–1487.

Betz, M.; Kulozik, U. Microencapsulation of Bioactive Bilberry Anthocyanins by Means of Whey Protein Gels. *Procedia Food Sci.* **2011**, *1*, 2047–2056.

Bhosale, P.; Bernstein, P. S. Vertebrate and Invertebrate Carotenoid Binding Proteins. *Arch. Biochem. Biophys.* **2007**, *458*, 121–127.

Borkar, A.; Rout, M. K.; Hosur, V. Denaturation of HIV-1 Protease (PR) Monomer by Acetic Acid: Mechanistic and Trajectory Insights from Molecular Dynamics Simulations and NMR. *J. Biomol. Struct. Dyn.* **2012**, *29*, 893–903.

Brownlow, S.; Morais Cabral, J. H.; Cooper, R.; Flower, D. R.; Yewdall, S. J.; Polikarpov, I.; North, A. C. T.; Sawyer, L. Bovine β-Lactoglobulin at 1.8 A Resolution—Still an Enigmatic Lipocalin. *Structure* **1997**, *5*, 481–495.

Carvalho, E.; Póvoas, M. J.; Mateus, N.; de Freitas, V. Application of Flow Nephelometry to the Analysis of the Influence of Carbohydrates on Protein-Tannin Interactions. *J. Sci. Food Agric.* **2006**, *86*, 891–896.

Chaudhuri, A.; Haldar, S.; Chattopadhyay, A. Organization and Dynamics of Tryptophans in the Molten Globule State of Bovine α-Lactalbumin Utilizing Wavelength-Selective Fluorescence Approach: Comparisons with Native and Denatured States. *Biochem. Biophys. Res. Commun.* **2010**, *394*, 1082–1086.

Chen, L.; Remondetto; G. E. Subirade, M. Food Protein-Based Materials as Nutraceutical Delivery Systems. *Trends Food Sci. Technol.* **2006**, *17*, 272–283.

Chrysina, E. D.; Brew, K.; Acharya, K. R. Crystal Structures of Apo- and Holo-Bovine α-Lactalbumin at 2.2-A Resolution Reveal an Effect of Calcium on Inter-Lobe Interactions. *J. Biol. Chem.* **2000**, *275*, 37021–37029.

Chung, C.; Rojanasasithara, T.; Mutilangi, W.; McClements, D. J. Enhanced Stability of Anthocyanin-Based Color in Model Beverage Systems through Whey Protein Isolate Complexation. *Food Res. Int.* **2015**, *76*, 761–768.

Collini, M.; D'Alfonso, L.; Baldini, G. New Insight on β-Lactoglobulin Binding Sites by 1-anilinonaphtalene-8-sulfonate Fluorescence Decay. *Protein Sci.* **2000**, *9*, 1968–1974.

Cook, N.; Samman, S. Flavonoids-Chemistry, Metabolism, Cardioprotective Effects, and Dietary Sources. *J. Nutr. Biochem.* **1996**, *7*, 66–76.

De Freitas, V.; Carvalho, E.; Mateus, N. Study of Carbohydrate Influence on Protein-Tannin Aggregation by Nephelometry. *Food Chem.* **2003**, *81*, 503–509.

Deming, D. M.; Baker, D. H.; Erdman, Jr., J. W. The Relative Vitamin A Value of 9-cis β-Carotene is Less and that of 13-cis β-Carotene may be Greater than the Accepted 50% that of All-trans β-Carotene in Gerbils. *J. Nutr.* **2002**, *132*, 2709–2712.

Divsalar, A.; Saboury, A. A.; Moosavi-Movahedi, A. A. Conformational and Structural Analysis of Bovine β-Lactoglobulin-A Upon Interaction with Cr+3. *Protein J.* **2006**, *25*, 157–165.

Dumitraşcu, L.; Ursache, F. M.; Stănciuc, N.; Aprodu, I. Studies on Binding Mechanism Between Carotenoids from Sea Buckthorn and Thermally Treated α-Lactalbumin. *J. Mol. Struct.* **2016,** *1125,* 721–729. DOI: 10.1016/j.molstruc.2016.07.070.

Dupont, D.; Arnould, C.; Rolet-Repecaud, O.; Duboz, G.; Faurie, F.; Martin, B.; Beuvier, E. Determination of Bovine Lactoferrin Concentrations in Cheese with Specific Monoclonal Antibodies. *Int. Dairy J.* **2006,** *16,* 1081–1087.

Esfanjani, A. E.; Jafari, S. M.; Assadpoor, E.; Mohammadi, A. Nano-Encapsulation of Saffron Extract Through Double-Layered Multiple Emulsions of Pectin and Whey Protein Concentrate. *J. Food Eng.* **2015,** *165,* 149–155

Fang, B.; Zhang, M.; Tian, M.; Jiang, L.; Guo, H. Y.; Ren, F. Z. Bovine Lactoferrin Binds Oleic Acid to form an Anti-Tumor Complex Similar to HAMLET. *Biochim. Biophys. Acta* **2014,** *1841,* 535–543.

Farnaud, S.; Evans, R. W. Lactoferrin-A Multifunctional Protein with Antimicrobial Properties. *Mol. Immunol.* **2003,** *40,* 395–405.

Fickel, J.; Pitra, C. H.; Joest, B. A.; Hofmann, R. R. A Novel Method to Evaluate the Relative Tannin-Binding Capacities of Salivary Proteins. *Comp. Biochem. Physiol. Part C: Toxicol. Pharmacol.* **1999,** *122,* 225–229.

Frazier, R. A.; Papadopoulou, A.; Mueller-Harvey, I.; Kissoon, D.; Green, R. J. Probing Protein-Tannin Interactions by Isothermal Titration Microcalorimetry. *J. Agric. Food Chem.* **2003,** *51,* 5189–5195.

Gholami, S.; Bordbar, A. K. Exploring Binding Properties of Naringenin with Bovine β-Lactoglobulin: A Fluorescence, Molecular Docking and Molecular Dynamics Simulation Study. *Biophys. Chem.* **2014,** *187–188,* 33–42.

Giancola, C.; De Sena, C.; Fessas, D.; Graziano, G.; Barone, G. DSC Studies on Bovine Serum Albumin Denaturation. Effects of Ionic Strength and SDS Concentration. *Int. J. Biol. Macromol.* **1997,** *20,* 193–204.

Gustafsson, L.; Leijonhufvud, I.; Aronsson, A.; Mossberg, A. K.; Svanborg, C. Treatment of Skin Papillomas with Topical Alpha-Lactalbumin-Oleic Acid. *N. Engl. J. Med.* **2004,** *350,* 2663.

He, X. M.; Carter, D. C. Atomic Structure and Chemistry of Human Serum Albumin. *Nature* **1992,** *358,* 209–215.

Heim, K. E.; Tagliaferro, A. R.; Bobilya, D. J. Flavonoid Antioxidants: Chemistry, Metabolism and Structure-Activity Relationships. *J. Nutr. Biochem.* **2002,** *13,* 572–584.

Horne, J.; Hayes, J.; Lawless, H. T. Turbidity as a Measure of Salivary Protein Reactions with Astringent Substances. *Chem. Senses* **2002,** *27,* 653–659.

Humphrey, W.; Dalke, A.; Schulten, K. VMD: Visual Molecular Dynamics. *J. Mol. Graphics* **1996,** *14,* 33–38.

Jöbstl, E.; Howse, J. R.; Fairclough, A.; Williamson, M. P. Noncovalent Cross-Linking of Casein by Epigallocatechin Gallate Characterized by Single Molecule Force Microscopy. *J. Agric. Food Chem.* **2006,** *54,* 4077–4081.

Joye, I. J.; Davidov-Pardo, G.; Ludescher, R. D.; McClements, D. J. Fluorescence Quenching Study of Resveratrol Binding to Zein and Gliadin: Towards a More Rational Approach to Resveratrol Encapsulation Using Water-Insoluble Proteins. *Food Chem.* **2015,** *185,* 261–267.

Kanakis, C. D.; Hasni, I.; Bourassa, P.; Tarantilis, P. A.; Polissiou, M. G.; Heidar-Ali, T-R. Milk β-Lactoglobulin Complexes with Tea Polyphenols. *Food Chem.* **2011,** *127,* 1046–1055.

Knockaert, G.; Pulissery, S. K.; Colle, I.; van Buggenhout, S.; Hendrickx, M.; van Loey, A. Lycopene Degradation, Isomerization and in vitro Bioaccessibility in High Pressure Homogenized Tomato Puree Containing Oil: Effect of Additional Thermal and High Pressure Processing. *Food Chem.* **2012,** *135,* 1290–1297.

Lakowicz, J. R. *Principles of Fluorescence Spectroscopy,* 2nd ed.; Kluwer/Plenum: New York, 1999.

Lakowicz, J. R. *Principles of Fluorescence Spectroscopy,* 3rd ed.; Springer: New York, 2006.

Laskowski, R. A.; Swindells, M. B. LigPlot+: Multiple Ligand–Protein Interaction Diagrams for Drug Discovery. *J. Chem. Inf. Model.* **2011,** *51,* 2778–2786.

Latruffe, N.; Menzel, M.; Delmas, D.; Buchet, R.; Lançon, A. Compared Binding Properties between Resveratrol and Other Polyphenols to Plasmatic Albumin: Consequences for the Health Protecting Effect of Dietary Plant Microcomponents. *Molecules* **2014,** *19,* 17066–17077.

Lätti, A.; Riihinen, K.; Kainulainen, P. Analysis of Anthocyanin Variation in Wild Populations of Bilberry (*Vaccinium myrtillus* L.) in Finland. *J. Agric. Food Chem.* **2008,** *56,* 190–196.

Lättia, A. K.; Riihinena, K. R.; Jaakola, L. Phenolic Compounds in Berries and Flowers of a Natural Hybrid Between Bilberry and Lingonberry (Vaccinium × intermedium Ruthe). *Phytochemistry* **2011,** *72,* 810–815

Le Bourvellec, C.; Renard, C. M. Interactions Between Polyphenols and Macromolecules: Quantification Methods and Mechanisms. *Crit. Rev. Food Sci. Nutr.* **2012,** *52*(3), 213–48.

Legrand, D.; Elass, E.; Pierce, A.; Mazurier, J. Lactoferrin and Host Defence: An Overview of its Immuno-Modulating and Anti-Inflammatory Properties. *Biometals* **2004,** *17,* 225–229.

Li, M.; Maa, Y.; Ngadi, M. O. Binding of Curcumin to β-Lactoglobulin and its Effect on Antioxidant Characteristics of Curcumin. *Food Chem.* **2013,** *141,* 1504–1511.

Livney, Y. D. Milk Proteins as Vehicles for Bioactives. *Curr. Opin. Colloid Interface Sci.* **2010,** *15,* 73–83.

Loch, J.; Polit, A.; Gerocki, A.; Bonarek, P.; Kurpiewska, K.; Dziedzicka-Wasylewska, M.; Lewinski, K. Two Modes of Fatty Acid Binding to Bovine Beta-Lactoglobulin-Crystallographic and Spectroscopic Studies. *J. Mol. Recognit.* **2011,** *24*(2), 341–349.

Manach, C.; Williamson, G.; Morand, C.; Scalbert, A.; Rémésy, C. Bioavailability and Bioefficacy of Polyphenols in Humans. I. Review of Bioavailability Studies. *Am. J. Clin. Nutr.* **2005,** *81,* 230S–242S.

Marengo, M.; Miriani, M.; Ferranti, P.; Bonomi, F.; Iametti, S.; Barbiroli, A. Structural Changes in Emulsion-Bound Bovine Beta-Lactoglobulin Affect its Proteolysis and Immunoreactivity. *Biochim. Biophys. Acta* **2016,** *1864,* 805–813.

McClements, D. J. *Food Emulsions: Principles, Practice, and Techniques,* 3rd ed.; CRC Press: Boca Raton, FL, 2015.

McClements, D. J.; Li, F.; Xiao, H. The Nutraceutical Bioavailability Classification Scheme: Classifying Nutraceuticals According to Factors Limiting their Oral Bioavailability. *Ann. Rev. Food Sci. Technol.* **2015,** *6,* 299–327.

Middleton, E.; Kandaswami, C.; Theoharides, T. C. The Effects of Plant Flavonoids on Mammalian Cells: Implications for Inflammation, Heart Disease, and Cancer. *Pharmacol. Rev.* **2000,** *52,* 673–751.

Mocz, G.; Ross, J. A.; Fluorescence Techniques in Analysis of Protein–Ligand Interactions. In *Protein-Ligand Interactions: Methods and Applications, Methods in Molecular Biology;* Williams, M. A., Daviter, T., Eds.; Springer Science-Business Media: New York, 2013; Vol. 1008, pp 169–210.

Mohammadi, F.; Bordbar, A. K.; Divsalar, A.; Mohammadi, K.; Saboury, A. A. Interaction of Curcumin and Diacetylcurcumin with the Lipocalin Member β-Lactoglobulin. *Protein J.* **2009,** *28,* 117–123.

Møller, A. P.; Biard, C.; Blount, J. D.; Houston, D. C.; Ninni, P.; Saino, N.; Surai, P. F. Carotenoid-Dependent Signals: Indicators of Foraging Efficiency, Immunocompetence or Detoxification Ability? *Avian Poult. Biol. Rev.* **2000,** *11,* 137–159.

Munin, A.; Edwards-Lévy, F. Encapsulation of Natural Polyphenolic Compounds: A Review. *Pharmaceutics* **2011,** *3,* 793–829.

Nandave, M.; Ojha, S.; Arya, D. Protective Role of Flavonoids in Cardiovascular Diseases. *Natural Product Radiance* **2005,** *4,* 166–176.

Oancea, A. M.; Aprodu, I.; Râpeanu, G.; Bahrim, G.; Stănciuc. Binding Mechanism of Anthocyanins from Sour Cherries (*Prunuscerasus* L) Skin to Bovine β-Lactoglobulin: A Fluorescence and in Silico Based Approach. *Int. J. Food Prop.* **2017.** DOI: http://dx.doi.org/10.1080/10942912.2017.1343347.

Papadopoulou, A.; Frazier, R. A. Characterization of Protein-Polyphenol Interactions. *Trends Food Sci. Technol.* **2004,** *15,* 186–190.

Parker, R. S. Carotenoids in Human Blood and Tissues. *J. Nutr.* **1989,** *119*(1), 101–104.

Pérez, M. D.; Calvo, M. Interaction of β-Lactoglobulin with Retinol and Fatty Acids and its Role as a Possible Biological Function for this Protein. *J. Dairy Sci.* **1995,** *78,* 978–988.

Pike, A. C.; Brew, K.; Acharya, K. R. Crystal Structures of Guinea-Pig, Goat and Bovine α-Lactalbumin: Highlight the Enhanced Conformational Flexibility of Regions that are Significant for its Action in Lactose Synthase. *Structure* **1996,** *4,* 691–703.

Pop, R. M.; Weesepoel, Y.; Socaciu, C.; Pintea, A.; Vincken; J. P. Gruppen, H. Carotenoid Composition of Berries and Leaves from Six Romanian Sea Buckthorn (*Hippophaerhamnoides* L.) Varieties. *Food Chem.* **2014,** *147,* 1–9.

Prigent, S. V. E.; Gruppen, H.; Visser, A. J. W. G.; Van Koningsveld, G. A. H. D.; Alfons, G. J. V. Effects of Noncovalent Interactions with 5-o-Caffeoylquinic Acid (CGA) on the Heat Denaturation and Solubility of Globular Proteins. *J. Agric. Food Chem.* **2003,** *51,* 5088–5095.

Rodrigues, R. M.; Martins, A. J.; Ramos, O. L.; Malcata, F. X.; Teixeira, J. A.; Vicente, A. A.; et al. Influence of Moderate Electric Fields on Gelation of Whey Protein Isolate. *Food Hydrocolloids* **2015,** *43,* 329–339.

Ross, P. D.; Subramanian, S. Thermodynamics of Protein Association Reactions: Forces Contributing to Stability. *Biochemistry* **1981,** *26,* 3096–3102.

Roufik, S.; Gauthier, S. F.; Leng, X.; Turgeon, S. L. Thermodynamics of Binding Interactions Between Bovine β-Lactoglobulin A and the Antihypertensive Peptide β-Lg f142–148. *Biomacromolecules* **2006,** *7*(2), 419–426.

Sahihi, M.; Heidari-Koholi, Z.; Bordbar, A. K. The Interaction of Polyphenol Flavonoids with β-Lactoglobulin: Molecular Docking and Molecular Dynamics Simulation Studies. *J. Macromol. Sci. Part B* **2012**, *51,* 2311–2323.

Sánchez, L.; Peiro, J. M.; Castillo, H.; Perez, M. D.; Ena, J. M.; Calvo, M. Kinetic Parameters for Denaturation of Bovine Milk Lactoferrin. *J. Food Sci.* **1992,** *57,* 873–879.

Šaponjac, V. T.; Ćetković, G.; Čanadanović-Brunet, J.; Pajin, B.; Djilas, S.; Petrović, I.; Lončarević, I.; Stajčić, S.; Vulic, J. Sour Cherry Pomace Extract Encapsulated in Whey and Soy Proteins: Incorporation in Cookies. *Food Chem.* **2016,** *207,* 27–33.

Sarkar, M.; Shankar Paul, S.; Mukherjea, K. K. Interaction of Bovine Serum Albumin with a Psychotropic Drug Alprazolam: Physicochemical, Photophysical and Molecular Studies. *J. Lumin.* **2013,** *142,* 220–230.

Schantz, M.; Mohn, C.; Baum, M.; Richling, E. Antioxidative Efficiency of an Anthocyanin Rich Bilberry Extract in the Human Colon Tumor Cell Lines Caco-2 and HT-29. *J. Berry Res.* **2010,** *1,* 25–33.

Sellappan, S.; Akoh, C. C.; Krewer, G. Phenolic Compounds and Antioxidant Capacity of Georgia-Grown Blueberries and Blackberries. *J. Agric. Food Chem.* **2002,** *50,* 2432–2438.

Shahidi, F.; Naczk, M. *Phenolics on Food and Nutraceuticals;* CRC Press: Boca Raton, FL, 2004.

Simion, A. M.; Aprodu, I.; Dumitraşcu, L.; Bahrim, G. E.; Alexe, P.; Stănciuc, N. Probing Thermal Stability of the β-Lactoglobulin–Oleic Acid Complex by Fluorescence Spectroscopy and Molecular Modeling. *J. Mol. Struct.* **2015,** *1095,* 26–33.

Soares, S.; Mateus, N.; De Freitas, V. Interaction of Different Polyphenols with Bovine Serum Albumin (BSA) and Human Salivary r-Amylase (HSA) by Fluorescence Quenching. *J. Agric. Food Chem.* **2007,** *55,* 6726–6735.

Stănciuc, N.; Aprodu, I.; Râpeanu, G.; Bahrim, G. Fluorescence Spectroscopy and Molecular Modeling Investigations on the Thermally Induced Structural Changes of Bovine β-Lactoglobulin. *Innovative Food Sci. Emerging Technol.* **2012a,** *15,* 50–56.

Stănciuc, N.; Râpeanu, G.; Bahrim, G.; Aprodu, I. pH and Heat-Induced Structural Changes of Bovine Apo-α-Lactalbumin. *Food Chem.* **2012b,** *131*(3), 956–963.

Stănciuc, N.; Aprodu, I.; Râpeanu, G.; Van der Plancken, I.; Bahrim, G.; Hendrickx, M. Analysis of the Thermally Induced Structural Changes of Bovine Lactoferrin. *J. Agric. Food Chem.* **2013,** *61*(9), 2234–2243.

Steijns, J. M.; van Hooijdonk, A. C. M. Occurrence, Structure, Biochemical Properties and Technological Characteristics of Lactoferrin. *Br. J. Nutr.* **2000,** *84,* S11–17.

Sugio, S.; Kashima, A.; Mochizuki, S.; Noda, M.; Kobayashi, K. Crystal Structure of Human Serum Albumin at 2.5 A Resolution. *Protein Eng.* **1999,** *12,* 439–446.

Sui, Q.; Roginski, H.; Williams; R. P.; Versteeg, C.; Wan, J. Effect of Pulsed Electric Field and Thermal Treatment on the Physicochemical Properties of Lactoferrin with Different Iron Saturation Levels. *Int. Dairy J.* **2010,** *20*(10), 707–714.

Svensson, M.; Hakansson, A.; Mossberg; A. K.; Linse, S.; Svanborg, C. Conversion of Alpha-Lactalbumin to a Protein Inducing Apoptosis. *Proc. Natl. Acad. Sci.* **2000,** *97,* 4221.

Tang, L.; Zuo, H.; Li, S. Comparison of the Interaction Between Three Anthocyanins and Human Serum Albumins by Spectroscopy. *J. Lumin.* **2014,** *153,* 54–63.

Tavares, G. M.; Croguennec, T.; Carvalho, A. F.; Bouhallab, S. Milk Proteins as Encapsulation Devices and Delivery Vehicles: Applications and Trends. *Trends Food Sci. Technol.* **2014,** *37,* 5–20.

Teng, Z.; Xu, R.; Wang, Q. Beta-Lactoglobulin-Based Encapsulating Systems as Emerging Bioavailability Enhancers for Nutraceuticals: A Review. *RSC Adv.* **2015,** *5,* 35138–35154.

Togashi, D. M., Ryder, A. G. A Fluorescence Analysis of ANS Bound to Bovine Serum Albumin: Binding Properties Revisited by Using Energy Transfer. *J. Fluoresc.* **2008,** *18,* 519–526.

Ulrih, N. P. Analytical Techniques for the Study of Polyphenol-Protein Interactions. Critical Reviews in Food Science and Nutrition, **2015,** http://dx.doi.org/10.1080/1040 8398.2015.1052040.

Ursache, F. M.; Aprodu, I.; Nistor, O. V.; Bratu, M.; Botez, E.; Stănciuc, N. Probing the Heat-Induced Structural Changes in Bovine Serum Albumin by Fluorescence Spectroscopy and Molecular Modelling. *Int. J. Dairy Technol.* **2016.** DOI; 10.1111/1471–0307.12351.

Ursache, F. M.; Dumitrascu, L.; Aprodu, I.; Stănciuc, N. Screening the Thermal Stability of Carotenoids-α Lactalbumin Complex by Spectroscopic and Molecular Modeling Approach. *J. Macromol. Sci. Pure Appl. Chem.* **2017,** *54,* 316–322 (accepted for publication).

Vachali, P.; Bhosale, P.; Bernstein, P. S. Microbial Carotenoids. In *Microbial Carotenoids from Fungi: Methods and Protocols;* Barredo, J. L., Ed.; Springer: New York, 2012; pp 41–59.

Van de Weert, M.; Stella, L. Fluorescence Quenching and Ligand Binding: A Critical Discussion of a Popular Methodology. *J. Mol. Struct.* **2011,** *998,* 144–150.

Verma, D. K.; Srivastav, P. P. *Microorganisms in Sustainable Agriculture, Food and the Environment;* (as part of book series on *Innovations in Agricultural Microbiology);* Apple Academic Press: USA, 2017.

Verma, D. K.; Mahato, D. K.; Billoria, S.; Kapri, M.; Prabhakar, P. K.; Ajesh Kumar, V.; Srivastav, P. P. Microbial Spoilage in Milk Products, Potential Solution, Food Safety and Health Issues. In *Microorganisms in Sustainable Agriculture, Food and the Environment;* Verma, D. K., Srivastav, P. P., Eds.; (as part of book series on *Innovations in Agricultural Microbiology);* Apple Academic Press: USA, 2017; pp 171–196.

Vetri, V.; Militello, V. Thermal Induced Conformational Changes Involved in the Aggregation Pathways of β-Lactoglobulin. *Biophys. Chem.* **2005,** *113,* 83–91.

Wiese, S.; Gärtner, S.; Rawel; H. M.; Winterhalter, P.; Kulling, S. B. Protein Interactions with Cyanidin-3-Glucoside and Its Influence on α-Amylase Activity. *J. Sci. Food Agric.* **2009,** *89,* 33–40.

Xiong, S.; Melton; L. D.; Easteal, A. J.; Siew, D. Stability and Antioxidant Activity of Black Currant Anthocyanins in Solution and Encapsulated in Glucan Gel. *J. Agric. Food Chem.* **2006,** *54,* 6201–6208.

Xue, Y.; Miao, Q.; Zhao, A.; Zheng, Y.; Zhang, Y.; Wang, P.; Kallio, H.; Yang, B. Effects of Sea Buckthorn (*Hippophaë rhamnoides*) Juice and L-Quebrachitol on Type 2 Diabetes Mellitus in db/db Mice. *J. Funct. Foods* **2015,** *16,* 223–233.

Yan, Y.; Hu, J.; Yao, P. Effects of Casein, Ovalbumin, and Dextran on the Astringency of Tea Polyphenols Determined by Quartz Crystal Microbalance with Dissipation. *Langmuir* **2009,** *25,* 397–402.

Zhang, L.; Mou, D.; Du, Y. Procyanidins: Extraction and Microencapsulation. *J. Sci. Food Agric.* **2007,** *87,* 2192–2197.

Zhang, Y.; Luo, J.; Bi, J.; Wang, J.; Sun, L.; Liu, Y.; Zhang, G.; Ma, G.; Su, Z. Efficient Separation of Homologous α-Lactalbumin from Transgenic Bovine Milk Using

Optimized Hydrophobic Interaction Chromatography. *J. Chromatogr A* **2010,** *1217,* 3668–3673.

Zhu, J.; Sun, X.; Wang, S.; Xu, Y.; Wang, D. Formation of nanocomplexes Comprising Whey Proteins and Fucoxanthin: Characterization, Spectroscopic Analysis, and Molecular Docking. *Food Hydrocolloids* **2017,** *63,* 391–403.

Zorilla, R.; Liang, L.; Remondetto, G.; Subirade, M. Interaction of Epigallocatechin-3-Gallate with β-Lactoglobulin: Molecular Characterization and Biological Implication. *Dairy Sci. Technol.* **2011,** *91,* 629–644.

INDEX